Quantitative Treatments of
Solute/Solvent Interactions

THEORETICAL AND COMPUTATIONAL CHEMISTRY

SERIES EDITORS

Professor P. Politzer
Department of Chemistry
University of New Orleans
New Orleans, LA 70418, U.S.A.

Professor Z.B. Maksic
Ruder Boskovic Institute
P.O. Box 1016,
41001 Zagreb, Croatia

VOLUME 1
Quantitative Treatments of Solute/Solvent Interactions
P. Politzer and J.S. Murray (Editors)

1

THEORETICAL AND COMPUTATIONAL CHEMISTRY

Quantitative Treatments of Solute/Solvent Interactions

Edited by

P. Politzer

J.S. Murray

Department of Chemistry
University of New Orleans
New Orleans, LA 70148, USA

1994

ELSEVIER

Amsterdam – Lausanne – New York – Oxford – Shannon – Tokyo

ELSEVIER SCIENCE B.V.
Sara Burgerhartstraat 25
P.O. Box 211, 1000 AE Amsterdam, The Netherlands

Library of Congress Cataloging-in-Publication Data

Quantitative treatments of solute/solvent interactions / edited by P.
  Politzer, J.S. Murray.
       p.   cm. -- (Theoretical and computational chemistry ; 1)
    Includes bibliographical references and index.
    ISBN 0-444-82054-X
    1. Solvation.   I. Politzer, Peter, 1937-   . II. Murray, J. S.
  (Jane S.) III. Series.
  QD541.Q36 1994
  541.3'4--dc20
                                                        94-31159
                                                            CIP

ISBN: 0 444 82054 X

Printed and bound by Antony Rowe Ltd, Eastbourne
Transferred to digital print on demand, 2005

FOREWORD

The emphasis in this volume is on theoretical and computational treatments of experimental observations. The motivation of the former is to correlate, explain and be able to predict the latter. This philosophy has also been evident in a series of excellent annual meetings dealing with solute/solvent interactions; the earlier ones were organized by Mortimer Kamlet and Bob Taft, in Laguna Beach, California, and the more recent (after an interruption of several years) by George Famini at the Aberdeen Proving Grounds in Maryland. These relatively small and informal gatherings have been very intense and stimulating, and from them evolved the idea for this book. Indeed, most of the contributors have also been participants at one or more meetings. Our objective in this volume has been to provide a wider audience with a concentrated look at what has been achieved and what is being done in this very active area.

ACKNOWLEDGMENTS

We greatly appreciate the very efficient assistance
of Ms. Anita Buckel and Ms. Elizabeth Blankenship
in putting together this volume, and especially
in the tedious task of preparing the index.

TABLE OF CONTENTS

P. Politzer and J.S. Murray
Quantitative Treatments of Solute/Solvent Interactions
Theoretical and Computational Chemistry, Vol. 1

Perspectives and Overview

Peter Politzer and Jane S. Murray

Department of Chemistry, University of New Orleans, New Orleans, Louisiana, 70148, USA

How can we understand, describe and predict the ways in which solutes and solvents interact with each other, and the effects that they have upon each other? Our primary objective in putting together this volume has been to survey some effective approaches to answering these questions. Such answers can of course be obtained on various levels, some more empirical and others more rigorous. We have sought to be at least representative of this range of options; however our main emphasis will be somewhere in the middle, perhaps leaning to the less rigorous side.

One important approach to the problem of solute/solvent interactions has been based on *solvatochromism*. This involves determining the effects of various solvents upon the frequency corresponding to a specific electronic transition in a particular probe solute molecule. Solvatochromism can be described theoretically in terms of a perturbation treatment using a reaction field model [1-3], and it has been a useful tool for studying solvent effects; for reviews, see Kamlet *et al* [4] and Abraham [5]. Particularly successful and very extensive applications of solvatochromism have evolved from a "solvatochromic comparison" approach used by Kamlet and Taft initially to develop quantitative scales of the tendencies of solvents to accept and donate hydrogen bonds (hydrogen bond basicity and acidity, respectively) [6, 7]. Subsequent papers in this series addressed the problem of quantifying solvent polarity, and led to a second series, entitled "Linear Solvation Energy Relationships," which began in 1979 [8-10]. Taft, Kamlet and their collaborators have developed a number of quantitative relationships between a variety of physicochemical solute properties (not just spectral transitions) and certain solvent parameters; the latter include hydrogen-bond-accepting and -donating tendencies, polarity/polarizability, and a "cavity term" related to the size of the solute molecule [4, 11, 12]. These have all been described as solvatochromic equations, even though the solute properties being represented have included formation constants, solvolysis and reaction rates, enthalpies of solution, solubilities, partition coefficients and others [4, 12].

The original approach, which examined the effects of different solvents upon a given solute, was eventually complemented by its inverse: studying variations in the properties of a series of solutes in a specified solvent. This is indeed the focus of most current work in this area. Abraham and his collaborators developed scales of solute hydrogen bonding tendencies [13-16]; however these were based not upon solvatochromism but rather upon the formation of 1:1 complexes between the solute molecule and a given reference acid or base in CCl_4. A continuing theme in this work has been the objective of separating hydrogen-bonding from other types of solute/solvent interactions.

The development and applications of linear solvation energy relationships, which have been only sketchily introduced in the preceding paragraph, have had a considerable and widespread impact. Solute/solvent interactions have been partitioned into contributions from certain key factors, e.g. hydrogen bonding vs. other dipolar effects, and scales have been devised for quantitatively relating these to individual solutes and solvents. The applications have gone beyond ordinary solutions; for example there has been extensive use of solvatochromic equations in analyzing the properties of supercritical solutions [17-24].

The continuing development and use of linear solvation energy relationships, in various forms, is the theme of a considerable portion of this volume. However we felt that the chapter by Cramer and Truhlar would be a very appropriate one with which to begin (Chapter 2). They examine in detail the problems associated with trying to treat solvent effects in a rigorous manner. A relatively successful approach has been to treat the solvent as a continuum, characterized only by one or more of its bulk properties, e.g. the dielectric constant. This leads to the concept of the reaction field, which is the electric field felt by the solute due to the polarization that it has induced in the solvent. The solute molecule is now viewed as residing within a cavity of parametrically-defined size. Cramer and Truhlar discuss and analyze various approaches to approximately solving Poisson's equation for the system, to obtain the electrostatic portion of the free energy of solvation. They also examine contributions arising from other factors, such as the creation of the cavity, dispersion forces and structural changes that the solute causes in the surrounding solvent.

Cramer and Truhlar summarize their own procedures for evaluating the various quantities that comprise the total free energy of solvation, using semi-empirical computational techniques, and then present several applications to biological systems. They report studies of the aqueous solvation of the nucleic acid bases and some of their methyl derivatives, including the effects upon the dipole moments of the former. They also describe the effects of aqueous solvation upon the conformations of dopamine, ethylene glycol and glucose. Extensions of the approach to other solvents are discussed.

Finally, in a very useful introduction to the chapters that will follow, Cramer and Truhlar mention the development and use of linear solvation energy relationships. They comment on the solvatochromic equations which have evolved from the work of Kamlet and Taft , as well as some variations and extensions, two of which (due to Famini, Wilson and collaborators and Politzer, Murray and collaborators) are the subjects of chapters in this volume.

A very common and extremely important form of solute/solvent interaction is hydrogen bonding, and it has accordingly been the focus of a great deal of attention in this context. As mentioned above, the initial applications of solvatochromism by Kamlet and Taft in 1976 had the objective of developing quantitative measures of the hydrogen bonding tendencies of solvents [6, 7], and Abraham and his collaborators later established analogous scales for solutes [13-16], using a different approach. (Of course hydrogen bonding is not the only kind of electrostatic interaction that can occur between a solute and solvent; for example, the significance of "halogen bonding" between an electron-pair donor and the positive tip of a chlorine, bromine or iodine atom should not be overlooked [25, 26].) In recognition of the importance of hydrogen bonding in many of the systems of present interest, it is the subject of Chapters 3-5.

The first portion of Chapter 3, by Taft and Murray, reviews the development of these solvent and solute scales of hydrogen-bond-donating and -accepting abilities. A problem arises when the molecule contains several possible hydrogen-bonding sites. How does one then obtain a value for the overall tendency to accept or donate a proton? It is not necessarily a simple sum of the individual contributions. Taft and Murray present a technique, based upon an analysis of partition coefficients, for determining the total hydrogen-bond-donating tendency of a multi-site donor. Values are listed for about 30 compounds.

There has been an extensive effort by Politzer, Murray and collaborators to develop analytical representations of various properties related to molecular interactions, in which these properties are expressed in terms of quantities computed for the isolated molecules. This is discussed in detail in Chapter 8, but it is already introduced by Taft and Murray in Chapter 3. They proceed to show that the solvent and solute hydrogen-bonding parameters can be represented by such relationships, as can the experimentally-determined infra-red O–H frequency shifts of methanol complexed to various hydrogen-bond acceptors. Thus, these solution properties can be predicted by means of computations made for the isolated molecules.

In chapter 4, Abraham describes further efforts to achieve improved scales of solute hydrogen bonding tendencies, both donating and accepting. Partition coefficients are also the key to his approach, and he begins with a background discussion of these that leads him to linear free energy relationships. Abraham obtains solute parameters for both hydrogen bond acidity and basicity, and he addresses the problem of multi-site systems. For the latter, he contrasts his approach to that of Carr [27, 28], which is based on gas/liquid chromatography. Abraham also examines the relationship between his treatment and the computational ones of Famini, Wilson and collaborators (Chapter 7) and Politzer, Murray and collaborators (Chapter 8).

A large portion of Chapter 4 deals with the use of Abraham's hydrogen bonding parameters in two linear free energy relationships, one for gas/condensed phase processes and the other for processes within condensed phases. A number of applications are presented and discussed in detail, involving gas/liquid and gas/solid chromatographic data in the first case and partition coefficients, high pressure liquid chromatography and aqueous anesthesia in the second.

The discussions by Taft and Murray and by Abraham deal only with solutions. Abboud, Notario and Botella, in Chapter 5, seek to go beyond this and to present a unified treatment of free energy relationships for 1:1 hydrogen-bonded complexes in solution and also in the gas phase, for both neutral and ionic species. They begin with a review of experimental techniques used to study hydrogen bonding, and then survey some of the methods that have been developed to analyze and interpret the data obtained by these techniques. Equilibrium constants have long provided a convenient route to quantifying hydrogen bond acidity and basicity, and are the basis for much of the discussion in this chapter. Abboud, Notario and Botella show that the formalism and parameter values that are applicable in solution can be extended to the gas phase.

Chapters 3-5 focused on hydrogen bonding, and emphasized solute behavior. Chapter 6, by Fawcett, deals with a wider range of interactions but in terms of the properties of the solvent, specifically its ability to accept or donate a pair of electrons (Lewis acidity and basicity). He has proposed solvent acidity and basicity scales based on the free energies of solvation of alkali metal cations and halide anions [29], and he compares these to others that have been introduced, including those of Kamlet and Taft for hydrogen-bond-donating and -accepting tendencies [6, 7]. (Hydrogen bonding is a special case of a Lewis acid-base interaction.) In general, solvent polarity/polarizability also needs to be taken into account in linear solvation energy relationships [4, 11, 12, 30]. Fawcett analyzes the roles of solvent acidity, basicity, polarity and polarizability in several applications, involving electrolyte solutions, electrode reactions and solute infra-red frequency shifts.

The preceding four chapters have described efforts to relate experimental data to parameters defined as representing certain factors believed to be important in solute/solvent interactions. Among the more important of these factors are hydrogen bond acidity and basicity, and polarity/polarizability. The values of such parameters for different solutes and solvents have been obtained from the experimental data. However one could also adopt a theoretical/computational approach. For example, correlations could be sought between known parameter values and calculated molecular quantities, and used to evaluate the former in cases where experimental data are not available. Work along these lines is reported in Chapters 7-9. One possibility is to represent the same or similar physical factors by means of new parameters, or descriptors, which can be determined by theoretical means. This is the approach of Famini and Wilson (Chapter 7). Their descriptors have interpretations qualitatively similar to those in the equations of Kamlet, Taft and collaborators [4, 12], but they are formulated in terms of, for example, molecular orbital energies, atomic charges and van der Waals volumes [31, 32]. Famini and Wilson discuss applications of their theoretical linear solvation energy relationships in representing toxicity data, spectral peak positions and decarboxylation rates.

Whereas Famini and Wilson's procedure is firmly based in linear solvation energy equations, a significantly different approach is presented by Murray and Politzer in Chapter 8. They and their collaborators were originally interested in developing correlations between solvatochromic parameters (hydrogen bond acidity and basicity, and polarity/polarizability) and quantities (electronic and structural) computed for the respective molecules [33, 34]. These results are reviewed. However this evolved into a general unified treatment whereby macroscopic properties determined by molecular interactions in fluid media are represented by relationships involving some subset of a group of key computed molecular quantities [35]. The latter are evaluated on the molecular surface, through *ab initio* calculations; most of them are related to the electrostatic potential of the molecule. There is no direct correspondence with the solvatochromic parameters, and hydrogen bonding is not distinguished from any other type of interaction. Murray and Politzer discuss applications of this general interaction properties function to a wide variety of properties: critical constants, boiling points, heats of vaporization, supercritical solubilities, partition coefficients, aqueous and gas phase acidities, and hydrogen bonding parameters. The Famini/Wilson and Murray/Politzer approaches are similar in that each uses a limited set of descriptors for all of the relationships, thus maintaining a common thread between them.

A third computational procedure for correlating and predicting a wide variety of physical and chemical properties is presented by Hilal, Carreira and Karickhoff in Chapter 9. They have designed a program that is highly user-friendly and is intended to mimic the reasoning of an expert chemist. It involves contributions from a number of factors that are presumed to be significant in each particular case. These may represent structural units, some of which are treated as perturbations which produce electrostatic, field, induction, resonance, mesomeric, solvation and hydrogen bonding effects. These are evaluated empirically, and/or through semi-empirical quantum chemical calculations. The specific nature of the treatment depends upon the property being represented. In Chapter 9, Hilal, Carreira and Karickhoff describe in detail the procedures used to calculate pK_a's, electron affinities and chromatographic retention times.

The treatments of solute/solvent interactions that are presented in this volume emphasize a synergism between theory and experiment. Data obtained experimentally are used as a basis for developing quantitative theoretical models that permit the correlation and interpretation of the data, and also provide a predictive capability. The latter is of course a key motivation for these efforts. Linear solvation energy relationships have been quite successful in these respects and accordingly receive considerable attention in the chapters that follow. As has already been indicated, there is a continuing focus upon improving the parameters used in these equations. Other effective approaches, including computational ones, are also being pursued, and are discussed in several chapters. This is an area that is continually evolving, and we hope that the present volume will convey a sense of its dynamic nature.

REFERENCES

1. Y. Ooshika, J. Phys. Soc. Japan, 9 (1954) 594.
2. E. G. McRae, J. Phys. Chem., 61 (1957) 562.
3. W. Liptay, in *Modern Quantum Chemistry, Part II: Interactions*, O. Sinanoglu, ed., (Academic Press, New York, 1965), ch. B5.
4. M. J. Kamlet, J.-L. M. Abboud and R. W. Taft, in *Progress in Physical Organic Chemistry*, vol. 13, R. W. Taft, ed., (Wiley-Interscience, New York, 1981), ch. 6.
5. M. H. Abraham, Pure Appl. Chem., 57 (1985) 1055.
6. M. J. Kamlet and R. W. Taft, J. Am. Chem. Soc., 98 (1976) 377.
7. R. W. Taft and M. J. Kamlet, J. Am. Chem. Soc., 98 (1976) 2886.
8. M. J. Kamlet and R. W. Taft, J. Chem. Soc., Perkin Trans. 2, (1979) 337.
9. M. J. Kamlet, M. E. Jones, J.-L. M. Abboud and R. W. Taft, J. Chem. Soc., Perkin Trans. 2, (1979) 342.

6

10. M. J. Kamlet and R. W. Taft, J. Chem. Soc., Perkin Trans. 2, (1979) 349.
11. M. J. Kamlet, J.-L. M. Abboud, M. H. Abraham and R. W. Taft, J. Org. Chem., 48 (1983) 2877.
12. M. J. Kamlet, R. M. Doherty, J.-L. M. Abboud, M. H. Abraham and R. W. Taft, Chemtech, 16 (1986) 566.
13. M. H. Abraham, P. P. Duce, P. L. Grellier, D. V. Prior, J. J. Morris and P. J. Taylor, Tetrahedron Letters, 29 (1988) 1587.
14. M. H. Abraham, P. L. Grellier, D. V. Prior, P. P. Duce, J. J. Morris and P. J. Taylor, J. Chem. Soc., Perkin Trans. 2, (1989) 699.
15. M. H. Abraham, P. L. Grellier, D. V. Prior, J. J. Morris, P. J. Taylor, C. Laurence and M. Berthelot, Tetrahedron Letters, 30 (1989) 2571.
16. M. H. Abraham, P. L. Grellier, D. V. Prior, J. J. Morris and P. J. Taylor, J. Chem. Soc., Perkin Trans. 2, (1990) 521.
17. M. E. Sigman, S. M. Lindley and J. E. Leffler, J. Am. Chem. Soc., 107 (1985) 1471.
18. C. R. Yonker, S. L. Frye, D. R. Kalkwarf and R. D. Smith, J. Phys. Chem., 90 (1986) 3022.
19. R. D. Smith, S. L. Frye, C. R. Yonker and R. W. Gale, J. Phys. Chem., 91 (1987) 3059.
20. S. L. Frye, C. R. Yonker, D. R. Kalkwarf and R. D. Smith, in *Supercritical Fluids*, T. G. Squires and M. E. Paulaitis, eds., *ACS Symp. Ser 329*, (American Chemical Society, Washington, 1987), ch. 3.
21. C. R. Yonker and R. D. Smith, J. Phys. Chem., 92 (1988) 235. 2374.
22. J. P. Blitz, C. R. Yonker and R. D. Smith, J. Phys. Chem., 93 (1989) 6661.
23. K. P. Johnston, S. Kim and J. Combes, in *Supercritical Fluid Science and Technology*, K. P. Johnston and J. M. L. Penninger, eds., *ACS Symp. Ser. 406*, (American Chemical Society, Washington, 1989), ch. 5.
24. M. L. O'Neill, P. Kruus and R. C. Burk, Can. J. Chem., 71 (1993) 1834.
25. T. Brinck, J. S. Murray and P. Politzer, Int. J. of Quantum Chem., Quantum Biol. Symp., 19 (1992) 57.
26. P. Politzer and J. S. Murray, in *Supplement D: Chemistry of the Halides*, S. Patai and Z. Rappoport, eds., (Wiley, Chichester, England, in press).
27. J. Li, Y. Zhang, A. J. Dallas and P. W. Carr, J. Chromatogr., 550 (1991) 101.
28. J. Li, Y. Zhang, H. Ouyang and P. W. Carr, J. Am. Chem. Soc., 114 (1992) 9813.
29. W. R. Fawcett, J. Phys. Chem., 97 (1993) 9540.
30. I. A. Koppel and V. A. Palm, in *Advances in Linear Free Energy Relationships*, N. B. Chapman and J. Shorter, eds., (Plenum Press, London, 1972), Ch. 5.
31. G. R. Famini, Using Theoretical Descriptors in Quantitative Struture Activity Relationships 1989, V.CRDEC-TR-085, US Army Chemical, Research, Development and Engineering Center, Aberdeen Proving Ground, MD
32. G. R. Famini, C. A. Penski and L. Y. Wilson, J. Phys. Org. Chem., 5 (1992) 395.

33. J. S. Murray and P. Politzer, J. Chem. Res., S (1992) 110.
34. T. Brinck, J. S. Murray and P. Politzer, Mol. Phys., 76 (1992) 609.
35. J. S. Murray, T. Brinck, P. Lane, K. Paulsen and P. Politzer, J. Mol. Struct. (Theochem), 307 (1994) 55.

P. Politzer and J.S. Murray
Quantitative Treatments of Solute/Solvent Interactions
Theoretical and Computational Chemistry, Vol. 1
© 1994 Elsevier Science B.V. All rights reserved.

DEVELOPMENT AND BIOLOGICAL APPLICATIONS OF QUANTUM MECHANICAL CONTINUUM SOLVATION MODELS

Christopher J. Cramer and Donald G. Truhlar

Department of Chemistry and Supercomputer Institute, University of Minnesota, Minneapolis, MN 55455.

1. INTRODUCTION

The development of computational methods for condensed phases that have predictive powers equivalent to those available for gas-phase systems has been a long-standing goal of theoretical chemists. The complementary nature of theory and experiment has become strikingly apparent for the study of molecular structure and dynamics in the gas-phase, and indeed we have reached the point where many gas-phase observables are more accurately and efficiently predicted by theoretical methods than they may be measured. However, the corresponding role of theory in *solution*-phase chemistry is nowhere near so well established. Since the bulk of preparative organic chemistry and all of biological chemistry occur in condensed phases, the importance of further progress in this area should be clear.

The transition from the gas phase to solution is by no means a small perturbation; often the effects of solvation on issues of structure and reactivity are extremely large [1]. For example, the Menschutkin reaction of ammonia and chloromethane is illustrated in Figure 1. The reaction involves nucleophilic displacement of chloride by ammonia and thereby converts the neutral reactants into a pair of charged products, methylammonium cation and chloride anion, a so-called Type II S_N2 reaction. This reaction is exothermic and proceeds readily in aqueous solution. Modeling efforts in the gas phase are unable to provide much information about this process, since the product ions are very high in energy in the absence of solvation—in fact the corresponding reaction path is entirely uphill in the gas phase. The reaction is made possible by virtue of aqueous solvation, which is roughly 150 kcal/mol more favorable for the products than for the reactants [2,3].

While the above example illustrates the effect of solvent on the equilibrium between reactants and products, another interesting case is the effect of solvent on the activation energy, i.e., the differential solvation of the transition state relative to the reactants. This effect has been well studied for S_N2 reactions of anions [4-11]. Figure 2 illustrates this effect for a more complex organic reaction, namely the Claisen rearrangement, an electrocyclic reaction which converts an allyl vinyl ether to a γ,δ-unsaturated carbonyl compound. The Claisen rearrangement has been demonstrated to be accelerated on the order of 1000-fold for the unsubstituted parent ether on going from the gas phase to aqueous

10

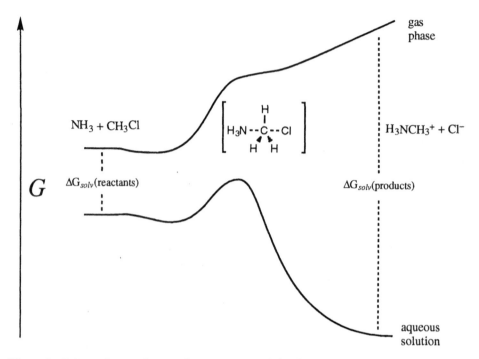

Figure 1. Schematic gas-phase and aqueous potentials of mean force for the Menschutkin reaction projected onto a single generalized reaction coordinate.

solution at room temperature [12-16]. The Claisen rearrangement is a particularly interesting case, insofar as there are two distinct transition state structures, a chair and a boat form, each of which leads to stereochemically distinct products when the reactive termini are appropriately substituted (illustrated with deuterium substitution in Figure 2). In addition to the differential solvation effect between reactants and transition states for this reaction, there are additional solvation differences between the two transition states, giving rise to the possibility that in specific instances one might tune the stereoselectivity of the reaction by judicious choice of solvent [14,17].

These kinds of effects also occur in a more biologically focused paradigm, which is illustrated in Figure 3. The top portion of the illustrated cycle corresponds to the unimolecular, gas-phase reaction of some molecule A. It is convenient to continue using the Claisen rearrangement as an example, since this appears to be the mechanism by which chorismate is converted into prephenate in vivo, an important step in the shikimic acid pathway [18,19] by which a family of sugars is transformed into aromatic amino acids. The middle portion of the cycle is similar to what we have discussed above, i.e., the effect of aqueous solvation has been included. The bottom portion of the cycle, on the other hand,

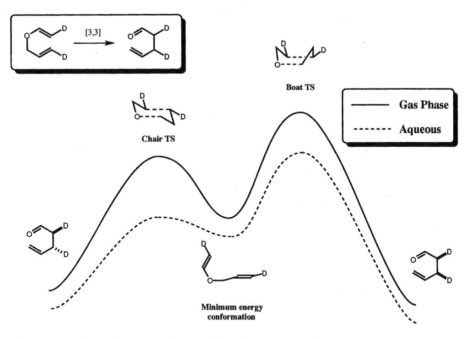

Figure 2. Schematic comparison of the differential effect of aqueous solvation on competing chair and boat transition states for the Claisen rearrangement.

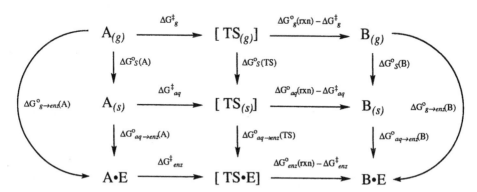

Figure 3. Thermodynamic cycles for a unimolecular reaction showing the relationship between the gas-phase, aqueous solution, and the enzyme-mediated reaction.

represents a reaction that is aqueous overall, but the reaction now proceeds within an enzyme active site (e.g., chorismate dismutase). The cycle illustrates that the manner in which an enzyme influences reaction rates is very similar to the manner in which solvent does [20-22]. Thus, enzyme catalysis involves the competition between solvation of the substrate by the bulk solvent and "solvation" by the enzyme (the latter more commonly associated with terms like "complexation," "binding," or "stabilization") [23].

Section 2 reviews theories of aqueous solvation based on combining quantum mechanics with continuum models, including both a continuum representation of the bulk solvent and also the continuum approach (based on atomic surface tensions) to first hydration shell effects. Section 3 reviews selected applications to biomolecules. Section 4 considers other solvents. Section 5 presents a brief comparison of the present approach to the method of theoretical linear solvation energy relationships, which also involves both quantum mechanics and a continuum treatment of the solvent.

2. MODELING AQUEOUS SOLVATION

The most obvious way to account for solvation in a theoretical calculation is to surround a substrate of interest with sufficient solvent molecules to mimic the effects of bulk solvation. Regrettably, the number of solvent molecules which is required to mimic bulk solvent is usually quite large. Moreover, the supersystem in general has a very large number of energetically accessible states (differing by the individual orientations of solvent molecules, for example) which must be statistically sampled to obtain thermodynamically averaged information [24-30]. This sampling may be performed using either probabilistic methods like Monte Carlo [21,25,27,28,30] or by following molecular dynamics trajectories in phase space [20-22,26-28,31,32], but the net result is that converged treatment of the explicit quantum mechanical representation of the entire system is effectively impossible. As a result, simulations along the lines described above are typically carried out with classical mechanical Boltzmann factors replacing the quantum mechanical density operators for atomic coordinates. Furthermore, the atomic potential energy function which, according to the Born-Oppenheimer separation of electronic and nuclear motions, is actually governed by the quantum mechanical adiabatic evolution of the electronic states, is often replaced by a set of pairwise interactions governed by a molecular mechanics force field. Even in such classical simulations, convergence with respect to long-range forces [33-35] and multidimensional sampling [36-42] remains problematic. In the latter respect, there is no general way to insure the adequate sampling of phase space during a simulation. Thus, in systems with hundreds to thousands of position and momentum coordinates, phase-space bottlenecks may prevent the system from exploring regions which are important for the accurate prediction of observed properties.

Classical simulations suffer from other drawbacks as well. Inevitably, force fields are derived from fitting functional forms for interaction energies to experimental or theoretical data [43,44], and the extension of such approaches to transition states is a severe difficulty because of the limited database available. Thus, the development of force fields capable of handling transition states is technically challenging [45].

In order to alleviate the problems associated with solvent sampling, it proves useful to simplify the microscopic representation of the solvent. For instance, Warshel and coworkers pioneered replacement of individual water molecules with Langevin dipoles [46-50]. One critical advantage of eliminating some of the technical difficulties associated with carrying out the solvent sampling is that it permits a more accurate representation of the solute, e.g., it makes it much more practical to introduce a quantum mechanical treatment of the solute [10,46,51-75]. Another approach to simplifying the problem is to replace hydrogen bonding interactions in the first hydration shell by an effective potential method [76]. An older and more radical approach to simplifying the treatment of the solvent is to remove all microscopic coordinates and leave behind only a continuum characterized by one or more of its bulk properties. The simplest such solvation models retain no information about the solvent beyond its bulk dielectric constant. This level of solvent detail is all that is required for the cavity-based electrostatic approximations of Born [77], Kirkwood [78], and Onsager [79] which capture the essential physics of an ion or small dipolar molecule in a dipolar solvent. These classical treatments lead to simple analytic expressions for the solvation free energy when the solute can be assumed to be spherical, and these analytical expressions provide the essential scaling laws for qualitative understanding of how solvation energies vary with solute radii.

Three key elements have been involved in extending the usefulness of continuum solvation models to essentially arbitrarily complicated solutes: (1) By using computers, one can eliminate the restriction to spherical solutes. The modern approach is to represent the solute shape in terms of overlapping spheres centered at the atomic nuclei. Such space-filling models are now recognized to provide realistic models of the shape of essentially all molecules and molecular fragments, and even transition states. (2) By using molecular orbital theory, the charge distribution of any solute can be modeled reasonably well by a series of partial charges at the atomic centers; this is called the distributed monopole model. This allows the treatment of solutes that contain more than a single dipolar center. Of course, the more complicated models of solutes as arbitrarily large numbers of spheres, not necessarily arranged symmetrically, and each with an arbitrary partial charge at the center, means that analytic solutions are no longer available; however, the cost of computer solutions (with varying degrees of numerical complexity and auxiliary approximations) is a rapidly decreasing function of time. (3) Finally, modern extensions of the continuum solvent model have benefited from the realization that the main defect of the continuum treatment of the solvent is the breakdown of bulk properties in the first solvation shell. In early work this concern was expressed in attempts to define effective solute radii that encompass a portion of the solvent where "dielectric saturation" occurs. This saturation of the solvent orientational polarizability in the vicinity of a large electric field is accompanied by other first solvation shell effects, for example, "electrostriction", which refers to the loss of motional freedom experienced by dipolar solvent molecules in that same region. In modern work, one focuses more directly on the first hydration shell and explicitly accounts for its effects in addition to the bulk electrostatic effects.

This section will continue with more details of these modern extensions. Section 2.1.1 reviews the essentials of classical theory. Sections 2.1.2 and 2.1.3 review the classical analytical theories. Sections 2.1.4 and 2.1.5 review the elements of extensions (1) and (2). Section 2.2 reviews extension (3).

2.1. Electrostatic components of solvation free energy
We begin with a discussion of that portion of the solvation free energy which arises from solute-solvent, solvent-solvent, and internal solute electrostatic interactions.

2.1.1. The Poisson Equation
The key reason for choosing to characterize the solvent continuum by its dielectric constant is that it allows one to use the power of classical electrostatics. When the solute is represented explicitly, and the solvent is treated as a continuum, the Laplacian of the electrostatic potential, $\phi(\mathbf{r})$, is related to the free charge density (i.e., the charge density due exclusively to the solute), $\rho(\mathbf{r})$, by Poisson's equation [80-82],

$$\nabla^2\phi = -\frac{4\pi\rho(\mathbf{r})}{\varepsilon},$$

(1)

where ε is the homogeneous dielectric constant (relative permittivity of the solvent, e.g., 78.3 for water at 298 K), $\mathbf{r}$ denotes the position in space, and the equation is written in gaussian units. If ε depends on $\mathbf{r}$, one can replace equation 1 by the slightly more complicated [81]

$$\nabla\cdot\varepsilon(\mathbf{r})\nabla\phi = -4\pi\rho(\mathbf{r}).$$

(2)

Assuming that thermal equilibrium is maintained by an external heat bath, the free energy of solvation G, which is the maximum work which may be extracted from the solvation process, is then obtained from [81,82]:

$$G = -\frac{1}{2}\int d^3\mathbf{r}\,\frac{1}{4\pi}\phi(\mathbf{r})\,\nabla\cdot\varepsilon(\mathbf{r})\nabla\phi(\mathbf{r})$$

(3)

or equivalently

$$G = \frac{1}{2}\int d^3\mathbf{r}\,\frac{1}{4\pi}\phi(\mathbf{r})\,\nabla\cdot\varepsilon(\mathbf{r})\,\mathbf{E}(\mathbf{r})$$

(4)

where $\mathbf{E}(\mathbf{r})$ is the electric field given by the gradient of the electrostatic potential and is given by

$$\mathbf{E}(\mathbf{r}) = -\nabla\phi(\mathbf{r}).$$

(5)

In solution, the electric field contains contributions from both the intrinsic solute charges and from the polarization *induced* by the solute in the solvent. The latter contribution is called the reaction field. Equations 3 and 4 can also be written as

$$G = \frac{1}{2} \int d^3r \, \mathbf{D}(\mathbf{r}) \cdot \mathbf{E}(\mathbf{r}), \tag{6}$$

where $\mathbf{D}(\mathbf{r})$ is the electric displacement due to the solute charges. This illustrates how the solute interacts with its own reaction field, which has a significant effect on the energy of the system. For conceptual purposes, equation 6 may be thought of as arising from an iterative process. That is, the gas-phase solute is placed into solution, inducing a reaction field determined from the gas-phase charge distribution. The interaction of the reaction field with the solute charge distribution in general induces a relaxation of the gas-phase nuclear and electronic structure in order to minimize the free energy of the whole system. Of course, as relaxation proceeds it changes the reaction field quantitatively, such that additional changes in the solute charge distribution may be favorable. The entire procedure reaches its terminus when the internal cost of additional change in the charge distribution of the solute and solvent begins to exceed the resultant gain in their interaction free energy.

We will have cause to refer to this electrostatic portion of the free energy of solvation repeatedly in later discussion. We label it ΔG_{ENP} to indicate that it includes the work required to distort the electronic (E) and nuclear (N) structure (i.e., the molecular electronic wave function and the geometry, respectively) of the solute from their optimal gas-phase values, and this is driven by the gain in polarization (P) free energy, which is the net gain in solute-solvent interaction free energy minus the cost in solvent internal free energy.

2.1.2. The Born equation

In actual practice, analytical solutions to the Poisson equation exist only for rather simple cases. One example is a charge q on a conducting sphere of radius α. Since a charge on a metallic sphere is spread uniformly over its surface, but the effect of this outside the sphere is the same as for a point charge at the sphere center, this is a simple model for a monatomic ion. Solution of Poisson's equation in this instance for the gas-phase ($\varepsilon = 1$) and the dielectric medium ($\varepsilon > 1$) gives Born's formula [77,83,84] for the free energy of transfer of the charged sphere from a medium with a dielectric constant of unity (vacuum or sufficiently dilute gas phase) to a solvent of dielectric constant ε:

$$\Delta G_S^0 = -\frac{1}{2} \left(1 - \frac{1}{\varepsilon} \right) \frac{q^2}{\alpha}. \tag{7}$$

Of course, the meaning of α is somewhat less clear for a monatomic ion than for the ideal case of the conducting sphere [84-88]. That is, the surrounding solvent dielectric is not homogeneous all the way up to the ionic "surface". Many models have been proposed for relating α to the electronic structure of the monatomic ion, but ΔG_S^0 is very sensitive to the precise value of α, so these models must be used with caution. We prefer the empirical route where α is determined from equation 7 using an experimental value for the free energy of solvation. Such an α is called an effective ionic radius, or a Born radius. Based on analysis of classical simulations employing explicit solvent representation, Jayaram et al. have concluded that dielectric saturation does not affect the quadratic dependence of ΔG_S^0 on charge for spheres with a charge less than 1.1 electronic units [87]; in such instances, the ionic radii in equation 7 which reproduce the simulation free energies of solvation are generally in reasonable agreement with other standard measures of atomic radii, such as the van der Waals radii suggested by Bondi [89] (although, of course, they are not exactly the same). Hirata et al. [86] and Roux et al. [90] have separately discussed this result in terms of extended reference interaction site method (RISM) calculations [91,92], suggesting that sensible Born radii may be derived from analysis of the first peak in the solute-solvent radial distribution function, again in concert with a simple spherical model for the ion. Furthermore, it has been shown that the average distance between ions and nearest neighbor water molecules is about 1.4 Å larger than the ionic crystal radius for typical ions [93].

As mentioned at the end of Section 2.1.1, the electrostatic energies which we have so far discussed include not only the interaction of the solute with the solvent but also the change in solvent-solvent interactions when the solute is inserted. Under the mild assumptions of linear response theory, it can be shown that the latter increase in intrasolvent energy cancels half the favorable solute-solvent interactions, which is one way to think about the factors of $\frac{1}{2}$ appearing in equations 3, 4, 6, and 7 [63,67,69,72,86,87,90,94-100].

2.1.3. The reaction field approach

Clearly only monatomic ions may be unambiguously regarded as spherical. However, at large enough distance, the interaction of any multipole distribution with a surrounding field is dominated by its leading term. In other words, at long enough range the remaining interaction of an ion with a surrounding continuum is well approximated by the Born equation. Indeed, in molecular simulations of ions which *include* explicit solvent molecules, the calculation of electrostatic interactions is typically truncated at some maximal distance from the solute to maintain computational efficiency, and the remaining interactions out to infinite distance are often approximated using the Born equation [33-35].

The leading multipole moment for *uncharged* solutes is usually the dipole term. Once again, an analytic solution to Poisson's equation exists. For a point dipole of magnitude μ in a sphere of radius α, the result, derived in different ways by Kirkwood and Onsager, is [78,79]

$$\Delta G_S^0 = -\frac{(\varepsilon - 1)\mu^2}{(2\varepsilon + 1)\alpha^3} \tag{8}$$

where ε is again the dielectric constant of the solvent modeled as a continuum. It should be emphasized that the spherical radius α appearing in equations 7 and 8 is only defined insofar as it represents the point at which a discontinuity in the dielectric constant occurs in the integral of equations 3 and 4. In other words, although it has clear mathematical utility in the solution of the Poisson equation, there is no prescription for its determination in molecular cases. We note at this point that analytical solutions to the Poisson equation for a point dipole inside an ellipsoid are also available, and implementations of this approximation have also appeared [71].

As mentioned above, α may be assigned on the basis of fitting the Born equation to experimental data for ions, but for uncharged molecules it would be a much more severe and ambiguous approximation to determine α in the same way from the Kirkwood-Onsager equation because of the spherical cavity approximation; it is clearly impossible to construct an uncharged, spherically symmetric molecule which has a dipole moment. Furthermore, little work has appeared establishing whether or not there are any clear connections between optimal α values and known physical properties for atoms and molecules. In this regard, an obvious approach would be to choose α as the radius of a sphere whose volume matches the cavity enclosed by the solvent-excluding surface of the solute, also called the molecular surface [101-105]. This approach yields similar results to those obtained using the van der Waals surface to divide the regions of unequal dielectric constant [106]. Wong et al. [107] have used a quantum mechanical approach in which the van der Waals surface is replaced by an isodensity surface and empirical scale factors are employed. Since the choice of surface determines the value of α, and the electrostatic free energy of solvation depends on the third power of α in equation 8, the calculations are quite sensitive to the choice of surface, and some nonphysical results have been reported in the literature when insufficient care was taken in assigning a value to α. (Further discussion of alternative definitions of solute surface is provided in Section 2.2.)

One of the key differences between the Born and Kirkwood-Onsager equations is that the former involves the solute charge, which is unchanged by the polarization field induced in the dielectric (we neglect charge transfer to or from the solvent); the latter, however, involves the molecular dipole moment, which may readily increase by relaxation of the electronic structure as described in Section 2.1, thereby contributing more polarization free energy at the cost of reorganizational free energy. Since the polarization field now affects the solvated electronic structure, it should be treated self-consistently in any quantum mechanical calculations designed to incorporate the effects of continuum solvation.

Within the Kirkwood-Onsager approximation, the quantum mechanical Hamiltonian operator that includes reaction field effects for neutral solutes is

$$(H_0 - \lambda g \, \mu \cdot < \psi | \mu | \psi >) \, | \psi > = E \, | \psi > \tag{9}$$

where $\lambda = 0.5$, $g = 2 \, (\varepsilon - 1) / (2 \, \varepsilon + 1) \, \alpha^3$, α is the solute cavity radius, and H_0 is the gas-phase Hamiltonian constructed in the usual manner. The corresponding Hartree-Fock equations are then [57,66,71,107-123]

$$(F_0 - \lambda g \, \mu \cdot < \psi | \mu | \psi >) \, | \phi_i > = \varepsilon_i \, | \phi_i > \tag{10}$$

where F_0 is the usual gas-phase Fock operator [124,125], and the ε_i are the one-electron orbital energies associated with the molecular orbitals ϕ_i. Note that these equations must be solved self-consistently insofar as the Fock operator, the one-electron density matrix involved in the solution of the Hartree-Fock equations, and the molecular dipole moment are all mutually interdependent. It is easily seen that these equations capture the phenomenon of increased charge separation being favored in solvents of increasing dielectric constant. The electrostatic portion of the free energy of solvation, ΔG_{ENP}, is then simply the energy calculated from the orbitals obtained from equation 10 minus the gas-phase energy. Note that this will generally be less negative than the energy calculated from equation 8, which is the portion associated with ΔG_P, since the costs of distorting the electronic and molecular structure of the *solute* have been included in the calculations implicitly by the SCF formalism.

In addition, as discussed in Section 2.1.2, the cost of distorting the *solvent* structure is taken into account via the factor λ of 0.5 preceding the reaction field portion of the solution-phase Fock operator. This formalism thus treats all three effects—favorable solute-solvent interactions (twice ΔG_P), cost of solute reorganization (ΔE_{EN}), and cost of solvent reorganization (minus ΔG_P)—mutually self-consistently; changes in solute electronic and molecular structure cease when additional gain in ΔG_P fails to be larger than the costs of distortion, ΔE_{EN}. An alternative approach has been considered, however, which minimizes the solute energy plus solute-solvent interaction, rather than the system energy. Formally, this convention involves taking $\lambda = 1$ in equations 9 and 10; therefore, *solute* electronic and structural reorganization proceeds until that portion of ΔG_P associated *only with the solute* no longer exceeds any increase in ΔE_{EN}. Clearly, the solute distortion for the latter approach must be greater than that for the former. However, when optimization is carried out with $\lambda = 1$, the *total* energy of the system can only be obtained by adding back a factor of $0.5 g \mu^2$, where μ is the relaxed dipole moment. In other words, if the cost of solvent reorganization is not accounted for self-consistently, it is included *ex post facto*.

Both methods, employing $\lambda = 0.5$ [71,107,109-113,116-123] and $\lambda = 1$ [57,66,108,114,115,122,126,127], have seen extensive use in the literature. Regrettably, it is not always clear in certain instances which method has been employed, and careful analysis of the equations in individual papers even reveals cases where the mathematical derivation switches haphazardly from one approach to the other! Not much work has been done to assess which, if either, of these two methods is to be preferred. Szafran et al. [122] compared the two approaches for the prediction of solvent effects on tautomeric equilibria in 2-hydroxypyridine $\rightleftharpoons$ 2-pyridone and 4-hydroxypyridine $\rightleftharpoons$ 4-pyridone and found little difference between them in comparison to experimental results. However, any comparison of the methods is somewhat ambiguous insofar as the spherical cavity radius α appearing in the coupling factor g has no obvious physical interpretation (vide supra). Treating the radius as a free parameter makes it fairly simple to obtain reasonable results with either value of λ. What is *unambiguous*, however, is that optimization of solute geometries is much more straightforward when one takes $\lambda = 0.5$. The alternative, $\lambda = 1$, requires solution of a coupled-perturbed Hartree-Fock equation each time analytic gradients are required, since the total energy of the system includes the term $0.5 g \mu^2$ non-self consistently, i.e., the term depends on the dipole moment, and therefore the density matrix, but it is not accounted for in the Fock operator since it is added *ex post facto*. This latter consideration is significant enough to suggest that, from a purely practical standpoint, it would be better to adopt

$\lambda = 0.5$, regardless of the competing arguments over which is to be preferred based on fundamental principles.

In any case, the great simplicity of the Kirkwood-Onsager approach has prompted its incorporation into a number of quantum mechanical electronic structure programs, both at the ab initio and semiempirical levels [108,111,128-131]. Considerable caution should be exercised in its application, however. For instance, for charged solutes one should include an ionic Born term derived from equation 7, which will be a *considerably* larger contributor to the solvation free energy than the corresponding Kirkwood-Onsager term. However, for at least one commonly used electronic structure program this term is *not* included [130], perhaps because it does not require a self-consistent treatment in the Hartree-Fock equations and is thus easily added on post facto. Another consideration for ionic systems is that only the leading molecular multipole moment is independent of the origin of the chosen molecular coordinate system. Thus, Born-Kirkwood-Onsager calculations of such systems require a choice of where to evaluate the molecular dipole moment, e.g., at the molecular center of mass, center of charge, center of the encompassing sphere, etc. The final result will be correct only if a consistent choice of coordinates is used throughout the derivation and application. The two most critical considerations, however, are that (1) application of the model is justified only when higher-order multipole moments are negligible, and (2) application of the model is justified only for nearly spherical solutes. These issues are addressed in the next two sections.

2.1.4. Truncated single-center multipolar expansions

As discussed above, the Born-Kirkwood-Onsager approach includes only the solute's monopole and dipole interactions with the continuum. That is, the full classical multipolar expansion of the total solute charge distribution is truncated at the dipole term. Although these terms dominate at very large distances, one may imagine evaluating the electrostatic potential and the polarization contributions to the free energy of solvation while approaching more and more closely to the solute. Eventually, the contributions from higher-order moments cease to be negligible. The importance of such higher order moments is most obvious for neutral molecules whose dipole moments vanish as a result of symmetry. The Born-Kirkwood-Onsager model would require the electrostatic portion of the free energy of solvation for these molecules to be identically zero.

Generalization of the Born-Kirkwood-Onsager approach may be accomplished by solution of the Poisson equation for a single-center multipolar expansion to arbitrarily high order. This approach yields [60,78,98,132-135]

$$\Delta G_S^o = -\frac{1}{2} \sum_{l=0}^{\infty} \sum_{m=-l}^{+l} \sum_{l'=0}^{\infty} \sum_{m'=-l'}^{+l'} M_l^m f_{ll'}^{mm'} M_{l'}^{m'} \tag{11}$$

where each component $M_{l'}^{m'}$ of every multipole moment interacts with all of the reaction field multipole moments induced by the solute multipoles via a set of coupling factors $f_{ll'}^{mm'}$, called the reaction field factors. The assumption that the cavity is a sphere leads to the coupling factor being non-zero only for $l = l'$ and being independent of m and m'. The Born-Kirkwood-Onsager model is then seen as a special case involving only the net charge ($l = 0$)

and dipole moment ($l = 1$) terms. The assumption of a spherical or ellipsoidal solute cavity actually permits analytical determination of *all* of the reaction field factors in equation 11 [60,71,78,108]. When implemented into the SCF equations in a manner analogous to that described in the previous section with λ taken as 0.5, this simplification additionally permits efficient optimization of solvated geometries [71,132,133,135-137]. Even for more general cavities, e.g., cavities which more resemble a molecular van der Waals surface, it is possible to determine the reaction field factors numerically, since they appear in an overdetermined system of linear equations [133,134]. Both of these approaches have seen increasing use [134,138,139]. Although Tapia has discussed the competing derivations which yield $\lambda = 1$ (solvent as isothermal bath) and $\lambda = 0.5$ (work of solvent polarization included self-consistently) in the equations analogous to equations 9 and 10 for this more general multipole approach [66], it appears that implementations appearing in the recent literature have exclusively used $\lambda = 0.5$, probably to take advantage of the simplified geometry optimizations for this case [71,132-139].

A point of obvious interest is how fast the electrostatic portion of the solvation free energy converges with respect to multipole order. Interestingly, this convergence can be quite slow, even for fairly simple molecules. A specific example is Z-3-aminoacrylonitrile, which was studied by Pappalardo and co-workers [140]. Immersion of this solute in a continuum with a dielectric constant of 38.8 yields a total electrostatic polarization free energy of −13.2 kcal/mol. Decomposition of this energy finds 66% contained in the dipole term, 22% in the quadrupole, and the remaining 12% in the terms up through 2^6-pole, which was the highest multipole considered. Moreover, this slow convergence was still more pronounced for the transition state for rotation about the carbon-carbon double bond, where the polarization free energy for this nearly zwitterionic structure is −44.8 kcal/mol, now consisting of 64% dipole, 18% quadrupole, and 19% in the higher order terms. The convergence of the multipole expansion is also very dependent on the shape of the employed cavity [134,138].

Although typically the multipolar expansion of the electronic structure is performed at only a single point, e.g., the center of mass of the molecule, this is not a requirement. Instead, an arbitrary number of *distributed* multipoles may be placed at multiple points, e.g., the atomic coordinates or the atomic coordinates and bond midpoints [72,134,141-144]. Numerical fitting of the multipoles and reaction field factors proceeds equivalently. One of the simplest approaches is to use only atomic monopoles (i.e., partial charges); this may be made equivalent to a single-center expansion up through $l = N - 1$ where N is the number of atoms. Indeed, this is quite similar to the generalized Born approximate solution to Poisson's equation which is discussed in greater detail in Section 2.3.

As expected, this distributed approach is much more rapidly convergent in terms of the multipole order required at each center. In the modeling of formamide, for instance, a one-center expansion in a generalized cavity still has 1% fluctuations by the 2^6-pole term [134]. The distributed expansion at the atomic positions in the same cavity, on the other hand, has only a 1% contribution from the quadrupole, and is effectively converged after this point [134]. Moreover, it is generally more efficient computationally to describe the molecular electronic structure as a set of N distributed monopoles rather than a single multipolar expansion of order N. Of course, the precise method for determining the magnitude of the distributed monopoles (partial charges) remains controversial insofar as atomic partial charges are not physical observables; a large number of models and algorithms is available [144-166]. Although including higher multipoles at *every* center clearly increases the flexibility of the approach, the cost is considerable in terms of computational effort and so far has limited its application to fairly simple systems, e.g., the NH$_3$-HCl complex [167].

2.1.5. Generalized reaction fields from surface charge densities

If the multipolar expansion of the molecular charge density implicit in equation 11 were carried out to infinite order, the resulting equation would be a complete solution to the volume integral expression of Poisson's equation discussed so far. An alternative approach is application of Green's theorem to convert the volume integral of equation 6 to an integral over the molecular cavity surface S. In particular, the effect of the reaction field may be modeled by a continuous polarization charge density spread over that surface, where this virtual charge density, $\sigma(r)$, is in Gaussian units

$$\sigma(r) = \frac{1-\varepsilon}{4\pi\varepsilon} \frac{\partial}{\partial n} \left[\phi_\rho(r) + \phi_\sigma(r) \right]_{S_-} \tag{12}$$

with $\phi_\rho(\mathbf{r})$ being the electrostatic potential due to the solute charge distribution, and $\phi_\sigma(\mathbf{r})$ being the potential due to the virtual charges [65,74,168-171]. The derivative is with respect to an outward surface normal evaluated on the solute side (indicated by the S_- subscript) of the dielectric interface. The potential created by the surface virtual charge density is

$$\phi_\sigma(\mathbf{r}) = \int_S \frac{\sigma(\mathbf{r}')}{|\mathbf{r}-\mathbf{r}'|} d^2\mathbf{r}', \tag{13}$$

and it must be added to the potential due to the solute charge distribution to obtain the total electrostatic potential at $\mathbf{r}$. The electrostatic portion of the free energy of solvation is then defined as

$$\Delta G_S^o = <\psi | H_0 + \frac{1}{2}\phi_\sigma | \psi> - G_g^o. \tag{14}$$

In practice, the surface charge density is approximated by a discrete set of point charges which are distributed as uniformly as possible, and the appropriate integrals are then replaced by summations. This model is often referred to as the Polarized Continuum Model or PCM, and it saw most of its early development by Tomasi and co-workers [65,74,169-171]. More recently, it has been implemented in a variety of ab initio and semiempirical quantum chemistry programs [128,172-179]. Of all of the models discussed so far, PCM has seen the most effort spent upon the development of prescriptions for choosing the optimal cavity surface, to include as a function of basis set at the ab initio level [180].

It is probably worth emphasizing here, in case it isn't obvious, that the multipole methods of Section 2.1.4 and the surface-charge-density methods discussed in this section are physically identical, and they will yield the same result if (1) the same molecular surface is used in both methods (e.g., the same solute atomic radii if the surface is constructed from overlapping spheres), (2) the expansion of equation 11 is well-converged, (3) the numerical representation of the surface charge density is well converged in the PCM, and (4) the factor λ used in the development of the SCF equations is identical. With respect to the last point, equation 14 explicitly takes $\lambda = 0.5$, as is most common in the multipolar reaction field schemes discussed previously. However, again a choice of $\lambda = 1$ may be made (i.e., assuming the solvent to be an isothermal bath) in which case, in exact analogy to the discussion in Section 2.1.3, the free energy of solvation must include the addition of a work of solvent polarization term *ex post facto*. While this latter approach has been pursued by Tomasi and co-workers [65,74,169-171], all other implementations of the PCM approach have used the formalism of equation 14 [128,172-179,181]. To our knowledge, no comparison between the two implementations has appeared.

The broadly general nature of the PCM technique makes it uniquely attractive, especially as it is somewhat more straightforward to implement than the analogous truncated multipolar expansion method taken to arbitrarily high order. Nevertheless, it remains extremely demanding in computational resources, primarily because of the time required to generate the surface virtual charges. Klamt and Schüürmann [178] have presented a particularly efficient algorithm [182] for accomplishing this at the Neglect of Diatomic Differential Overlap (NDDO) [183] semiempirical level of theory.

While the scope of this chapter is intended only to cover quantum mechanical continuum solvation models, we mention in passing that classical approaches to solving the Poisson equation also exist which are similar in spirit to the PCM model but involve representation of the solute charge density as a discrete, grid-mapped set of charges [184-198]. These models fail to allow for self-consistent relaxation of the molecular electronic structure, although obviously they are considerably faster than fully quantal models as a result. To make up for the lack of self-consistency, the dielectric constant is sometimes set equal to a value in the range of 2 to 4 in the non-self-consistent Poisson approaches, whereas it is properly set equal to 1 in the solute when solute polarization is included explicitly.

2.2. Non-electrostatic components of solvation free energy

So far in this chapter, we have concerned ourselves only with the bulk electric polarization of the *volume* surrounding the solute. However, there are other effects that are more specifically associated with the *surface* layer of solvent, i.e., the first solvation shell. One example is the free energy required to create a solute-sized vacuum within the solvent. This cavitation energy, which is approximately proportional to the surface area of the created cavity, is quite dependent on the particular solvent. Additional components at solvent-solute interfaces include the attractive dispersion forces between the solute and the nearby solvent molecules and local structural changes in the surrounding solvent as a result of the insertion of the solute. A key example of the latter effect is found in water, where the solute may induce solute-solvent hydrogen bonding and/or cause especially significant changes in solvent-solvent hydrogen bonding in the first solvation shells [199,200]. Although continuum solvation models arguably include the electrostatic component of hydrogen bonding to some degree in the dielectric polarization term, short-range components cannot be fully modeled by a uniform dielectric constant. Considering the other extreme, for solutes that do *not* hydrogen bond to an acceptor/donor solvent, the solvent structural change may be unfavorable due to the loss of solvent orientational entropy, i.e.,

23

the hydrophobic effect. We will refer to the sum of these effects as the CDS term, for Cavitation, Dispersion, and solvent-Structural rearrangement.

The difference between the electrostatic effect calculated using the bulk dielectric constant and that calculated taking account of local structural factors may be viewed as arising from an inhomogeneous dielectric constant. In the absence of a microscopic model for the solvent, one approach to incorporating the CDS term would be to allow the dielectric constant of the surrounding medium to take on different values at different locations. However, no clear prescription for accomplishing this in a physically meaningful way is available. An attractive alternative is to assume that the approximate proportionality of the cavitation term to the cavity surface area extends to the remaining terms as well. This seems intuitively reasonable for dispersion, which operates over so short a range that one expects it to be proportional to the number of molecules in the first solvation shell, which is clearly dependent on the solvent accessible surface area of the solute [201]. There is a key distinction to be made between cavitation and dispersion, however, and that is to note that cavitation is independent of the solute, while dispersion is expected to depend on the local polarizability of the solute in any given region. Thus, one might model the C and D terms by assigning a surface tension σ_i to each atom i in the solute and calculating:

$$\Delta G^o_{CD} = \sum_i \sigma_i^{CD} A_i \tag{15}$$

where A_i is the solvent accessible surface area of atom i. The surface tension will contain a constant component which is independent of i, while the remainder will be associated with dispersion and will be dependent on the atomic polarizability of the individual atom. The solvent accessible surface area is most readily calculated following the definition of Lee and Richards [101,202], and it is calculated as the surface mapped out by the center of a solvent-sized ball rolling over the molecular van der Waals surface.

Finally, the sum of the solvent structural rearrangement free energy and the free energy due to specific electrostatic and hydrogen bonding effects in the first solvation shell (including the non-homogeneity of the dielectric constant) can also be assumed to be proportional to a cavity surface area, although it need not be the case that the CS cavity will be defined in the same way as the CD cavity. We include the C term in both cases because it is not rigorously clear how it might be separated from either dispersion or structural rearrangement when the CDS term is simply known in its entirety. Of course, if the two cavities are identical, then equation 15 may be used for the entire term ΔG^o_{CDS} with a single set of atom-specific surface tensions which would now be called σ_i^{CDS}, i.e., [203-209]

$$\Delta G^o_{CDS} = \sum_i \sigma_i^{CDS} A_i . \tag{16}$$

It is worth noting that in cases where the CDS term dominates the entire free energy of solvation, it may be possible to calculate the full ΔG^o_S using the formalism of equation 16.

Such an approach works reasonably well for estimating the solvation free energies of hydrocarbons in water, where the ENP terms are very small [203-209]. Scheraga and co-workers had moderate success applying this idea to the twenty biologically important amino acids, although this is admittedly a crude approximation for these much more polar molecules [207].

Very few quantum mechanical solvation models have attempted to incorporate simultaneously both the ENP and some or all of the CDS components of solvation. One common approach is to model cavitation free energies using the scaled-particle theory of Pierotti [210], and some attempts to model dispersion have also appeared [58,67,135,177,211-220]. Rigorous quantum mechanical calculation of this term can be quite costly, and one must question whether the inherent accuracy of a continuum model, with its strong dependence on solute radii, warrants such an approach, as compared to a simplified approximation such as that offered by equation 15 or 16

One particularly interesting attempt to address both the ENP and hydrophobic CDS components of solvation has been described by Still and co-workers within the framework of molecular mechanics [221]. In their model the ENP term is arrived at non-self-consistently by using a generalized Born formalism [10,11,51-54,61-63,66,69,70,221,222] to approximately solve the Poisson equation, and the hydrophobic portion of the CDS term is calculated using an analog of equation 16 where all atoms have the same surface tension, which is a good start but is not flexible enough to model all CDS effects quantitatively. We will devote Section 2.3 to describing our own extension of these ideas to a quantum chemical implementation at the NDDO level, in which we attempt to include all of the important solvation terms in σ_i^{CDS} while simultaneously permitting self-consistent relaxation of molecular electronic structure as a function of solvation for one particular solvent, water.

Before closing this section, it is useful to comment in more detail on the precise interpretation of A_i in equations 15 and 16. Lee and Richards [101,202] and Pascual Ahuir, Silla, and Tuñon [104,105] have carefully distinguished three definitions of the surface area of a solute or its associated cavity. The three definitions will be given here for the case where the solute is modeled as a set of overlapping "atomic" spheres, one representing every atom i (or a group i consisting of a nonhydrogenic atom and its attached hydrogens), with radii R_i, and the solvent molecules are modeled as spheres of radius R_S. The van der Waals surface, also called WSURF [105], is composed of those surfaces of the atomic spheres that are not encompassed by any other sphere. The solvent-accessible surface [202], also called the cavity surface [203] or ASURF [105], is the surface traced out by the center of a solvent sphere rolling on the van der Waals surface. The third surface, originally called the Molecular Surface [101], but more clearly labeled the solvent-excluded surface [105] and also called ESURF [105], is the surface traced out by the points of contact of a solvent sphere rolling on the van der Waals surface in those regions where the solvent sphere touches only one solute sphere; in regions where the solvent sphere simultaneously contacts two or more solute spheres, it is defined to be the surface traced by the inward surface of the solvent sphere. Note that WSURF is independent of R_S, ASURF depends strongly on R_S, and ESURF has an intermediate dependence. We believe that the ASURF definition is most physical for equations 15 and 16. This is because the ASURF surface passes through the center of the first solvation shell and thus, when the solvent is modeled by a continuum, it is proportional to the number of solvent molecules in the first solvation shell. Since dispersion, solvent-structural free-energy changes, and the effect of nonhomogeneous dielectric constant in the first solvation shell are all approximately proportional to the number of solvent molecules in that shell, this definition is ideal for equation 16.

2.3. The SMx models and absolute free energies of solvation

In this section we review our own SMx models, based on the generalized Born model [10,11,51-54,61-63,66,69,70,221,222] for electrostatic effects and a generalization of equation 16 for first-solvation-shell effects. Models AM1-SM1 [223], AM1-SM1a [223], AM1-SM2 [224], and PM3-SM3 [225] are named to indicate their extension of an underlying gas-phase semiempirical NDDO Hamiltonian, either Austin Model 1 (AM1) [226] or Parameterized Model 3 (PM3) [227,228], with a particular solvation model (SM) where the numbering of the solvation model is primarily chronological in nature. Two new models, AM1-CM1A-SM4A and PM3-CM1P-SM4P, are under development [229] and will be published soon. We have extensively reviewed the theory behind these models elsewhere [75,230], and we present here a less exhaustive description for purposes of completeness.

In the SMx models, the standard-state free energy of solvation, ΔG_S^0, is calculated from

$$\Delta G_S^0 = G_S^0 - E_{EN}(g), \tag{17}$$

where $E_{EN}(g)$ is the gas-phase electronic kinetic and electronic and nuclear coulombic energy, and G_S^0 is that part of the aqueous free energy given by

$$G_S^0 = E_{EN}(aq) + G_P(aq) + G_{CDS}^0(aq), \tag{18}$$

where $E_{EN}(aq)$ is equivalent to $E_{EN}(g)$ except now calculated in the presence of solvent, i.e., including distortion energy, $G_P(aq)$ is the electric polarization free energy in solution, and G_{CDS}^0 is the cavitation-dispersion-structural free energy summarized in the previous section. Other contributions to the total free energy of the solute in solution (e.g., vibrational) are assumed (in work carried out so far) to make an identical contribution to the gas-phase and aqueous free energy, and thus not to affect ΔG_S^0. Clearly they would have to be added to G_S^0 in order to obtain the total free energy in solution.

The polarization free energy is calculated from the generalized Born approximation [10,11,51-54,61-63,66,69,70,221,222] to the solution of the Poisson equation,

$$G_P = \frac{1}{2}\left(1 - \frac{1}{\varepsilon}\right) \sum_{k,k'} q_k q_{k'} \gamma_{kk'} \tag{19}$$

where q_k is the atomic partial charge on atom k, where k and k' run over all atoms ($k = 1,2,...,k_{max}$), and where $\gamma_{kk'}$ is a coulomb integral. We have adopted the form for the coulomb integrals proposed by Still *et al.* [221]

$$\gamma_{kk'} = \{r^2_{kk'} + \alpha_k \alpha_{k'} C_{kk'}(r_{kk'})\}^{-1/2}, \tag{20}$$

where α_k is the Born radius of atom k, $r_{kk'}$ is the interatomic distance between atoms k and k', and $C_{kk'}$ is in general given by

$$C_{kk'} = \exp(-r^2_{kk'} / d^{(0)} \alpha_k \alpha_{k'}), \tag{21}$$

with $d^{(0)}$ being an empirically optimized constant equal to 4. Equation 20 is designed so that G_P behaves properly in three important limits for a pair of atoms k and k': infinite separation of atoms k and k', where it yields a sum of Born monatomic ion energies, coalescence of identical atoms, where it again yields the monatomic Born formula, and close approach of dissimilar atoms, where it approaches the Kirkwood-Onsager result.

For the monatomic case ($k = k'$ and $k_{max} = 1$), α_k is set equal to

$$\rho_k = \rho_k^{(0)} + \rho_k^{(1)} \left[-\frac{1}{\pi} \arctan \frac{q_k + q_k^{(0)}}{q_k^{(1)}} + \frac{1}{2} \right] \tag{22}$$

where $\rho_k^{(0)}$ and $\rho_k^{(1)}$ are empirically optimized parameters corresponding to positive and negative ionic radii, $q_k^{(0)}$ is the charge about which the switch is centered, q_k is the calculated partial charge, and $q_k^{(1)}$ has been fixed at 0.1 for all atoms. In the multicenter case, α_k is determined numerically so that the G_P which would be derived for the monatomic case (i.e., as if using equation 7) is equal to the G_P determined via a numerical integration [221,223,230,231] that corresponds to dividing the solute from solvent with a WSURF-type surface based on a set of spheres centered at nuclear location $\mathbf{x}_{k'}$ with radii $\rho_{k'}$, where $k' = 1, 2, ..., N$, and N is the number of atoms in the solute. In particular,

$$\frac{1}{\alpha_k} = \int_{\rho_k}^{\infty} \frac{dr}{r^2} \, a(r, \{\mathbf{x}_{k'}, \rho_{k'}\}_{k' \neq k}) \tag{23}$$

where $a(r, \{\mathbf{x}_{k'}, \rho_{k'}\}_{k' \neq k})$ is the fraction of the surface area of a sphere of radius r at the origin that is not contained in any of the $N - 1$ spheres specified by the set $\{\mathbf{x}_{k'}, \rho_{k'}\}_{k' \neq k}$ when the system is translated so that atom k is the sphere at the origin. The integration is performed by precisely specified (and hence reproducible) quadrature rules in the SM1, SM1a, SM2, and SM3 models, and it is converged in SM4A and SM4P. The integral in equation 23 accounts for screening of the solute from the solvent by other parts of the solute; this solute screening effect is critical to solvation modeling [17,75,221,223,232] but is often neglected or—when included—often unappreciated.

At the valence-electron NDDO level, the electrostatic terms are calculated using the density matrix $\mathbf{P}$ of the aqueous-phase SCF calculation as

$$G_{ENP} = \frac{1}{2}\sum_{\mu\nu} P_{\mu\nu}\left(H_{\mu\nu} + F_{\mu\nu}\right) + \frac{1}{2}\sum_{k,k'\neq k}\frac{Z_k Z_{k'}}{r_{kk'}} - \frac{1}{2}\left(1 - \frac{1}{\varepsilon}\right)\sum_{k,k'}Z_k q_{k'}\gamma_{kk'} \tag{24}$$

where $\mathbf{H}$ and $\mathbf{F}$ are, respectively, the one-electron and Fock matrices [233-235], μ and ν run over basis orbitals, Z_k is the effective (valence) nuclear charge of atom k. We point out here that both equation 5 of reference 225 and equation 19 of reference 230 (both being analogs of equation 24 above) are incorrect: the former is missing the last two terms of equation 24, and the latter is missing the final term and has "=" in the summation index instead of "$\neq$".

The critical point in implementing the generalized Born model at this semiempirical level is that the Fock matrix is related to the energy functional of equation 18 as its partial derivative with respect to the density matrix. The partial charges which appear in the definition of G_P (equation 19) are themselves derived from the density matrix. In the SM1, SM1a, SM2, and SM3 models, this is accomplished by a simple Mulliken population analysis [147] under the assumptions of zero overlap

$$q_k = Z_k - \sum_{\mu \in k} P_{\mu\mu}. \tag{25}$$

A more complicated dependence of atomic partial charges on the density matrix elements is used in SM4A and SM4P, which we call Charge Model 1A and 1P (CM1A or CM1P) [166]. In any case, using either formalism to define q_k, the partial differentiation of equation 18 is straightforward, and it delivers a solution-phase Fock matrix which *self-consistently* includes polarization effects. Thus, as for the other quantum models, self-consistency is required in these calculations: in particular, the Fock matrix, the density matrix, and the interacting solvent field are made self-consistent.

The last term in equation 18, which is required to accurately calculate absolute free energies of solvation, is calculated by a version of equation 16 that is modified in a way that depends on which of the SMx models is involved. For example, in SM2 and SM3, we use

$$G_{CDS}^0 = \sum_{k'}\left\{\sigma_{k'}^{(0)} + \sigma_{k'}^{(1)}\left[f(B_{k'H}) + g(B_{k'H})\right]\right\}A_{k'}(\beta_{k'},\{\beta_k\}) \tag{26}$$

where the $\sigma_{k'}$ are atomic surface tension parameters, and $A_{k'}(\beta_{k'},\{\beta_k\})$ is the solvent-accessible surface area for *non-hydrogen* atoms k'. The latter is defined as that portion of

ASURF associated with atom k' when the set of radii for all the atoms is $\{\beta_k\}$. We set $R_S = 1.4$ Å, and we set $\beta_k = 0$ for hydrogen atoms.

In the remainder of equation 26, B_{kH} is the sum of the bond orders, defined more specifically as covalent bond indices [236], of atom k to all hydrogen atoms in the solute, i.e.,

$$B_{kH} = \sum_{\mu \in k, \nu \in H} P_{\mu\nu}^2 \tag{27}$$

where μ runs over the atomic orbitals of atom k, and ν runs over all hydrogen orbitals. The hydrogen atom is specifically defined to have zero solvent-accessible surface area, and moreover not to block the accessible area of the atom(s) to which it is attached. Finally

$$f(B_{kH}) = \tan^{-1}(\sqrt{3}\, B_{kH}) \tag{28}$$

and

$$g(B_{kH}) = \begin{cases} a_k \exp\left\{-b_k \big/ 1 - \left[(B_{kH} - c_k)/w_k\right]^2\right\}, & |B_{kH} - c_k| < w_k \\ 0, & \text{otherwise.} \end{cases} \tag{29}$$

This more complicated expression of G_{CDS}^o was found to be required in water because hydrogen atoms in that solvent interact with the first solvation shell differently depending on the heavy atom to which they are attached. For example, an alkane hydrogen is hydrophobic but an amine hydrogen is hydrophilic. To maintain flexibility and accuracy in the model, it was convenient to make the heavy atom surface tension be a function of the number of attached hydrogen atoms. We emphasize that the zero-radius treatment of hydrogen occurs only for the CDS term in SM2 and SM3, not for the other models, and never for the ENP term.

The various parameters (van der Waals radii, surface tensions, etc.) were fit to reproduce experimental aqueous solvation data. In practice, one begins the parameterization process by focusing on ions, for which most of the solvation free energy is found in the ENP term. This being the case, the ENP parameters are effectively "decoupled" from the CDS parameters, and indeed in the first pass through the ions, all surface tensions are set to zero. Once cavity radii have been defined for the ions, they are then used in neutral molecule calculations, with the difference between the calculated and experimental free energies being assigned to the CDS term. Surface tensions and radii are then fit using equation 26 in a multilinear regression to minimize residual error compared to experiment. These CDS terms are then used in another pass through the set of ions, and the ENP parameters are allowed to relax accordingly to maximize agreement with experiment. This iterative procedure is continued until the parameters converge, usually by the second or third pass through the data. Throughout this process the parameters are monitored so as to ensure the location of a physically meaningful local minimum in parameter space.

As a consequence of this type of parameterization, the CDS parameters do more than account for cavitation, dispersion, and structural rearrangement. In particular, they correct for the impact on the ENP terms of errors in the NDDO wave function, in the representation of a continuous charge distribution by a set of partial charges on the atoms, and in the generalized Born approximation to solving the Poisson equation. In addition in models SM1, SM1a, SM2, and SM3, they also account for systematic numerical errors in the quadrature of equation 23.

We have reviewed the performance of the AM1-SM2, AM1-SM1a, and PM3-SM3 models elsewhere [75,230] and the SM4A and SM4P models will be described in a forthcoming paper [229]. We note that the parameterizations ultimately permit the prediction of absolute free energies of solvation for a large set of neutral molecules (> 150) with a mean unsigned error of less than 1 kcal/mol (where the data span a range from roughly +5 to –10 kcal/mol) and for about 30 ions with a mean unsigned error of about 3-4 kcal/mol. The latter number is well within experimental error since the measurements require the completion of thermodynamic cycles which include gas-phase deprotonation enthalpies. The models are all available in the semiempirical package AMSOL [237], and they have also been implemented in commercial software packages. All of the calculations discussed below were performed with various versions of AMSOL. The remainder of this contribution will focus on the application of these models to systems of biological interest.

2.4. Nonequilibrium solvation

The above considerations all apply to the solvent being equilibrated to the solute and vice versa. This is a reasonable assumption for free energies of solvation of solutes executing small-amplitude vibrations around a single equilibrium structure. For dynamics problems, though, one must sometimes consider non-equilibrium solvation. The theoretical treatment of nonequilibrium solvation involves a careful consideration of time scales [238], and it is much less well understood than equilibrium solvation. Nonequilibrium solvation effects can be included in dynamics calculations by treating solvent degrees of freedom explicitly, or they can be incorporated as corrections to transition state theory [239,240] The reader is directed elsewhere for recent literature on continuum models of nonequilibrium solvation [241-245].

Another area where nonequilibrium solvation is important is electronic spectroscopy. To a first approximation, excitation from an electronic ground state into an excited state occurs much more rapidly than reorganization of the structure of the surrounding solvent shells. As such, the solvent reaction field acting upon the excited state at the instant of excitation is just that field that was derived from the ground state. Over a longer time period, solvent relaxes to a new equilibrium reaction field, and if radiative emission occurs to create the ground state, again the instantaneous reaction field experienced by the ground state will be that for the excited state. This differential solvation can lead to a change in Franck-Condon factors and significantly shifted absorption maxima in solution compared to the gas phase. There is great interest in the theoretical prediction of the effects of solvation on spectroscopy, and again we will simply refer the reader to recent discussions of these issues [126,177,179,246-252].

3. APPLICATIONS TO BIOLOGICALLY INTERESTING SYSTEMS

A major motivation for the development of aqueous solvation models is the modeling of biological and pharmacological structure and reactivity. For instance, it is clear that the development of the tertiary structure of non-membrane-associated proteins is driven in part by the energetically favorable tendency for them to bury hydrophobic amino acid residues in the interior of the protein while leaving hydrophilic residues exposed to the surrounding aqueous environment [104,209,253-257]. Moreover, the energetics of substrate-enzyme or substrate-receptor binding may be viewed as a differential solvation effect between the active site and the bulk solvent [23,258], i.e., electrostatic and binding stabilization of the substrate at the active site is in competition with solvation of the free substrate in aqueous solution. The references in this paragraph provide only a small sampling of the wide activity in this field.

Sections 3.1 and 3.2 are dedicated to specific examples of the effects of aqueous solvation on systems of biological import. Section 3.1 illustrates for the nucleic acids how the electronic structure changes as a result of solvent-induced polarization, and section 3.2 provides examples of how aqueous solvation influences the equilibrium population of conformers for flexible biomolecules.

3.1. Aqueous solvation effects on electronic structure — the nucleic acids

Quantum mechanical studies of the nucleic acids [113,129,259-262] and classical mechanical studies of their molecular dynamics, both in the gas phase and when surrounded by hundreds or thousands of water molecules [26,263-269], have done much to advance our understanding of these important molecules. As discussed in section 2, classical simulations involve molecular mechanics force fields which are parameterized by electronic structure calculations and/or semiempirical fitting to experimental data. The charges built into such force fields are not subject to solvent-induced polarization. That is, empirically optimized partial atomic charges on the individual atoms are taken to be constant and are not self-consistently determined with respect to either molecular conformation or environment. However, quantum mechanical calculations on solutes using the self-consistent models detailed in section 2 nearly always indicate electronic polarization to be a non-negligible component of the overall solvation free energy. In very polarizable systems, like heterocycles, the effect can in fact be quite large [259,270,271].

Such a finding raises interesting questions about the requirements for realistic force fields for molecular mechanics simulations of polarizable solutes in aqueous media. Potential functions optimized for liquid-phase simulations [267,269,272] must necessarily have larger partial charges and dipole moments than would be appropriate for the gas-phase molecules. One way to accomplish this is to parameterize the force field based on calculations on hydrogen bonded or ion-water complexes. A popular alternative has been to base the partial charges on ab initio Hartree-Fock wave functions derived using the 6-31G* basis set [273], since calculations with this basis set are known to overestimate molecular dipole moments by about 10% in many cases [125]—this appears to be roughly the contribution of polarization to the overall electrostatic component of the free energy of solvation. It is clear, however, that such an approach will be of limited value in more polarizable systems or in systems where it is difficult to account for the bulk of the solvent-induced polarization by performing electronic structure calculations for the solute when it is complexed to only one or two water molecules. The nucleic acids appear to serve as a particularly important example of this phenomenon. We have reported AM1-SM2 and PM3-SM3 calculations for these solutes [259], and the two methods are in reasonable agreement for the absolute free energies of solvation. Both models predict that solvent induced polarization of the solute accounts for 23–34% of ΔG_S^o. Since experimental free energies of solvation remain to be measured for the nucleic acid bases, it is most instructive to compare diverse theoretical models in order to assess the relative importance of polarization to the aqueous solvation of these molecules.

One pertinent study is that of Gao and Xia [274], who estimated polarization effects on solute-solvent interaction energies by a combined quantum/classical mechanical approach which included 260 molecular mechanics water molecules in the AM1 Hamiltonian with the nucleic acid bases constrained to their gas-phase geometries. Although calculations of the relative free energies of hydration are expensive by this method and were therefore not carried out, canonical ensemble averages are less expensive, and average values of the molecular dipole moments were computed for each solute. Katritzky and Karelson [113] have also studied the nucleic acid bases, employing the Kirkwood-Onsager approximation described in section 2.1.3 as implemented into the AM1 Hamiltonian. Since both of these studies, as well as our own AM1-SM2 calculations, start from the AM1 gas-phase Hamiltonian, it is particularly interesting to compare them. Table 1 lists the dipole

Table 1.
Dipole moments of the nucleic acid bases in aqueous solution compared to the gas phase (Debyes)

		In solution			
		At gas-phase geometries			At optimized solution geometries
	In the gas phase	Katritzky and Karelson	Gao and Xia	AM1-SM2	AM1-SM2
cytosine	6.3	7.4	9.9	9.0	10.0
thymine	4.2	5.2	5.9	6.2	6.6
uracil	4.3	5.0	5.9	6.4	6.9
adenine	2.2	2.9	3.8	3.1	3.1
guanine	5.9	6.6	8.5	8.5	9.3

Table 2.
Absolute free energies of solvation for the methylated nucleic acid bases (kcal/mol)

	AMBER [275]	AMBER [276]	OPLS [277]	OPLS [278]	6-31G* [279]	AM1-SM2 frozen	AM1-SM2 relaxed
1-Methylthymine	−7.7	−9.4	−10.4	−9.9	−8.6	−10.8	−13.3
9-Methyladenine	−12.8	−14.9	−10.8	−9.4	−6.5	−16.7	−20.9
1-Methylcytosine	−12.9		−16.8	−17.9	−13.0	−14.4	−18.7
9-Methylguanine	−19.8		−19.7	−19.5	−16.0	−18.2	−24.3

moments found for the nucleic acid bases in the gas phase, and with each of the solvation models, for the gas-phase geometries [i.e., only the electronic structure (not the geometry) has been permitted to relax in the presence of solvent]. We have also optimized the solutes at the AM1-SM2 level in order to study the additional effects of geometric relaxation.

Three points are especially worthy of note. First, the increase in the molecular dipole moment is smallest for the Kirkwood-Onsager model, as expected given the severely limiting assumptions of that model (as discussed in section 2.1.3). Second, the results of Gao and Xia are remarkably consistent with our own, especially considering the different representations of the solvent either as explicit and classical or as a continuum dielectric. Finally, the additional effects of geometric relaxation are not trivial for several of these solutes, illustrating the importance of reoptimizing geometries in solution.

To illustrate the importance of the polarization contribution to the absolute free energy of solvation, Table 2 presents the calculated free energies of solvation for the methylated nucleic acid bases of DNA as calculated by several different methods [259,275-279]. Elcock and Richards [278] calculated only relative free energies of solvation, and for comparison purposes their results have been arbitrarily normalized to sum to the same value as those of Mohan et al. [277] who used the same charge model. The results of Elcock and Richards and of Bash et al. [275] are from classical simulations employing the Optimized Parameters for Liquid Simulations (OPLS) [267,272] and AMBER [269] charge sets, respectively, the results of Mohan et al. [277] are from a numerical solution of the classical Poisson equation (i.e., non-self-consistent charges are employed) using the OPLS charge set, and the results of Young and Hillier [279] are based on a multipole expansion through $l=7$, with the 6-31G* basis set. The results of Ferguson et al. [276] are based on repeating the calculations of Bash et al. [275] with new algorithms. AM1-SM2 results are provided as calculated both for the frozen gas-phase molecules (i.e., no relaxation of either electronic structure or molecular geometries) and for self-consistently optimized solutes. Although it will require experimental data to establish which model is most consistently accurate for these molecules, it is clear that consideration of polarization dramatically increases the

AM1-SM2 calculated absolute free energies of solvation, which are otherwise in reasonable agreement with all of the other non-self-consistent models, with the exception of 9-methyladenine. The wide variation between even the different classical models illustrates the need for continued study of the solvation free energies of nucleic-acid bases.

3.2. Aqueous solvation effects on molecular conformation

The nucleic acids are examples of systems where solvation induces large changes in electronic structure. In addition to the change in *electronic* structure, there is an additional small change in molecular geometry, as evidenced by comparing the last two columns of Table 1. Further inspection, however, reveals that really there have been only very small changes in bond lengths and bond angles, the net result of which has been to permit additional electronic relaxation with minimal geometric distortion. Put another way, bond stretching potentials and bond angle bending potentials are very steep for the nucleic acids compared to the additional solvation free energy which may be gained by distortion from the gas-phase equilibrium point for any given degree of freedom. In order to observe a significant effect of solvation on molecular geometry, one of two situations must hold: either the gain in solvation free energy to be had by geometric distortion must be quite large, so that it competes with such steep potentials as those mentioned above or alternatively the geometric potential itself must be rather shallow or involve more than one minimum, in which case solvation may be instrumental in determining the shape of the potential. A good example of the latter situation is a torsional coordinate in a flexible molecule. Torsions often exhibit multiple minima and are characterized by fairly low-energy barriers separating the minima. In section 3.2.1, we discuss this situation in more detail for the neurotransmitter dopamine. In section 3.2.2, we expand the level of complexity to consider a number of conformational issues relevant to polyalcohols, to include ethylene glycol and the two anomers of D-glucopyranose.

3.2.1. Dopamine

At physiological pH, the neurotransmitter dopamine exists predominantly in its protonated form [280]. As a consequence, dielectric shielding may play a significant role in the stabilization/destabilization of particular conformers [281]. We focus in particular on the torsional isomerism about the carbon-carbon single bond which is defined by dihedral angle ϕ_1 in Figure 4. In general, torsion about the other carbon-carbon single bond, denoted by dihedral angle ϕ_2, is not particularly affected by solvation, and we restrict our discussion to cases where that torsion has been fixed to its near optimal value of 90°. Figure 4 illustrates that in the absence of solvation, the global minimum occurs for $\phi_1 = 60°$, i.e., when the ammonium group is *gauche* to the aromatic ring (the catechol hydroxyl groups introduce a slight asymmetry into the system, so that $\phi_1 = -60°$ is not quite isoenergetic). This is consistent with numerous theoretical [282,283] and experimental [284,285] studies which illustrate that, in the absence of solvation, aromatic pi clouds interact very favorably with ammonium cations, the most notable examples being interactions in enzyme binding sites [282,285].

When aqueous solvation was included by using the original AM1-SM1 method, the situation changed markedly [281]. The *trans* isomer ($\phi_1 = 180°$) is considerably stabilized relative to the *gauche* rotamers. As a result, the *trans* rotamer became the global minimum

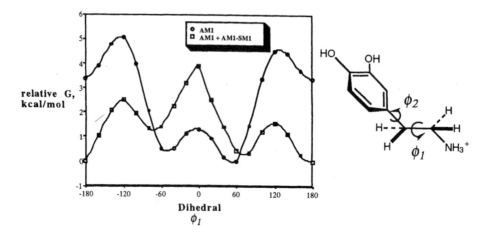

Figure 4. Electronic-nuclear energy of dopamine in the gas phase and free energy of dopamine in aqueous solution as a function of the torsion about the sp^3-sp^3 carbon-carbon single bond. For all points on the torsional coordinate, the dihedral angle ϕ_2 is fixed at 90°, which is the value illustrated in the molecular structure.

on the rotational coordinate because of the considerably smaller dielectric shielding experienced by the positively charged amino group when it is more distant from the bulky aromatic ring. Expressedly differently, the hydrophilic ammonium group is more accessible to solvent in the extended *trans* conformation.

In order to assess the quantitative accuracy of the model, it is useful to compare to rotameric populations determined from aqueous nuclear magnetic resonance (NMR) studies [286]. In particular, the gauche:anti ratio about ϕ_1 has been observed by NMR to be 58:42. Of the two theoretical models, AM1 (gas phase) calculations predict a >99:1 ratio and AM1 + AM1-SM1 (implying a standard-state free energy in aqueous solution arrived at by adding the AM1 gas-phase energy to the AM1-SM1 free energy of solvation) predicts 37:63. The latter represents an error of only 0.5 kcal/mol. We have here repeated these calculations using the more recent AM1-SM2 model, and the results are similar: the predicted aqueous gauche:anti ratio is 12:88 which corresponds to an error of 1.4 kcal/mol. It is worth noting that the calculated ratios rely in part on the relative accuracy of the AM1 gas-phase surface to which the solvation free energies are being added in order to arrive at Boltzmann-averaged equilibrium populations. Thus, it is impossible to judge which model, SM1 or SM2, is actually predicting the differential free energies of solvation more accurately. The fact that they are within one kcal/mol of each other reflects the similar way in which the two models treat the ENP portion of the solvation free energy, which is the dominant term for the dopamine cation. We will not devote extensive discussion to ϕ_2, however we do note that AM1-SM1 predicts aqueous solvation to lower the rotational barrier about the indicated bond [281], and this result is consistent with the rapid rotation observed experimentally by NMR [286].

3.2.2. Ethylene glycol and glucose

Accurately modeling the aqueous solvation of 1,2-ethanediol (ethylene glycol) provides a very challenging test for continuum solvation models insofar as much of the favorable solvation of this molecule results from numerous hydrogen bonding interactions with the water molecules found in the first solvation shell. In the SMx models, such interactions are accounted for by parameterization in the surface tension terms, and it is of considerable interest to analyze the robustness of this scheme. Ethylene glycol is additionally challenging because its equilibrium population at room temperature is composed of numerous isomers. Figure 5 illustrates the ten possible symmetry-unique isomers for this system. The nomenclature refers to the torsion angles about the leftmost C-O bond (lower case), the central C-C bond (upper case), and the rightmost C-O bond (lower case). A "g" (or "G") indicates a gauche torsion angle (either positive or negative, as marked) while a "t" (or "T") indicates the torsion to be trans. Although the presence of three torsions, each of which is characterized by three rotational minima, would normally give rise to $3^3 = 27$ separate minima, symmetry in this system causes a number of these cases to be degenerate. The numbers in parentheses beneath the molecular structures in figure 5 indicate that degeneracy, and, as required, these sum to 27.

Experimentally, it has been established by a number of methods that isomer tG^+g^- predominates, there is a smaller fraction of $g^+G^+g^-$, and there is a barely detectable fraction of $g^-G^+g^-$ [287-295]. High level ab initio calculations which employ very large basis sets

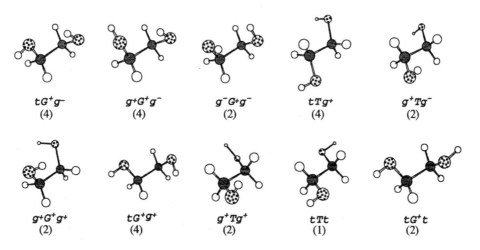

| tG^+g^- | $g^+G^+g^-$ | $g^-G^+g^-$ | tTg^+ | g^+Tg^- |
| (4) | (4) | (2) | (4) | (2) |

| $g^+G^+g^+$ | tG^+g^+ | g^+Tg^+ | tTt | tG^+t |
| (2) | (4) | (2) | (1) | (2) |

Figure 5. The ten unique structures of 1,2-ethanediol. Geometries were optimized at the MP2/cc-pVDZ level of theory.

(the correlation-consistent polarized valence triple-ζ basis of Dunning [296]) and account for correlation at sophisticated levels (coupled cluster analysis including all single and double excitations with a perturbative treatment of triples [297]) reproduce these trends nicely [298]. Results are summarized in Table 3. It is noteworthy that lower levels of theory, especially semiempirical models, do very poorly at reproducing these results.

Table 3 also presents the equilibrium population predicted by addition of PM3-SM3 solvation free energies for structures fully optimized in aqueous solution to the ab initio gas-phase energies. Comparison may again be made to experiment, in this case referring to aqueous NMR measurements which estimated the percentage of conformers *gauche* about the C–C bond to be 88±3% [299]. This is in good agreement with the predicted value of 92%. Results based on adding AM1-SM2 solvation free energies to the ab initio solute energies were found to be quite similar, although in that instance the gas-phase geometries were employed since AM1 geometries are qualitatively incorrect [298].

Table 3.
Equilibrium populations of ethylene glycol conformers in the gas phase and in aqueous solution (%)

Isomer	Triple-ζ[a] $G^o_{298}(gas)$	Triple-ζ[a] + PM3-SM3 $G^o_{298}(aqueous)$
tG^+g^-	55.9	45.8
$g^+G^+g^-$	27.4	31.0
$g^-G^+g^-$	13.4	14.0
tTg^+	1.3	4.2
g^+Tg^-	0.6	2.5
$g^+G^+g^+$	0.4	0.4
tG^+g^+	0.3	0.6
g^+Tg^+	0.2	0.9
tTt	0.2	0.6
tG^+t	0.2	not stationary
Total C–C gauche	97.6	91.8
Total C–C trans	2.4	8.2
Internal H-bond	83.4	76.8
No internal H-bond	16.7	23.2

[a] Ab initio Hartree-Fock calculations.

Finally, the population-averaged absolute free energy of solvation may be calculated from [300]

$$\exp\left[-\Delta G_S^o/RT\right] = \sum_C P_C \, \exp\left[-\Delta G_S^o(C)/RT\right] \qquad (30)$$

where P_C is the equilibrium mole fraction of conformation C in the gas phase. Following this procedure, we find the PM3-SM3 value for ΔG_S^o to be –9.4 kcal/mol, in outstanding agreement with a very recent experimental measurement of –9.6 kcal/mol [301].

The solvation of glucose, a monosaccharide, is expected to be dominated by the same effects as those which are important for ethylene glycol. Of course, the degree of complexity present in the sugar molecule is considerably greater than that observed for the simple diol! Since thorough discussions of the energetics of glucose are available [302-306], especially as regards the influence of aqueous solvation, we focus here on the contributions of the SMx models to further illustrate the applicability of these continuum solvation models [307].

Several noteworthy conformational equilibria have been experimentally determined. NMR experiments have been interpreted to indicate that only the $\overline{G}$ and G

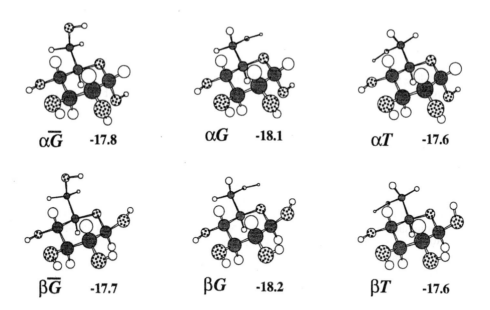

Figure 6. The important conformers for D-glucose. Under each structure is the nomenclature designating the anomer (α or β 2-hydroxyl group), the hydroxymethyl conformer ($\overline{G}$, G, or T), and the AM1-SM2 value predicted for ΔG_S^o in kcal/mol.

hydroxymethyl conformers are present in aqueous solution (Figure 6) [308]. Consistent with this observation, we find the alternative T conformer to be the highest in energy in aqueous solution. This is due in part to a differential solvation effect: the T conformer is 0.6 kcal/mol less well solvated than the lowest energy G conformer for both anomers. This preferential solvation of the G conformer is in good agreement with results from explicit-solvent molecular simulations [304,305], once again illustrating the ability of the SMx models to accurately reflect first-shell solvation effects. This agreement extends to consideration of the effect of solvation on the anomeric equilibrium, which is known to be 36:64 α:β in aqueous solution [309]. The effect of solvation on the anomeric equilibrium, as calculated from the Boltzmann-averaged equilibrium populations using equation 34, is predicted to be zero. This may be compared to the results of Ha et al. [305] and van Eijck et al. [302] who used explicit-solvent aqueous simulations to predict a differential solvation effect for the β anomer relative to the α of only 0.6 ± 0.5 kcal/mol and –0.5 ± 0.2 kcal/mol, respectively. The SMx prediction is thus halfway between the two simulation results. Moreover, when the AM1-SM2 solvation free energies are added to Hartree-Fock gas-phase relative free energies [306], the predicted anomeric equilibrium is within 0.8 kcal/mol of the experimental result. While more work clearly remains to be done with respect to converging the gas-phase relative free energies and examining in more detail the individual components of solvation, this is heartening agreement.

4. OTHER SOLVENTS

Section 4.1 discusses the development of solvation models for solvents other than water, and Section 4.2 briefly reviews empirical relationships for the prediction of solvent-dependent properties.

4.1. Modeling solvation in non-aqueous solvents

Although water is clearly the single most important solvent in which chemistry of biological relevance takes place, there are many solvents which find extensive use in other areas, e.g., organic synthesis, or which find use in the modeling of biological processes, e.g., octanol and hexadecane are sometimes used to model lipid membranes. For such solvents, all of the quantum mechanical continuum models discussed in Section 2 may be used for calculation of the ENP portion of the standard-state gas to condensed phase solvation free energy (i.e., free energy of transfer). One simply uses the dielectric constant appropriate for the solvent being modeled. The CDS portion of the solvation free energy, on the other hand, is not so straightforwardly addressed, and unfortunately very little work has been carried out incorporating the CDS effects into quantum mechanical continuum solvation models.

Within the SMx models, the CDS terms must be parameterized separately against available experimental data for each new solvent, and it is not clear, since they account for several physical effects as well as for errors in the ENP terms, that they will be related to any particular bulk property of the solvent in question. This presents the most significant hurdle to the development of new solvent parameter sets. Presently, the only published SMx models are specific for water. However, a n-hexadecane parameter set is in the final stages of completion, an octanol parameter set will be available shortly, and other ethereal, hydroxylic, and other solvents will be parameterized as well. Improved methods for arriving at the atomic partial charges (e.g., CM1A, CM1P) and/or solving the Poisson equation will reduce these "error-correcting" contributions, and more clear correlations between the remaining surface tensions and such bulk properties as macroscopic surface tension, viscosity, cohesive energy density, etc., may be discernible.

In the parameterization of the n-hexadecane model [310], we have noted one refinement which appears to be required for larger solvent molecules: significant dispersion interactions involve only that part of the solvent within a distance R_{CD} of the solute, where $R_{CD} < R_S$. (Recall that R_S is the solvent radius.) In such instances, it is useful to generalize the concept of solvent-accessible surface area. In particular, we separate the CDS effects, very approximately, into a CD part and a CS part. Then, for the former, we use the ASURF definition taking this smaller radius, R_{CD}, for the rolling probe—this might be called DSURF. For water, which is quite small, we have implicitly taken R_{CD} equal to R_S, which is 1.4 Å. This is a reasonable value for the "radius" of a water molecule in liquid water [93]. Since R_S for water is so small, it is reasonable to assume that dispersive interactions will operate over the entire molecule. For a molecule like n-hexadecane, on the other hand, an R_{CD} on the order of 1.7 Å works best [310]. This radius mimics the size of one of the methyl or methylene groups forming a portion of the n-hexadecane solvent molecule. This is quite a bit smaller than the R_S of about 5 Å which may be calculated either from the bulk density of liquid n-hexadecane by assuming a spherical molecule, or from taking the calculated volume of the molecule using overlapping spheres with van der Waals radii and choosing R_S to be the spherical radius providing the same volume. The former method is less ambiguous for a flexible molecule like n-hexadecane, although studies on both the aqueous ASURF of n-heptane [311] and the WSURF of n-decane [312] have found that the average molecular surface areas for the Boltzmann-weighted population of conformers at 298 K are only 3% and 4% smaller, respectively, than the surface area for the fully extended chain conformation—obviously the effect on the volume will be similarly slight. Although it is clearly possible that the remaining cavitation and structural rearrangement terms might also each require a different radius, i.e., an R_{CS} that is not simply equal to R_S, we did not observe that to be required for n-hexadecane.

We have illustrated in Section 3 a number of instances where molecular structures and energetics change significantly on going from the gas phase to aqueous solution. Of course, experimental observations in the gas phase are not generally possible for medium- to large-sized organic and biological molecules. However, there is considerable data available detailing the effects of *changing* solvents from one to another. Indeed, certain partition coefficients, which describe the equilibrium concentration of a solute in two different liquids, prove to be quite useful in drug design [313,314]. The most notable is the octanol-water partition coefficient [313-320]; the octanol-water system is considered to be a good model for biological membranes which are composed of amphiphiles like octanol.

The development of new semiempirical quantum chemical continuum solvation models will permit a comparison to more empirical models which exist to explain the effect of different solvents on various chemical properties. Those latter models are the subject of the next section.

4.2 Quantitative Structure-Activity Relationships and Linear Solvation Energy Relationships

Although the calculation of molecular properties from first principles is an esthetically appealing proposition, it is often impractical. Difficulties may arise for relatively simple reasons, e.g., the system of interest is simply too large to be tractable for trustworthy levels of theory. A more fundamental problem may be that no theoretical model exists to accurately predict the chemical property which is of interest. As a concrete example, consider that the antitumor activity of a series of structurally related organic molecules is known, and the objective is to predict the related activity of an as yet unsynthesized set of additional congeners. In such instances, it often proves useful to pursue an empirical approach which seeks to relate a variety of "simpler" molecular properties to the biological activity. Such an approach is embodied in the techniques of quantitative structure-activity relationships (QSAR) [321,322]. In essence, a QSAR analysis involves a regression equation that correlates microscopic features of a set of chemical compounds with some macroscopic property. This approach has demonstrated its utility repeatedly and it finds widespread use in medicinal and pharmaceutical chemistry.

A different application of regression analysis, which is nevertheless similar in spirit, is to be found in the so-called "linear solvation energy relationship" (LSER) formalism developed by Kamlet, Taft, Abraham, and co-workers [323-326]. Seeking to explain the effects of different solvents on chemical properties and reactivities, these researchers proposed an equation of the form

$$\log \gamma = c_0 + c_1 \alpha + c_2 \beta + c_3 \pi^* \tag{31}$$

where γ is the solvent-influenced property of interest for a particular solute (typically free-energy based, and hence the logarithm), and α, β, and π^* are solvent-specific constants which describe the solvent's hydrogen bond donating, hydrogen bond accepting, and dipolarity/polarizability properties, respectively. These constants are determined by measurement of some reference process in the solvent, e.g., the shift in the absorption maximum for a particular dye whose absorbance is sensitive to hydrogen bond interactions. As a result of choosing this kind of a reference process, these parameters have come to be known as "solvatochromic parameters". In practice, this equation finds use in the following fashion. Following a series of experimental measurements of $\log(\gamma)$ in a number of solvents for which the solvatochromic parameters are known, one obtains optimal constants c_i by regression analysis of the data. In principle, one may now predict $\log(\gamma)$ for *any* solvent for which the solvatochromic parameters have been tabulated. There are clearly dangers involved in extrapolating from a limited data set if the solvents selected for experimental measurement were not representative of those for which predictions are being made; nevertheless the method is quite powerful and has demonstrated itself to be reasonably robust.

It will be particularly interesting to explore the interplay of ENP and CDS effects in the parameters α, β, and π^* when continuum solvation models are available for comparison to LSER approaches.

5. PROPERTIES OF SOLVATED MOLECULES AS PREDICTORS FOR STRUCTURE-ACTIVITY RELATIONSHIPS

This section discusses how calculated molecular properties which take account of solvation may be used in empirical structure-activity relationships to provide guidance in molecular design. Although the formalism discussed in the preceding section was originally developed to model the effect of solvents on some specific solute property, the regression analysis has been generalized to allow prediction of some solute property based on *solute* solvatochromic parameters [325,327]. That is,

$$\log \Gamma = c_0 + c_1 \alpha_2 + c_2 \beta_2 + c_3 \pi_2^* \tag{32}$$

where Γ is upper case to indicate a general property exhibited by a large number of solutes (as opposed to γ, which may be unique to a single solute within the formalism of equation 31) and the subscript "2" indicates the solvatochromic parameters to be associated with the solute, not the solvent (which is kept constant in equation 32). Solute-specific solvatochromic parameters continue to be arrived at by measurement of some reference process. An important point to note is that this is analogous to a QSAR analysis, i.e., a chemical property is being correlated with some set of other, measurable properties, albeit the latter approach developed more from a background of linear free-energy relationships [328-330] than from a purely empirical impetus. Equations 31 and 32 can be combined in a very general form, viz.,

$$\log \Xi = \sum_i C_i^{\text{solvent}} d_i^{\text{solute}} \tag{33}$$

where the parameters C_i and d_i may be solvatochromic parameters, or other general parameters [324,331-340].

An interesting variation on this theme, which has been pursued extensively by Famini and Wilson [341-344], is to maintain the formalism of equation 32, but to replace the solute parameters α_2, β_2, and π_2^*, which must be experimentally determined, with other constants derived from theoretical calculations. The "theoretical" linear solvation free energy relationship (TLSER) ansatz is expressed as

$$\log \Gamma = c_0 + c_1 V_{\text{mc}} + c_2 \pi^* + c_3 \varepsilon_\alpha + c_4 \varepsilon_\beta + c_5 q^+ + c_6 q^- \tag{34}$$

where V_{mc} is the molecular van der Waals volume, and π^* is a polarizability term derived from the calculated polarization volume (the latter should not be confused with the solvatochromic descriptors π^* and π_2^*, which are experimentally measured quantities). Just as with LSER, hydrogen bonding is separated into donor and acceptor components. Since *all* intermolecular interactions can be considered to have varying degrees of covalent and electrostatic components, separate descriptors have been chosen to address each of these two limiting paradigms. The covalent contribution to Lewis basicity, ε_β, is taken as the difference in energy between the lowest unoccupied molecular orbital (E_{LUMO}) of the solute and the highest occupied molecular orbital (E_{HOMO}) of water, i.e., smaller values of ε_β denote a greater covalent basicity. The electrostatic basicity contribution is denoted by q^-, the magnitude of the most negative atomic partial charge in the molecule. Analogously, the hydrogen bonding acidity is divided into two components: ε_α is the energy difference between E_{HOMO} of the solute and E_{LUMO} of water, and q^+ is the magnitude of the most positively charged hydrogen atom in the molecule. Calculation of these descriptors may be performed at any level of theory, of course, but to date the emphasis has been on using the NDDO semiempirical level of theory (as mentioned in Section 2.3, this is the level used in the AM1 and PM3 models) so as to take advantage of the relative efficiency of this method for larger molecules.

Although equation 37 has been quite successful at predicting a number of interesting chemical and biological properties [341-346], it is by no means the only possible way to relate calculated molecular properties with activities/toxicities. Lewis has recently provided an extensive review detailing other descriptors which have found use in TLSER-like regression analyses [347]. Politzer and Murray and co-workers have also provided alternative formulations of this approach [348-350]. One additional point which should be mentioned is that it is perfectly logical to consider developing regression equations which consider *both* experimentally determined and calculated descriptors [351,352]. Of course, part of the attraction of the purely theoretical methodology is that it may be considerably simpler and more economical to perform the calculations rather than synthesize the solute in order to measure some parameter if it is not already known.

An interesting question is how solvation may influence the descriptors present in the regression models [353]. As has been discussed at length above, both electronic and geometric structure may change significantly for a solute in an aqueous environment (e.g., in vivo) and it seems clear that any TLSER-like regression equation being used to predict properties in such a situation should accurately take account of that. In particular, the free energy of solvation *itself* may be a particularly important descriptor. Activity in a biomolecule typically requires its interaction with a receptor and/or, as mentioned in Section 4, its crossing of a hydrophobic cell membrane; since these processes usually require desolvation which may or may not be balanced by specific interactions within a receptor site, the cost of desolvation can influence the overall activity.

One interesting example illustrating this point has been provided by Alkorta et al. [354], who have analyzed the affinity (expressed as $1/K_i$, where K_i is the dissociation constant of the enzyme-substrate complex) of a set of 3-benzazepine cations for the dopamine D_1 receptor. Figure 7 illustrates the gross structure of these substrates. In this case, the descriptors chosen were the AM1-SM1 calculated free energies of solvation (ΔG_S^0), the dot product of the solute dipole moments with the unit vector parallel to the activity-weighted sum of *all* of the solute dipole moments (μ^*, i.e., this descriptor measures both the magnitude of the molecular dipole moment and the degree to which it is aligned with the dipole moments of the compounds observed to be most active), and the molecular polarizability (α). Using this approach they obtained a regression equation of:

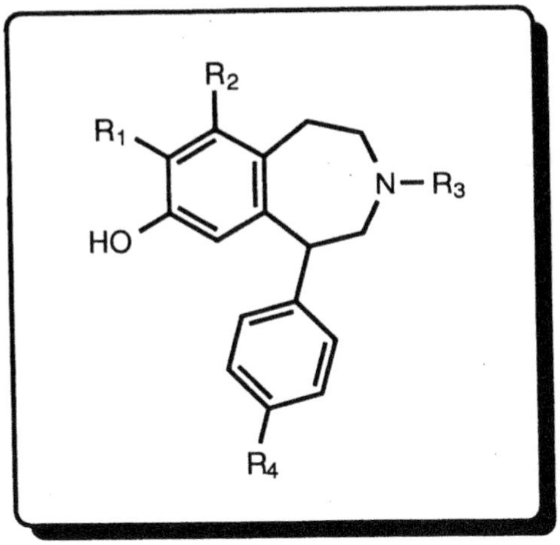

R_1 = H, F, Cl, Br, Me, OH, OMe, CN

R_2 = H, F, Cl

R_3 = H, Me

R_4 = H, Cl

Figure 7. 3-Benzazepines examined by Alkorta et al. Not all possible permutations of the listed R groups were explored.

$$\log (1/K_i) = -3.46 - 1.67\,\mu^* + 0.16\,\Delta G_S^0 + 0.07\,\alpha \tag{35}$$

with a multiple correlation coefficient R of 0.925 for a data set of 13 compounds. An analysis of the statistical significance of α suggests that it is the least useful parameter in the regression equation, which is perhaps not surprising since it represents a gas-phase property; its removal leads to only a slight drop in R (0.853). The most interesting aspects of this analysis, however, were the observations of Alkorta et al. [354] that (1) not even qualitative activities could be predicted from regression analyses performed for the neutral (i.e., non-protonated) 3-benzazepines, consonant with the expected protonation of these compounds in vivo, and (2) methylation of the azepine nitrogen leads to reduced solvation free energies as a result of dielectric shielding and loss of hydrogen bonding opportunities; this *increases* the affinity of the methylated compounds since it decreases the cost of their desolvation.

We anticipate that the continued development of rapid and accurate continuum solvation models will give rise to increasing use of such analyses in structure-activity predictions. The cost-effectiveness of the methodology for the pre-screening of potential synthetic targets gives it considerable practical importance.

6. CONCLUDING REMARKS

Quantum mechanical continuum solvation models span a wide range of complexity and utility in their ability to calculate electrostatic contributions to free energies of solvation.

When supplemented with models that also include those portions of the solvation free energy associated with the first surrounding shell of solvent, they become particularly powerful tools for studying complicated, condensed-phase systems. In particular, they can be used to provide insight into solutes that play important roles in biological systems. This information includes quantitative estimates of free energies of solvation, predictions of the detailed changes in electronic and geometric structure of biomolecules in solution, and calculation of micro- and macroscopic properties for use in structure-activity prediction. Continued comparison to explicit-solvent simulations and experimental results will be instrumental in improving the models.

ACKNOWLEDGMENTS

This work was supported in part by the National Science Foundation under grant no. CHE89-22048 and by the ILIR program of the United States Army.

REFERENCES

1. C. Reichardt, Solvents and Solvent Effects in Organic Chemistry, VCH, New York, 1990.
2. J.-L.M. Abboud, R. Notario, J. Bertrán, and M. Solà, Prog. Phys. Org. Chem., 19 (1993) 1.
3. J.L. Gao and X.F. Xia, J. Am. Chem. Soc., 115 (1993) 9667.
4. C.-C. Han, J.A. Dodd, and J.I. Brauman, J. Phys. Chem., 90 (1986) 471.
5. J. Chandrasekhar, S.F. Smith, and W.L. Jorgensen, 107 (1985) 154.
6. J. Chandrasekhar and W.L. Jorgensen, J. Am. Chem. Soc., 107 (1985) 2974.
7. P.A. Bash, M.J. Field, and M. Karplus, J. Am. Chem. Soc., 109 (1987) 8094.
8. J.-K. Hwang, G. King, S. Creighton, and A. Warshel, J. Am. Chem. Soc., 110 (1988) 5297.
9. S.E. Huston, P.J. Rossky, and D.A. Zichi, J. Am. Chem. Soc., 111 (1989) 5680.
10. T. Kozaki, M. Morihashi, and O. Kikuchi, J. Am. Chem. Soc., 111 (1989) 1547.
11. S.C. Tucker and D.G. Truhlar, Chem. Phys. Lett., 157 (1989) 164.
12. J.J. Gajewski, J. Jurayj, D.R. Kimbrough, M.E. Gande, B. Ganem, and B.K. Carpenter, J. Am. Chem. Soc., 109 (1987) 1170.
13. E. Brandes, P.A. Grieco, and J.J. Gajewski, J. Org. Chem., 54 (1989) 515.
14. P.A. Grieco, Aldrichim. Acta, 24 (1991) 59.
15. D.L. Severance and W.L. Jorgensen, J. Am. Chem. Soc., 114 (1992) 10966.
16. J. Gao, submitted for publication.
17. C.J. Cramer and D.G. Truhlar, J. Am. Chem. Soc., 114 (1992) 8794.
18. S. Sogo, T.S. Widlanski, J.H. Hoare, C.E. Grimshaw, G.A. Berchtold, and J.R. Knowles, J. Am. Chem. Soc., 106 (1984) 2701.
19. H. Dugas, Bioorganic Chemistry, 2 ed., Springer-Verlag, New York, 1989.
20. C.L. Brooks, M. Karplus, and B.M. Pettit, Adv. Chem. Phys., 71 (1989) 1.
21. P.A. Kollman and K.M. Merz, Acc. Chem. Res., 23 (1990) 246.

22. A. Warshel, Computer Modeling of Chemical Reactions in Enzymes and Solutions, Wiley-Interscience, New York, 1991.

23. A. Warshel, J. Åqvist, and S. Creighton, Proc. Natl. Acad. Sci., USA, 86 (1989) 5820.

24. T.L. Hill, An Introduction to Statistical Thermodynamics, Addison-Wesley, Reading, MA, 1960, 119.

25. M.P. Allen and D.J. Tildesley, Computer Simulations of Liquids, Oxford University Press, London, 1987.

26. J.A. McCammon and S.C. Harvey, Dynamics of Proteins and Nucleic Acids, Cambridge University Press, Cambridge, 1987.

27. W.L. Jorgensen, Acc. Chem. Res., 22 (1989) 184.

28. D.W. Heermann, Computer Simulation Methods in Theoretical Physics, 2nd ed., Springer-Verlag, Berlin, 1990.

29. P.M. King, C.A. Reynolds, J.W. Essex, G.A. Worth, and W.G. Richards, Mol. Sim., 5 (1990) 262.

30. K. Binder and D.W. Heermann, Monte Carlo Simulation in Statistical Physics, 2nd corrected ed., Springer-Verlag, Berlin, 1992.

31. D.L. Beveridge and F.M. DiCapua, Annu. Rev. Biophys. Biophys. Chem., 18 (1989) 431.

32. J.M. Haile, Molecular Dynamics Simulation, Wiley-Interscience, New York, 1992.

33. F.S. Lee and A. Warshel, J. Chem. Phys., 97 (1992) 3100.

34. K. Tasaki, S. McDonald, and J.W. Brady, J. Comp. Chem., 14 (1993) 278.

35. J. Guenot and P.A. Kollman, J. Comp. Chem., 14 (1993) 295.

36. C.B. Post, C.M. Dobson, and M.K. Karplus, Proteins, 5 (1989) 337.

37. D.A. Pearlman and P.A. Kollman, J. Chem. Phys., 94 (1991) 4532.

38. M. Mazor and B.M. Pettit, Mol. Sim., 6 (1991) 1.

39. M.J. Mitchell and J.A. McCammon, J. Comp. Chem., 12 (1991) 271.

40. C.A. Reynolds, J.W. Essex, and W.G. Richards, Chem. Phys. Lett., 199 (1992) 257.

41. J.E. Straub and D. Thirumalai, Proteins, 15 (1993) 360.

42. J.J. Urban and G.R. Famini, J. Comp. Chem., 14 (1993) 353.

43. J.P. Bowen and N.L. Allinger, in: Reviews in Computational Chemistry, Vol. 2, K.B. Lipkowitz and D.B. Boyd (eds.), VCH, New York, 1991, 81.

44. D.M. Ferguson, I.R. Gould, W.A. Glauser, S. Schroeder, and P.A. Kollman, J. Comp. Chem., 13 (1992) 525.

45. J.E. Eksterowicz and K.N. Houk, Chem. Rev., 93 (1993) 2439.

46. A. Warshel, J. Phys. Chem., 83 (1979) 1640.

47. V. Luzhkov and A. Warshel, J. Comp. Chem., 13 (1992) 199.

48. Y.W. Xu, C.X. Wang, and Y.Y. Shi, J. Comp. Chem., 13 (1992) 1109.

49. J. Åqvist and A. Warshel, Chem. Rev., 93 (1993) 2418.

50. F.S. Lee, Z.T. Chu, and A. Warshel, J. Comp. Chem., 14 (1993) 161.

51. F. Peradejordi, Cahiers Phys., 17 (1963) 343.

52. O. Chalvet and I. Jano, Compt. Rend. Acad. Sci. Paris, 259 (1964) 1867.

53. I. Jano, Compt. Rend. Acad. Sci. Paris, 261 (1965) 103.

54. R. Daudel, Adv. Quantum Chem., 3 (1967) 121.

55. S. Yomosa and M. Hasegawa, J. Phys. Soc. Japan, 29 (1970) 1329.

56. M. Newton, J. Phys. Chem., 79 (1975) 2795.

57.   O. Tapia and O. Goscinski, Mol. Phys., 29 (1975) 1653.

58.   J. Hylton-McCreery, R.E. Christofferson, and G.G. Hall, J. Am. Chem. Soc., 98 (1976) 7191.

59.   J.L. Burch, K.S. Raghuveer, and R.E. Christofferson, in: Environmental Effects on Molecular Structure and Properties, B. Pullman (ed.), Reidel, Dordrecht, 1976, 17.

60.   J.-L. Rivail and D. Rinaldi, Chem. Phys., 18 (1976) 233.

61.   I. Fischer-Hjalmars, I. Hendricksson-Entlo, and C. Hermann, Chem. Phys., 24 (1977) 167.

62.   R. Constanciel and O. Tapia, Theor. Chim. Acta, 48 (1978) 75.

63.   O. Tapia and B. Silvi, J. Phys. Chem., 84 (1980) 2646.

64.   G. Klopman and P. Andreozzi, Theor. Chim. Acta, 55 (1980) 77.

65.   S. Miertus, E. Scrocco, and J. Tomasi, Chem. Phys., 55 (1981) 117.

66.   O. Tapia, in: Quantum Theory of Chemical Reactions, Vol. 2, R. Daudel, A. Pullman, L. Salem, and A. Viellard (eds.), Reidel, Dordrecht, 1980, 25.

67.   P. Claverie, in: Quantum Theory of Chemical Reactions, Vol. 3, R. Daudel, A. Pullman, L. Salem, and A. Veillard (eds.), Reidel, Dordrecht, 1982, 151.

68.   O. Tapia, in: Molecular Interactions, Vol. 3, H. Rajaczak and W.J. Orville-Thomas (eds.), John Wiley & Sons, London, 1982, 47.

69.   R. Constanciel and R. Contreras, Theor. Chim. Acta, 65 (1984) 1.

70.   R. Contreras and A. Aizman, Int. J. Quant. Chem., 27 (1985) 293.

71.   J.L. Rivail, B. Terryn, and M.F. Ruiz-Lopez, J. Mol. Struct. (Theochem), 120 (1985) 387.

72.   O. Tapia, F. Colonna, and J.G. Angyan, J. Chem. Phys., 87 (1990) 875.

73.   O. Tapia, J. Mol. Struct. (Theochem), 226 (1991) 59.

74.   J. Tomasi, R. Bonaccorsi, R. Cammi, and F.J. Olivares del Valle, J. Mol. Struct. (Theochem), 234 (1991) 401.

75.   C.J. Cramer and D.G. Truhlar, in: Reviews in Computational Chemistry, Vol. 6, K.B. Lipkowitz and D.B. Boyd (eds.), VCH, New York, 1994, in press.

76.   J.H. Jensen, P.N. Day, M.S. Gordon, H. Basch, D. Cohen, D.R. Garmer, M. Krauss, and W.J. Stevens, ACS Monograph Series on Hydrogen Bonding, in press.

77.   M. Born, Z. Physik, 1 (1920) 45.

78.   J.G. Kirkwood, J. Chem. Phys., 2 (1934) 351.

79.   L. Onsager, J. Am. Chem. Soc., 58 (1936) 1486.

80.   N.H. Frank and W. Tobocman, in: Fundamental Formulas of Physics, Vol. 1, D.H. Menzel (ed.), Dover, New York, 1960, 307.

81.   R.K. Wangsness, Electromagnetic Fields, John Wiley & Sons, New York, 1979.

82.   D.R. Corson and P. Lorrain, Introduction to Electromagnetic Fields and Waves, W.H. Freeman, San Francisco, 1962.

83.   N. Hush, Aust. J. Sci. Res., Ser. A, 1 (1948) 480.

84.   A.A. Rashin and B. Honig, J. Phys. Chem., 89 (1985) 5588.

85.   A.A. Rashin and K. Namboodiri, J. Phys. Chem., 91 (1987) 6003.

86.   F. Hirata, P. Redfern, and R.M. Levy, Int. J. Quant. Chem. Symp., 15 (1988) 179.

87.   B. Jayaram, R. Fine, K. Sharp, and B. Honig, J. Phys. Chem., 93 (1989) 4320.

88.   H.-S. Kim and J.-J. Chung, Bull. Korean Chem. Soc., 14 (1993) 220.

89.   A. Bondi, J. Phys. Chem., 68 (1964) 441.

90.   B. Roux, H.-A. Yu, and M. Karplus, J. Phys. Chem., 94 (1990) 4683.

91. D. Chandler and H.C. Andersen, J. Chem. Phys., 57 (1972) 1930.
92. F. Hirata, B.M. Pettit, and P.J. Rossky, J. Chem. Phys., 77 (1982) 509.
93. Y. Marcus, J. Solution Chem., 12 (1983) 271.
94. C.J.F. Bottcher, O.C. van Belle, P. Bordewicijk, and A. Rip, Theory of Electric Polarization, 2nd ed., Elsevier, Amsterdam, 1973, 94.
95. A. Warshel and S.T. Russell, Quart. Rev. Biophys., 17 (1984) 283.
96. H.L. Friedman, Mol. Phys., 29 (1975) 29.
97. H.-A. Yu and M. Karplus, J. Chem. Phys., 89 (1988) 2366.
98. J.L. Rivail, in: New Theoretical Concepts for Understanding Organic Reactions, J. Bertrán and I.G. Czismadia (eds.), Kluwer, Dordrecht, 1989, 219.
99. F.S. Lee, Z.-T. Chu, M.B. Bolger, and A. Warshel, Protein Eng., 5 (1992) 215.
100. G. King and R.A. Barford, J. Phys. Chem., 97 (1993) 8798.
101. F.M. Richards, Annu. Rev. Biophys. Bioeng., 6 (1977) 151.
102. M.L. Connolly, Science, 221 (1983) 709.
103. J.L. Pascual-Ahuir and E. Silla, J. Comp. Chem., 11 (1990) 1047.
104. I. Tuñón, E. Silla, and J.L. Pascual-Ahuir, Protein Eng., 5 (1992) 715.
105. J.L. Pascual-Ahuir, E. Silla, and I. Tuñon, J. Comp. Chem., submitted for publication.
106. T. Furuki, A. Umeda, M. Sakurai, Y. Inoue, and R. Chûjô, J. Comp. Chem., 15 (1994) 90.
107. M.W. Wong, K.B. Wiberg, and M.J. Frisch, J. Am. Chem. Soc., 114 (1992) 1645.
108. M.M. Karelson, A.R. Katrizky, and M.C. Zerner, Int. J. Quant. Chem. Symp., 20 (1986) 521.
109. K.V. Mikkelson, H. Agren, H.J.A. Jensen, and T. Helgaker, J. Phys. Chem., 89 (1988) 3086.
110. M.M. Karelson, A.R. Katritzky, M. Szafran, and M.C. Zerner, J. Org. Chem., 54 (1989) 6030.
111. M.M. Karelson, T. Tamm, A.R. Katritzky, S.J. Cato, and M.C. Zerner, Tetrahedron Comput. Methodol., 2 (1989) 295.
112. M.M. Karelson, A.R. Katritzky, M. Szafran, and M.C. Zerner, J. Chem. Soc., Perkin Trans. 2, (1990) 195.
113. A.R. Katritzky and M. Karelson, J. Am. Chem. Soc., 113 (1991) 1561.
114. H.S. Rzepa, M.Y. Yi, M.M. Karelson, and M.C. Zerner, J. Chem. Soc., Perkin Trans. 2, (1991) 635.
115. H.S. Rzepa and M.Y. Yi, J. Chem. Soc., Perkin Trans. 2, (1991) 531.
116. M.W. Wong, M.J. Frisch, and K.B. Wiberg, J. Am. Chem. Soc., 113 (1991) 4776.
117. M.W. Wong, K.B. Wiberg, and M.J. Frisch, J. Chem. Phys., 95 (1991) 8991.
118. O.G. Parchment, I.H. Hillier, and D.V.S. Green, J. Chem. Soc., Perkin Trans. 2, (1991) 799.
119. M.W. Wong, K.B. Wiberg, and M.J. Frisch, J. Am. Chem. Soc., 114 (1992) 523.
120. L.C.G. Freitas, R.L. Longo, and A.M. Simas, J. Chem. Soc., Faraday Trans., 88 (1992) 189.
121. M.W. Wong, R. Leung-Toung, and C. Wentrup, J. Am. Chem. Soc., 115 (1993) 2465.
122. M. Szafran, M.M. Karelson, A.R. Katritzky, J. Koput, and M.C. Zerner, J. Comp. Chem., 14 (1993) 371.

123. P. Young, D.V.S. Green, I.H. Hillier, and N.A. Burton, Mol. Phys., 80 (1993) 503.

124. J. Avery, The Quantum Theory of Atoms, Molecules, and Photons, McGraw-Hill, London, 1972.

125. W.J. Hehre, L. Radom, P.v.R. Schleyer, and J.A. Pople, Ab Initio Molecular Orbital Theory, Wiley, New York, 1986.

126. M.M. Karelson and M.C. Zerner, J. Phys. Chem., 96 (1992) 6949.

127. A. Dega-Szafran, M. Gdaniec, M. Grunwald-Wyspianska, Z. Kosturkiewicz, J. Koput, P. Krzyzanowski, and M. Szafran, J. Mol. Struct., 270 (1992) 99.

128. M.F. Guest and J. Kendrick, GAMESS-UK, Daresbury Laboratory, 1986.

129. R.L. Longo and L.C.G. Freitas, Int. J. Quant. Chem., Quant. Biol. Symp., 17 (1990) 35.

130. M.J. Frisch, G.W. Trucks, M. Head-Gordon, P.M.W. Gill, M.W. Wong, J.B. Foresman, B.G. Johnson, H.B. Schlegel, M.A. Robb, E.S. Replogle, R. Gomperts, J.L. Andres, K. Raghavachari, J.S. Binkley, C. Gonzalez, R.L. Martin, D.J. Fox, D.J. Defrees, J. Baker, J.J. Stewart, and J.A. Pople, Gaussian 92, Revision B, Gaussian, Inc., Pittsburgh, PA, 1992.

131. D. Rinaldi, P.E. Hogan, and A. Cartier, QCPE Bull., in press.

132. J.L. Rivail and B. Terryn, J. Chem. Phys., 79 (1982) 1.

133. D. Rinaldi, J.-L. Rivail, and N. Rguini, J. Comp. Chem., 13 (1992) 675.

134. V. Dillet, D. Rinaldi, J.G. Angyán, and J.-L. Rivail, Chem. Phys. Lett., 202 (1993) 18.

135. J.-L. Rivail, in: Theoretical and Computational Models for Organic Chemistry, S.J. Formosinho, I.G. Czismadia, and L.G. Arnaut (eds.), Kluwer, Dordrecht, 1991, 79.

136. D. Rinaldi, M.F. Ruiz-Lopez, and J.-L. Rivail, J. Chem. Phys., 78 (1983) 834.

137. G.P. Ford and B. Wang, J. Comp. Chem., 13 (1992) 229.

138. M.J. Frisch, in: Abstracts of Papers, 205th National Meeting of the American Chemical Society, American Chemical Society, Washington, DC, 1993.

139. D. Rinaldi and R.R. Pappalardo, program to be submitted to the Quantum Chemistry Program Exchange.

140. R.R. Pappalardo, E.S. Marcos, M.F. Ruiz-Lopez, D. Rinaldi, and J.-L. Rivail, J. Am. Chem. Soc., 115 (1993) 3722.

141. A.J. Stone, Chem. Phys. Lett., 83 (1981) 233.

142. A.D. Buckingham and P.W. Fowler, J. Chem. Phys., 79 (1983) 6426.

143. A.J. Stone and M. Alderton, Mol. Phys., 56 (1985) 1047.

144. F. Colonna, E. Evleth, and J.G. Angyán, J. Comp. Chem., 13 (1992) 1234.

145. R.S. Mulliken, J. Chem. Phys., 3 (1935) 564.

146. C.A. Coulson and H.C. Longuet-Higgins, Proc. R. Soc. London, 191 (1947) 39.

147. R.S. Mulliken, J. Chem. Phys., 23 (1955) 1833.

148. F.L. Hirshfeld, Theor. Chim. Acta, 44 (1977) 129.

149. I. Mayer, Chem. Phys. Lett., 97 (1983) 270.

150. I. Mayer, Chem. Phys. Lett., 110 (1984) 440.

151. R.F. Bader, Acc. Chem. Res., 18 (1985) 9.

152. A.E. Reed, R.B. Weinstock, and F. Weinhold, J. Chem. Phys., 83 (1985) 735.

153. L.E. Chirlian and M.M. Francl, J. Comp. Chem., 8 (1987) 894.

154. A.E. Reed, L.A. Curtiss, and F. Weinhold, Chem. Rev., 88 (1988) 899.

155. J. Cioslowski, J. Am. Chem. Soc., 111 (1989) 8333.

156. B.H. Besler, J. K.M. Merz, and P.A. Kollman, J. Comp. Chem., 11 (1990) 431.

157. C.M. Breneman and K.B. Wiberg, J. Comp. Chem., 11 (1990) 361.

158. E.R. Davidson and S. Chakravorty, Theor. Chim. Acta, 83 (1992) 319.

159. K.M. Merz, J. Comp. Chem., 13 (1992) 749.

160. M.N. Ramos and B.d.B. Neto, Chem. Phys. Lett., 199 (1992) 482.

161. M.A. Spackman, Chem. Rev., 92 (1992) 1769.

162. T.K. Ghanty and S.K. Ghosh, J. Mol. Struct. (Theochem), 276 (1992) 83.

163. T.R. Stouch and D.E. Williams, J. Comp. Chem., 14 (1993) 858.

164. S.M. Bachrach, in: Reviews in Computational Chemistry, Vol. 5, K.B. Lipkowitz and D.B. Boyd (eds.), VCH, New York, 1993, 171.

165. G. Rauhut and T. Clark, J. Comp. Chem., 14 (1993) 503.

166. J.W. Storer, D.J. Giesen, C.J. Cramer, and D.G. Truhlar, J. Comp. Chem., submitted for publication.

167. C. Chipot, D. Rinaldi, and J.-L. Rivail, Chem. Phys. Lett., 191 (1992) 287.

168. J.L. Pascual-Ahuir, E. Silla, J. Tomasi, and R. Bonaccorsi, J. Comp. Chem., 8 (1987) 778.

169. R. Bonaccorsi, P. Palla, and J. Tomasi, J. Comp. Chem., 4 (1983) 567.

170. R. Bianco, S. Miertius, M. Persico, and J. Tomasi, Chem. Phys., 168 (1992) 281.

171. R. Bonaccorsi, R. Cammi, and J. Tomasi, J. Comp. Chem., 12 (1991) 301.

172. S. Woodcock, D.V.S. Green, M.A. Vincent, I.H. Hillier, M.F. Guest, and P. Sherwood, J. Chem. Soc., Perkin Trans. 2, (1992) 2151.

173. M. Peterson and R. Poirer, MONSTERGAUSS, (Department of Chemistry, University of Toronto, Toronto, Ontario, Canada,

174. M. Negre, M. Orozco, and F.J. Luque, Chem. Phys. Lett., 196 (1992) 27.

175. B. Wang and G.P. Ford, J. Chem. Phys., 97 (1992) 4162.

176. B. Wang and G.P. Ford, J. Am. Chem. Soc., 114 (1992) 10563.

177. G. Rauhut, T. Clark, and T. Steinke, J. Am. Chem. Soc., 115 (1993) 9174.

178. A. Klamt and G. Schüürmann, J. Chem. Soc., Perkin Trans. 2, (1993) 799.

179. T. Fox, N. Rösch, and R.J. Zaubar, J. Comp. Chem., 14 (1993) 253.

180. M.A. Aguilar and F.J. Olivares del Valle, Chem. Phys., 129 (1989) 439.

181. M. Orozco, W.L. Jorgensen, and F.J. Luque, J. Comp. Chem., 14 (1993) 1498.

182. H. Hoshi, M. Sakurai, Y. Inouye, and R. Chûjô, J. Chem. Phys., 87 (1987) 1107.

183. J.A. Pople and D.A. Beveridge, Approximate Molecular Orbital Theory, McGraw-Hill, New York, 1970.

184. R. Fowler and E.A. Guggenheim, Statistical Thermodynamics, Cambridge University Press, London, 1956.

185. M.K. Gilson and B. Honig, Proteins, 4 (1988) 7.

186. H. Nakamura, J. Phys. Soc. Japan, 57 (1988) 3702.

187. S.C. Harvey, Proteins, 5 (1989) 78.

188. M.E. Davis and J.A. McCammon, J. Comp. Chem., 10 (1989) 386.

189. M.E. Davis and J.A. McCammon, Chem. Rev., 90 (1990) 509.

190. D.A. Sharp and B. Honig, Annu. Rev. Biophys. Biophys. Chem., 19 (1990) 301.

191. B.J. Yoon and A.M. Lenhoff, J. Comp. Chem., 11 (1990) 1080.

192. A. Nicholls and B. Honig, J. Comp. Chem., 12 (1991) 435.

193. A. Jean-Charles, A. Nicholls, K. Sharp, B. Honig, A. Tempczyk, T.F. Hendrickson, and W.C. Still, J. Am. Chem. Soc., 113 (1991) 1454.

194. A.H. Juffer, E.F.F. Botta, B.A.M. van Keulen, A. van der Ploeg, and H.J.C. Berendsen, J. Comp. Phys., 97 (1991) 144.
195. B.A. Luty, M.E. Davis, and J.A. McCammon, J. Comp. Chem., 13 (1992) 768.
196. B.A. Luty, M.E. Davis, and J.A. McCammon, J. Comp. Chem., 13 (1992) 1114.
197. B. Honig, K. Sharp, and A.-S. Yang, J. Phys. Chem., 97 (1993) 1101.
198. A.A. Rashin, Prog. Biophys. Molec. Biol., 60 (1993) 73.
199. A.J. Hopfinger, Conformational Properties of Macromolecules, Academic Press, New York, 1973.
200. E.M. Huque, J. Chem. Ed., 66 (1989) 581.
201. G. Neméthy, W.J. Peer, and H.A. Scheraga, Annu. Rev. Biophys. Bioeng., 10 (1981) 459.
202. B. Lee and F.M. Richards, J. Mol. Biol., 55 (1971) 379.
203. R.B. Hermann, J. Phys. Chem., 76 (1972) 2754.
204. G.L. Amidon, S.H. Yalkowsky, S.T. Anik, and S.C. Valvani, J. Phys. Chem., 79 (1975) 2239.
205. S.C. Valvani, S.H. Yalkowsky, and G.L. Amidon, J. Phys. Chem., 80 (1976) 829.
206. G.D. Rose, A.R. Geselowitz, G.J. Lesser, R.H. Lee, and M.H. Zehfus, Science, 229 (1985) 834.
207. T. Ooi, M. Oobatake, G. Nemethy, and H.A. Scheraga, Proc. Natl. Acad. Sci., USA, 84 (1987) 3086.
208. D. Eisenberg and A.D. McLachlan, Nature, 319 (1986) 199.
209. B.v. Freyberg and W. Braun, J. Comp. Chem., 14 (1993) 510.
210. R.A. Pierotti, J. Phys. Chem., 67 (1963) 1840.
211. M.J. Huron and P. Claverie, J. Phys. Chem., 76 (1972) 2123.
212. P. Claverie, J.P. Daudey, J. Langlet, B. Pullman, D. Piazzola, and M.J. Huron, J. Phys. Chem., 82 (1978) 405.
213. B.T. Thole and P.T. van Duijnen, Theor. Chim. Acta, 55 (1980) 307.
214. D. Rinaldi, B.J. Costa Cabral, and J.-L. Rivail, Chem. Phys. Lett., 125 (1986) 495.
215. F. Floris and J. Tomasi, J. Comp. Chem., 10 (1989) 616.
216. J.G. Ángyán and G. Jensen, Chem. Phys. Lett., 175 (1990) 313.
217. J.A.C. Rullman and P.T. van Duijnen, Reports Molec. Theory, 1 (1990) 1.
218. F.M. Floris, J. Tomasi, and J.L. Pascual-Ahuir, J. Comp. Chem., 39 (1991) 784.
219. F.J. Olivares del Valle and M.A. Aguilar, J. Mol. Struct. (Theochem), 280 (1993) 25.
220. F.M. Floris, A. Tani, and J. Tomasi, Chem. Phys., 169 (1993) 11.
221. W.C. Still, A. Tempczyk, R.C. Hawley, and T. Hendrickson, J. Am. Chem. Soc., 112 (1990) 6127.
222. G.J. Hoijtink, E. de Boer, P.H. Van der Meij, and W.P. Weijland, Recl. Trav. Chim. Pays-Bas, 75 (1956) 487.
223. C.J. Cramer and D.G. Truhlar, J. Am. Chem. Soc., 113 (1991) 8305.
224. C.J. Cramer and D.G. Truhlar, Science, 256 (1992) 213.
225. C.J. Cramer and D.G. Truhlar, J. Comp. Chem., 13 (1992) 1089.
226. M.J.S. Dewar, E.G. Zoebisch, E.F. Healy, and J.J.P. Stewart, J. Am. Chem. Soc., 107 (1985) 3902.
227. J.J.P. Stewart, J. Comp. Chem., 10 (1989) 209.
228. J.J.P. Stewart, J. Comp. Chem., 10 (1989) 221.
229. J.W. Storer, D.J. Giesen, C.J. Cramer, and D.G. Truhlar, to be published.

230. C.J. Cramer and D.G. Truhlar, J. Comput.-Aid. Mol. Des., 6 (1992) 629.
231. D.A. Liotard, G.D. Hawkins, G.C. Lynch, C.J. Cramer, and D.G. Truhlar, to be published.
232. C.J. Cramer, G.D. Hawkins, and D.G. Truhlar, J. Chem. Soc., Faraday Trans., in press.
233. J.J.P. Stewart, in: Reviews in Computational Chemistry, Vol. 1, K.B. Lipkowitz and D.B. Boyd (eds.), VCH, New York, 1989, 45.
234. W. Thiel, Tetrahedron, 44 (1988) 7393.
235. M.C. Zerner, in: Reviews in Computational Chemistry, Vol. 2, K.B. Lipkowitz and D.B. Boyd (eds.), VCH, New York, 1990, 313.
236. D.R. Armstrong, P.G. Perkins, and J.J.P. Stewart, J. Chem. Soc., Dalton Trans., (1973) 838.
237. C.J. Cramer, G.C. Lynch, G.D. Hawkins, and D.G. Truhlar, QCPE Bull., 13 (1993) 78.
238. R.A. Marcus, J. Chem. Phys., 24 (1956) 979.
239. M.M. Kreevoy and D.G. Truhlar, in: Investigation of Rates and Mechanisms of Reactions, Part I, 4th ed., C.F. Bernasconi (ed.), Wiley, New York, 1986, 13.
240. B.C. Garrett and G.K. Schenter, Int. Rev. Phys. Chem., in press.
241. S. Lee and J.T. Hynes, J. Chem. Phys., 88 (1988) 6863.
242. J. Gehlen, D. Chandler, H.J. Kim, and J.T. Hynes, J. Phys. Chem., 96 (1992) 1748.
243. H.J. Kim and J.T. Hynes, J. Chem. Phys., 96 (1992) 5088.
244. D.G. Truhlar, G.K. Schenter, and B.C. Garrett, J. Chem. Phys., 98 (1993) 5756.
245. M.V. Basilevsky, G.E. Chudinov, and M.D. Newton, preprint.
246. J.S. Kwiatkowski and A. Tempczyk, Chem. Phys., 85 (1984) 397.
247. A. Razynska, A. Tempczyk, and Z. Grzonka, J. Chem. Soc., Faraday Trans. 2, 81 (1985) 1555.
248. A. Tempczyk, Z. Gryczynski, A. Kawski, and Z. Grzonka, Z. Naturforsch., 43a (1988) 363.
249. M. Karelson and M.C. Zerner, J. Am. Chem. Soc., 112 (1990) 9405.
250. T. Fox and N. Rösch, Chem. Phys. Lett., 191 (1992) 33.
251. T. Fox and N. Rosch, J. Mol. Struct. (Theochem), 276 (1992) 279.
252. M.A. Aguilar, F.J. Olivares del Valle, and J. Tomasi, J. Chem. Phys., 98 (1993) 7375.
253. G. Nemethy and H.A. Scheraga, J. Chem. Phys., 36 (1962) 3401.
254. T.J. Richmond, J. Mol. Biol., 178 (1984) 63.
255. J.T. Kellis, K. Nyberg, D. Sali, and A.R. Fersht, Nature, 788 (1988)
256. K.A. Dill, Biochemistry, 29 (1990) 7133.
257. K. Sharp, A. Nicholls, R. Friedman, and B. Honig, Biochemistry, 30 (1991) 9686.
258. J.W. Grate, R.A. McGill, and D. Hilvert, J. Am. Chem. Soc., 115 (1993) 8577.
259. C.J. Cramer and D.G. Truhlar, Chem. Phys. Lett., 198 (1992) 74 and 202 (1993) 567 (Erratum).
260. R. Rein, in: Intermolecular Interactions: From Diatomics to Biopolymers, B. Pullman (ed.), John Wiley & Sons, Chichester, 1978, 307.
261. R. Bonaccorsi, E. Scrocco, and J. Tomasi, Proc. Intl. Symp. Biomol. Struct. Interact., Suppl. J. Biosci., 8 (1985) 627.
262. R. Bonaccorsi, E. Scrocco, and J. Tomasi, Int. J. Quant. Chem., 29 (1986) 717.

263. M. Mezei, Mol. Sim., 10 (1993) 225.
264. A. Laaksonen, L. Nilson, B. Jönsson, and O. Teleman, Chem. Phys., 4 (1989) 81.
265. L. Nilsson and M. Karplus, J. Comp. Chem., 7 (1986) 59.
266. A. Pohorille, W.S. Ross, and J. I. Tinoco, Intl. J. Supercomput. Appl., 4 (1990) 81.
267. J. Pranta, S.G. Wierschke, and W.L. Jorgensen, J. Am. Chem. Soc., 113 (1991) 2810.
268. G.L. Seibel, U.C. Singh, and P.A. Kollmann, Proc. Natl. Acad. Sci., USA, 82 (1985) 6537.
269. S.J. Weiner, P.A. Kollman, D.T. Nguyen, and D.A. Case, J. Comp. Chem., 7 (1986) 230.
270. C.J. Cramer and D.G. Truhlar, J. Am. Chem. Soc., 113 (1991) 8552.
271. C.J. Cramer and D.G. Truhlar, J. Am. Chem. Soc., 115 (1993) 8810.
272. W.L. Jorgensen and J. Tirado-Rives, J. Am. Chem. Soc., 110 (1988) 1657.
273. W.J. Hehre, R. Ditchfield, and J.A. Pople, J. Chem. Phys., 56 (1972) 2257.
274. J. Gao and X. Xia, Biophys. Chem., submitted for publication.
275. P. Bash, U.C. Singh, R. Langridge, and P.A. Kollman, Science, 236 (1987) 564.
276. D.M. Ferguson, D.A. Pearlman, W.C. Swope, and P.A. Kollman, J. Comp. Chem., 13 (1992) 362.
277. V. Mohan, M.E. Davis, J.A. McCammon, and B.M. Pettitt, J. Phys. Chem., 96 (1992) 6428.
278. A.H. Elcock and W.G. Richards, J. Am. Chem. Soc., 115 (1993) 7930.
279. P.E. Young and I.H. Hillier, Chem. Phys. Lett., 215 (1993) 405.
280. P. Seeman, Pharmacol. Rev., 32 (1987) 229.
281. J.J. Urban, C.J. Cramer, and G.R. Famini, J. Am. Chem. Soc., 114 (1992) 8226.
282. D.A. Daugherty and D.A. Stauffer, Science, 250 (1990) 1558.
283. J. Gao, L.W. Chou, and A. Auerbach, Biophys. J., 65 (1993) 43.
284. M. Meot-Ner and C.A. Deakyne, J. Am. Chem. Soc., 107 (1985) 469.
285. M.F. Perutz, G. Fermi, D.J. Abraham, C. Poyart, and E. Bursaux, J. Am. Chem. Soc., 108 (1986) 1064.
286. P. Solmajer, D. Kocjan, and T. Solmajer, Z. Naturforsch., 38c (1983) 758.
287. O. Bastiansen, Acta Chem. Scand., 3 (1949) 415.
288. P. Buckley and P.A. Giguère, Can. J. Chem., 45 (1967) 397.
289. W. Caminati and G. Corbelli, J. Mol. Spectrosc., 90 (1981) 572.
290. H. Frei, T.K. Ha, R. Meyer, and H.H. Günthard, Chem. Phys., 25 (1977) 271.
291. T.K. Ha, H. Frei, R. Meyer, and H.H. Günthard, Theor. Chim. Acta, 34 (1974) 277.
292. L.P. Kuhn, J. Am. Chem. Soc., 74 (1952) 2492.
293. C.G. Park and M. Tasumi, J. Phys. Chem., 95 (1991) 2757.
294. H. Takeuchi and M. Tasumi, Chem. Phys., 77 (1983) 21.
295. E. Walder, A. Bander, and H.H. Günthard, Chem. Phys., 51 (1980) 223.
296. D.E. Woon and T.H. Dunning, J. Chem. Phys., 98 (1993) 1358.
297. K. Raghavachari, G.W. Trucks, J.A. Pople, and M. Head-Gordon, Chem. Phys. Lett., 157 (1989) 479.
298. C.J. Cramer and D.G. Truhlar, J. Am. Chem. Soc., submitted for publication.
299. K.G.R. Pachler and P.L. Wessels, J. Mol. Struct., 6 (1970) 471.
300. A. Ben-Naim, Solvation Thermodynamics, Plenum, New York, 1987.
301. D. Suleiman and C.A. Eckert, J. Chem. Eng. Data, submitted for publication.

302. B.P. van Eijck, R.W.W. Hooft, and J. Kroon, J. Phys. Chem., 97 (1993) 12093.
303. B.P. van Eijck, L.M.J. Kroon-Batenburg, and J. Kroon, J. Mol. Struct., 237 (1990) 315.
304. L.M.J. Kroon-Batenburg and J. Kroon, Biopolymers, 29 (1990) 1243.
305. S. Ha, J. Gao, B. Tidor, J.W. Brady, and M. Karplus, J. Am. Chem. Soc., 113 (1991) 1553.
306. P.L. Polavarapu and C.S. Ewig, J. Comp. Chem., 13 (1992) 1255.
307. C.J. Cramer and D.G. Truhlar, J. Am. Chem. Soc., 115 (1993) 5745.
308. Y. Nishida, H. Ohrui, and H. Meguro, Tetrahedron Lett., 25 (1993) 1575.
309. S.J. Angyal, Angew. Chem., Int. Ed. Engl., 8 (1969) 157.
310. D.J. Giesen, C.J. Cramer, and D.G. Truhlar, to be published.
311. R.B. Hermann, Proc. Natl. Acad. Sci., USA, 74 (1977) 4144.
312. I. Tuñón, E. Silla, and J.L. Pascual-Ahuir, Chem. Phys. Lett., 203 (1993) 289.
313. A. Leo, in: Environmental Health Chemistry, J.D. McKinney (ed.), Ann Arbor Science, Ann Arbor, MI, 1981, 323.
314. C. Hansch, J.P. Bjorkroth, and A. Leo, J. Pharm. Sci., 76 (1987) 663.
315. A. Leo, C. Hansch, and D. Elkins, Chem. Rev., 71 (1971) 525.
316. R. Balducci, A. Roda, and R.S. Pearlman, J. Solution Chem., 18 (1989) 355.
317. S.C. DeVito and R.S. Pearlman, Med. Chem. Res., 1 (1991) 461.
318. A.K. Debnath, R.L.d. Compadre, A.J. Shusterman, and C. Hansch, Envrion. Mol. Mutagen., 19 (1992) 53.
319. J. Grogan, S.C. DeVito, R.S. Pearlman, and K.R. Korsekwa, Chem. Res. Toxicol., 5 (1992) 548.
320. A.J. Leo, Chem. Rev., 93 (1993) 1281.
321. D. Hadzi and B. Jerman-Blazic (eds.), QSAR in Drug Design and Toxicology, Elsevier, Amsterdam, 1987 269.
322. C. Silipo and A. Vittoria (eds.), New Perspectives in QSAR, Elsevier, Amsterdam, 1991.
323. J.-L.M. Abboud, M.J. Kamlet, and R.W. Taft, J. Am. Chem. Soc., 99 (1977) 8325.
324. M.J. Kamlet and R.W. Taft, Prog. Org. Chem., 48 (1983) 2877.
325. M.J. Kamlet, M.H. Abraham, R.M. Doherty, and R.W. Taft, J. Am. Chem. Soc., 106 (1984) 464.
326. M.J. Kamlet, Prog. Phys. Org. Chem., 19 (1993) 295.
327. M.J. Kamlet and R.W. Taft, Acta Chem. Scand. B, 40 (1986) 619.
328. O. Exner, Prog. Phys. Org. Chem., 18 (1990) 129.
329. L.P. Hammett, Chem. Rev., 17 (1935) 125.
330. H.H. Jaffé, Chem. Rev., 53 (1953) 191.
331. F.W. Fowler, A.R. Katritzky, and R.J.D. Rutherford, J. Chem. Soc. B, (1971) 460.
332. I.A. Koppel and V.A. Palm, in: Advances in Linear Free Energy Relationships, N.B. Chapman and J. Shorter (eds.), Plenum Press, London, 1972, 203.
333. T.M. Krygowski and W.R. Fawcett, J. Am. Chem. Soc., 97 (1975) 2143.
334. R.C. Dougherty, Tetrahedron Lett., (1975) 385.
335. M.J. Kamlet, J.L. Abboud, and R.W. Taft, J. Am. Chem. Soc., 99 (1977) 6027.
336. M.J. Kamlet, J.L. Abboud, and R.W. Taft, J. Am. Chem. Soc., 99 (1977) 8325.
337. U. Mayer, Monatsh. Chem., 109 (1978) 421, 775.

338. C.G. Swain, M.S. Swain, A.L. Powell, and S. Alunni, J. Am. Chem. Soc., 105 (1983) 502.

339. R.W. Taft, J.-L.M. Abboud, M.J. Kamlet, and M.H. Abraham, J. Solution Chem., 14 (1985) 153.

340. R.S. Drago and A.P. Dadmun, J. Am. Chem. Soc., 115 (1993) 8592.

341. G.R. Famini, R.J. Kassel, J.W. King, and L.Y. Wilson, Quant. Struct.-Act. Relat., 10 (1991) 344.

342. L.Y. Wilson and G.R. Famini, J. Med. Chem., 34 (1991) 1668.

343. G.R. Famini, C.A. Penski, and L.Y. Wilson, J. Phys. Org. Chem., 5 (1992) 395.

344. G.R. Famini, W.P. Ashman, A.P. Mickiewicz, and L.Y. Wilson, Quant. Struct.-Act. Relat., 11 (1992) 162.

345. G.R. Famini and S.C. DeVito, in: Biomarkers of Human Exposure, J. Blancato and M. Saleh (eds.), American Chemical Society, Washington, DC, in press.

346. G.R. Famini, B.C. Marquez, and L.Y. Wilson, J. Chem. Soc., Perkin Trans. 2, (1993) 773.

347. D.F.V. Lewis, in: Reviews in Computational Chemistry, Vol. 3, K.B. Lipkowitz and D.B. Boyd (eds.), VCH, New York, 1992, 173.

348. J.S. Murray, P. Lane, T. Brinck, and P. Politzer, J. Phys. Chem., 97 (1993) 5144.

349. P. Politzer, J.S. Murray, P. Lane, and T. Brinck, J. Phys. Chem., 97 (1993) 729.

350. P. Politzer, J.S. Murray, M.C. Concha, and T. Brinck, J. Mol. Struct. (Theochem), 100 (1993) 107.

351. L.H. Hall and L.B. Kier, in: Reviews in Computational Chemistry, Vol. 2, K.B. Lipkowitz and D.B. Boyd (eds.), VCH, New York, 1991, 367.

352. I.B. Bersuker and A.S. Dimoglo, in: Reviews in Computational Chemistry, Vol. 2, K.B. Lipkowitz and D.B. Boyd (eds.), VCH, New York, 1991, 423.

353. C.J. Cramer, G.R. Famini, and A.H. Lowrey, Acc. Chem. Res., 26 (1993) 599.

354. I. Alkorta, H.O. Villar, and J.J. Perez, J. Comp. Chem., 14 (1993) 620.

P. Politzer and J.S. Murray
Quantitative Treatments of Solute/Solvent Interactions
Theoretical and Computational Chemistry, Vol. 1

Some Effects of Molecular Structure on Hydrogen-Bonding Interactions. Some Macroscopic and Microscopic Views from Experimental and Theoretical Results

R. W. Taft[1] and J. S. Murray[2]

[1]Department of Chemistry, University of California, Irvine, California, 92717
[2]Department of Chemistry, University of New Orleans, New Orleans,
  Louisiana, 70148, USA

1. INTRODUCTION

The understanding of how living organisms and nonliving matter are affected by interactions with organic chemicals is highly dependent upon an adequate knowledge of how molecular structures produce specific controllable properties, e.g. stable site-specific interactions classified as hydrogen bonds. Indeed, in the life sciences and in chemistry in general, the hydrogen-bond (HB) is perhaps the single most important kind of specific molecular interaction between molecular units. Its influence is exerted, for example, in drug solubility, bodily transport and docking to active sites. It was first recognized in the physical properties of liquids, and it was thought to involve only fluorine, oxygen or nitrogen atoms bridged by an intervening hydrogen [1]. Since then structural studies have indicated that atoms such as chlorine, bromine and sulfur can serve as hydrogen bond acceptors [2, 3], and that hydrogens bonded to carbon and sulfur can serve as hydrogen bond donors [3]. Another significant type of site-specific noncovalent interaction has been identified as "halogen bonding"; the latter occurs between an electron-pair donor (such as a hydrogen-bond acceptor) and the positive ends of the larger halogen atoms (chlorine, bromine and iodine) [4]. In this chapter we will focus primarily on the more traditional hydrogen bonding interactions.

Hydrogen bonded complexes between neutral molecules can be viewed as a union or partnership between two units, the hydrogen-bond-donor (HBD) and the hydrogen-bond-acceptor (HBA), in which the former retains the hydrogen atom covalently bonded (very much as in the uncomplexed form), whereas the

HBA utilizes an unshared electron pair to form a relatively long, largely electrostatic "bond". Strengths of hydrogen bonds for many neutral organic compounds range from 2 to 15 kcal/mole [3], with equilibrium constants of formation ranging from 10^1 to 10^4. In addition hydrogen bonds are generally very rapidly made and broken. The equilibrium constants of formation, also called stability constants, decrease in aqueous and other polar solvents compared to the corresponding values in relatively nonpolar solvents and the gas phase.

In this chapter we will begin with a brief history of the development of quantities reflecting the hydrogen-bond accepting and donating abilities of first solvent, and then solute molecules. This will be followed by recent developments in this area involving hydrogen bond complexes and polyfunctional hydrogen bond donors and acceptors, with an emphasis on biologically-interesting systems. Theoretical approaches for understanding and predicting fundamental monofunctional solute hydrogen bond acceptor and donor parameters will then be presented, followed by theoretical findings showing that shifts in vibrational frequencies upon complexation with methanol, $\Delta\nu_{OH}$, can also be correlated and predicted using a similar approach.

2. BACKGROUND AND DEVELOPMENT OF HYDROGEN BONDING PARAMETERS

2.1. The solvatochromic comparison method: LSER (linear solvation energy relationships)

In initiating studies of linear solvation energy relationships [5-7], Kamlet and Taft's goal was to explore how a variety of solvents influence a particular property of a single solute. Their first studies involved determining the effect of solvent on the wavelength of $\pi \rightarrow \pi^*$ and p–π^* electronic transitions of a carefully chosen set of solutes. For example, N,N-dimethyl-p-nitroaniline is an example of a molecule viewed as a non-HBD solute probe. Through a judicious choice of solutes the aim was to unravel the types and relative strengths of intermolecular interactions that were affecting these spectral transitions. The "solvatochromic" comparison method, as developed by Kamlet and Taft [8-10], is based on the idea that a solvent's ability to stabilize a solute by dipolar, hydrogen bond donor and hydrogen bond acceptor processes can be measured by a suitable choice of solute probe and reference molecules and spectroscopic measurement.

The work of Kamlet and Taft led to a set of three *solvent* parameters, π^*, α and β, of which the values for over two hundred solvents are compiled in a classic 1983 article [11]. A solvent's π^* value is viewed as a measure of its ability to stabilize a neighboring dipole by dipole-dipole and dipole-reduced dipole forces which exist between the solute probe and solvent. α and β are viewed as representing the ability of a solvent to donate or accept, respectively, a hydrogen bond from a solute. For various spectroscopic properties, XYZ, of a fixed solute in a series of solvents, the following equation was proposed:

$$XYZ = XYZ_0 + s\pi^* + a\alpha + b\beta \tag{1}$$

Equation 1 was found to hold for a select group of solvents, namely aliphatic, aprotic and those with only one basic functionality. However for correlating results with also aromatic and polyhalogenated solvents, it was necessary to introduce a correction factor $d\delta$ to π^*, as shown in equation 2, where δ is 0.5 for

$$XYZ = XYZ_0 + s(\pi^* - d\delta) + a\alpha + b\beta \tag{2}$$

halogen solvents and 1.0 for aromatics [11].

The various fitting coefficients in equations 1 and 2 are very important. As Kamlet and Taft pointed out very early in their work, both the signs and magnitudes were required to make chemical sense in order for a good regression to be accepted [5, 6]. In cases where hydrogen-bonding interactions are important the terms $a\alpha$ and $b\beta$ would be significant; in other instances, one and/or the other would drop out of the equation.

2.2 The emergence of solute parameters: LSER

Soon after the development and acceptance of the solvatochromic method as a means for defining solvent scales of polarity/polarizability, and hydrogen-bond-accepting and -donating parameters, the fundamental concept upon which these were based was inverted. The possibility that a conceptually similar approach could be used to assess solute-to-solute variations of some property in a fixed solvent was proposed. Clearly this seemed a reasonable approach to the development of chemically meaningful quantitative structure activity relationships (QSAR); the latter have found extensive use in the design of drugs, prediction of toxicity, biological activity, environmental transport chromatographic retention.

One of the earliest tests of this idea was its use in correlating retention in reversed phase liquid chromatography [12-14]. Soon thereafter it was used to study octanol-water partition coefficients, and solubility in water [15]. The general LSER used in these studies was of the general form shown in equation 3.

$$XYZ = XYZ_0 + mV_2 / 100 + s(\pi_2^* - d\delta) + a\alpha_2 + b\beta_2 \qquad (3)$$

The subscript 2 denotes a solute molecular property, with XYZ now being a solute macroscopic property. Although it was found that solvent parameters could be used in equation 3 as a first approximation to solute parameters, the concept expressed in equation 3 led to the development of separate scales of solute parameters.

Abraham et al, following upon work by Taft et al [16], developed *solute* scales of hydrogen-bond acidities (α_2^H) [17, 18] and basicities (β_2^H) [19, 20]. These were taken to be linear functions for the formation of 1:1 complexes between the solute molecule and a given reference base or acid, respectively, in CCl_4. These have been used extensively in the past few years and had led to their own refinements, as is discussed in Chapter 3 of this book.

3. HYDROGEN-BONDED COMPLEXES

The generalized basic equation for the equilibrium constant (K_c) for formation of a hydrogen bond complex is [21]:

$$\log K_c = -1.10 + (7.35)\alpha_2^H \beta_2^H \qquad (4)$$

where K_c is in units of l/mole or M^{-1}. In equation 4 the α_2^H and β_2^H are the scaled descriptor parameters for the relative HBD and HBA strengths of the combining neutral molecules. This equation was found to hold for over 1300 equilibrium constants with a precision of approximately 0.10 in log K [21]. There are known to be some systematic derivations for certain pairs, e.g. $HCCl_3$ or $(C_6H_5)_2NH$ with amines, anilines, ethers or sulfides [18, 22-24]. Equation 4 specifies that the relative orders of HBA ability are independent of the HBD ability and vice versa. Further, for hydrogen bonding to be observed

K_c cannot be less than 0.09 M^{-1}, which will occur if either α_2^H or β_2^H is zero. If, for example, the α_2^H and β_2^H values are both unity, $K_c = 1,780,000$ M^{-1} is predicted!

The coefficients -1.10 and 7.35 in equation 4 apply for dilute solutions in CCl$_4$ at 25° C, but values have also been found for Cl$_3$CCH$_3$ solutions [25], -1.13 and 6.84 at 25° C, and for the gas phase [26], -0.86 and 9.13 at 22° C. Equation 1 has been found [27] to be consistent with apparently reasonable accuracy to the formation constants of adenine-uracil and guanine-cytosine HB complexes in the gas phase and, with allowance for desolvation, in HCCl$_3$ solutions [28]. It is anticipated that molecular recognition and docking of drug molecules will also be within the utility of this equation.

Values for a great many α_2^H and β_2^H parameters (particularly the latter) have been made available for mostly singly-functionalized molecules [18, 22, 29, 30]. Laurence and Berthelot, from their IR spectroscopic studies, have determined quantitative family structure-HBA relationships for alchohols, phenols and nitriles, formamidines, amides, ureas, lactones, carbonates, nitro compounds, and lactams [31-37]. Using substituent constants [38] it has been possible to reliably predict the β_2^H values of drug substructures (fragments) containing many substituted functions. Furthermore, for many polyfunctional HBA's, the FTIR work has been used to distinguish the strongest HBA function from weaker ones [39]. A particularly important but simple rule has been generated:

> In polybasic (HBA) molecules, all functions can accept from excess HBD solvent except in push-pull systems (those with conjugated pi electron-pair donor and acceptor atoms) where only the (enhanced) pulling function is HBA.

The predictions made with this rule for simple drug molecules have been found to be consistent with observed partition constant behavior.

Table 1 gives some typical α_2^H and β_2^H values that include a number of functional groups found in drug molecules. The specificity of both parameters to molecular structures is illustrated by several examples.

For solubility and partitioning of solutes between immiscible liquids it has been found that HB formation between solute and solvent can be usefully treated by incorporation of the essence of equation 5 with other appropriate terms, where SP stands for any such solubility property [40-43].

$$SP = SP_0 + A(\text{cavity term})_{\text{solute}} + B\pi_1^*\pi_2^* + C\alpha_1\beta_2 + D\alpha_2\beta_1 \tag{5}$$

π_1^*, α_1 and β_1 are the solvent polarity/polarizablity, hydrogen-bond-donating and -accepting terms for the solvent, while π_2^*, α_2 and β_2 are the corresponding terms for the solute. This treatment applies to solubility in one solvent or to paritioning between two solvents with single site HBD and HBA solutes. Notice that one or more terms can easily drop out of equation 2; for example, if α_1 is essentially zero, then the $C\alpha_1\beta_2$ term goes to zero, and if β, is near zero, the $D\alpha_2\beta_1$ terms goes to zero. Thus there is a long way to go in extending equation 5 to drug-receptor complexes. In docking, desolvation of the drug and the receptor site is indicated to be very important. This depends in a potentially complex manner on the polysite nature of the drug and of the receptor. Furthermore, the structures of the receptor sites are generally in doubt.

The most successful predictive equations for solute solubility and partitioning are due to Abraham, (equation 6) [43] and to Taft (equation 7) [44].

$$\log P = c + vV_x + rR_2\, s\pi_2 + a\sum\alpha_2 + b\sum\beta_2 \tag{6}$$

$$\log P = c + vV_X + s(\mu_2)^2 + a\,\epsilon\,\alpha_2 + b\,\epsilon\,\beta_2 \tag{7}$$

The solute volume V in equations 6 and 7 is calculated by the McGowan and Abraham equation [45]. This molecular volume has been recently further confirmed by Monte Carlo integrative calculation (communicated to us by J. B. Foresman) that give the same volumes to an average deviation of 2% for a diverse set of 140 molecular structures [46]. London contributions are thought to be colinear with V. The parameters R and π of equation 6 are polarizability/dipolarity terms [43], and the μ^2 term of equation 7 is the molecular dipole moment squared. Taylor has previously utilized this parameter for nonspecific solute/solvent dipolar interaction [47]. Generally, the terms R, π and μ^2 are of second order importance compared to the HB terms in equations 6 and 7. Equations 6 and 7 were derived starting with known α_2^H and β_2^H values for monofunctional compounds, and by various procedures, the treatment was extended to relatively simple polyfunctional HBA and HBD solutes that can form complexes of 1:m stoichiometry with the large excess of HB forming solvents.

Table 1

Some typical values of α_2^H and β_2^H obtained directly from equilibrium constants for 1:1 complexes in CCl_4

Molecule	α_2^H	β_2^H	Molecule	α_2^H	β_2^H
$HC{\equiv}CC_6H_5$	.12	.16	$(C_2H_5)_2S$	.00	.26
[piperidine ring] NH	.08	.74	[tetrahydropyran ring] O	.00	.50
i-Pr_2NH	.08	.74	$CH_3CO_2C_2H_5$	.00	.47
O [morpholine ring] NH	.15	.64	CF_3CO_2H	.95	.26
$N(CH_3)_3$	.00	.72	$CF_3CO_2C_2H_5$	.00	.26
$N(C_3H_7)_3$	.00	.58	propylene carbonate	.00	.50
[quinuclidine ring] N	.00	.80	C_2H_5OH	.33	.44
[pyridine] N	.00	.64	CH_3CO_2H	.59	.43
$(CH_3)_2N$—[pyridine] N	.00	.84	C_6H_5CHO	.00	.41
[pyrimidine] N N	.00	.47	$HCONHCH_3$	.38	.66
N [pyrazine] N	.00	.44	CF_3CH_2OH	.57	.21
C_4H_9CN	.00	.44	$CF_3)_3COH$	.86	.02
C_6H_5CN	.00	.41	$(CH_3)_2NCO_2C_2H_5$	.00	.63
$(CH_3)_2N$—[benzene]—CN	.00	.48	$(C_2H_5)_2NCOCH_3$	.00	.77
$(CH_3)_2N$-C=C with H, CN	.00	.60	[benzene]—$NHCH_3$	.08	.38
CH_3N [imidazole] N	.00	.83	[benzene]—$SO_2N(CH_3)_2$	.00	.49
$[(CH_3)_2N]_2C{=}NH$	.44	.93	$(C_6H_5CH_2)_2SO$	.00	.74
DBU=diazobicycloundecene	.00	1.07	$(C_6H_5)_3PO$	.00	.92

In the structure of the antibacterial drug isoniazid (**I**), the six arrows are used to illustrate that here the value of m can be as large as its six available HB sites: two HBA functions (carbonyl and pyridine), two HBD functions (amide NH and amino NH_2), two sites on the carbonyl (one on pyridine N), two sites on N_2 (one on NH) [39]. The 1:2 complexes for the carbonyl and the amino functions are less stable than their respective 1:1 complexes. Stereoelectronic factors reduce the stabilities of HB complexes having greater than unit stoichiometry. For these reasons the overall molecular α or β values (given with prefix Σ in equation 6 and ∈ in equation 7 may be complex, i.e., not just a simple summation.

I

Taft et al are currently developing procedures for obtaining ∈α and ∈β values for polyfunctional hydrogen-bond-donating and -accepting solute molecules [44]. A discussion of his progress to date in determining ∈α values will be the topic of the next section.

4. RECENT DEVELOPMENTS IN THE DETERMINATION OF AN OVERALL HYDROGEN-BOND-DONATING SOLVATION ENERGY AND PARAMETER (∈α)

4.1 Methodology

The new methodology developed by Taft et al [48] for the determination of the overall HBD solvation energy and parameter ∈α involves several steps. These will be described below.

Measurements of log P(octanol/water) and log P(chloroform/water) allow one to calculate the value of log P for octanol/chloroform partitioning i.e. log P(octanol/chloroform) = log P(octanol/water) - log P(chloroform/water). In spite of possible error accumulation in the value of log P(octanol/chloroform), the following quite useful precise correlation equation has been found [44].

$$\log P \left(\text{octanol} / \text{chloroform} \right) = 3.23(\pm.05)\,\alpha_2^H - 1.00(\pm.06)\,V_x / 100 - .03(\pm.06) \quad (8)$$

The number of solutes used to determine equation 8 is 81; the linear correlation coefficient is 0.992 and the standard deviation is 0.12. 36 of the 81 solutes were single-site HBD's and 45 were non-HBD's. Equation 9 has been applied to the 48 HBD's in Table 2 to effectively calculate $\in \alpha$ values. These

$$\epsilon\alpha = \left[\log P \left(\text{octanol} / \text{chloroform} \right) + 100\,V_x / 100 + .03 \right] / 3.23 \quad (9)$$

should be nearly equal to the corresponding α_2^H values for monofunctional HBD's. Table 2 illustrates that this is indeed the case.

The application of equation 9 to polyfunctional HBD's allows $\in \alpha$ to be calculated for multi-site HBD's. Several requirements must be met for equations 8 and 9 to be valid: 1) it is required that there be a linear free energy relationship between the hydrogen-bonding of neutral single-site HBD solutes with the HBA solvent octanol and the α_2^H descriptors from the equilibrium constants for the 1:1 hydrogen bond complexes with CCl_4 and $CHCl_3$; 2) the V_x descriptor must satisfactorily describe the greater unfavorable free energy of cavity formation involved in octanol than in chloroform; 3) solute HBA and dipolar/polarizability interactions must be nearly equal in wet octanol and chloroform and therefore make little or no contribution to log P (octanol/chloroform) values. In other words, just as the differential solvation for octanol/water cancels the solute HBD effects, so also does the differential solvation for octanol/chloroform cancel the HBA effects.

4.2 Current Results

Table 3 lists $\in \alpha$ values for hydrocortisone and some salicylic derivatives, and Table 4 lists $\in \alpha$ for two polyfunctional HBD families of molecules, those containing –CONH and those containing $-CONH_2$ (amides).

It is interesting to examine critically the $\in \alpha$ values obtained for polyfunctional HBD molecules, such as are in Tables 3 and 4, to determine whether $\in \alpha = \Sigma \alpha_2^H$, where the α_2^H's involved in the summation are taken from appropriate molecules that could be considered as subunits of the larger molecule. For this additivity to prevail there must be very little intramolecular

Table 2.
Comparison of $\in \alpha$ and α_2^H values

Molecule	$\in \alpha$	α_2^H	Molecule	$\in \alpha$	α_2^H
methanol	0.29	0.36	bromoacetic acid	0.69	0.67
ethanol	0.33	0.33	p-Me benzoic acid	0.62	0.60
propanol	0.36	0.33	benzoic acid	0.62	0.59
s-propanol	0.32	0.32	p-Cl benzoic acid	0.62	0.65
butanol	0.38	0.33	o-Cl benzoic acid	0.69	0.64
i-butanol	0.37	0.33	o-NO$_2$ benzoic acid	0.76	0.68
s-butanol	0.33	0.32	o-NO$_2$ phenol	0.07	0.05
t-butanol	0.36	0.32	o-OMe phenol	0.20	0.26
benzyl alcohol	0.34	0.38	p-OMe phenol	0.56	0.57
2-Ph ethanol	0.35	0.33	m-OMe phenol	0.58	0.59
allyl alcohol	0.39	0.38	3,5-Me phenol	0.56	0.57
2,2,2-trifluoroethanol	0.58	0.57	2,5-Me phenol	0.57	0.54
N-Me Ph carbamate	0.18	0.25	2,4-Me phenol	0.58	0.53
N-Mc benzamide	0.31	0.25	phenol	0.58	0.60
N-Ph N',N'-Me$_2$ urea	0.33	0.40	p-Cl phenol	0.68	0.67
N-Me benzenesulfonamide	0.27	0.40	ethyl paraben	0.66	0.69
acetanilide	0.45	0.45	p-acetyl phenol	0.73	0.72
benzamide	0.47	0.49	m-NO$_2$ phenol	0.76	0.79
O-Ph carbamate	0.54	0.50	p-NO$_2$ phenol	0.84	0.82
benzene sulfonamide	0.53	0.60	1-naphthol	0.67	0.61
acetic acid	0.60	0.58	2-naphthol	0.66	0.61
propionic acid	0.60	0.54	2-aminopyrimidine	0.27	0.27
butanoic acid	0.57	0.54	carbazole	0.40	0.47
pentanoic acid	0.60	0.54			

ave. dev. = $\pm$ 0.035;   Ph = C$_6$H$_5$;   Me = CH$_3$

interaction between the HBD substructure as well as between these and any HBA substructures. The intramolecular interactions may take the form of (1) inductive effects (including perhaps σ bond resonance effects); (2) pi electron conjugative (push-pull) resonance effects; (3) steric effects; (4) tautomerism; (5) intramolecular HBD/HBA cyclic HB chelation; (6) some combination of these effects.

Hydrocortisone (Table 3) is a trihydroxysteroid with $\in \alpha$ = 1.12. The structure is sufficiently rigid to prevent intramolecular HB chelation. A value of α_2^H = 0.40 is reasonable for both of the two OH groups with the common intervening alpha keto function (cf. structure II in Table 3) so the observed value of $\in \alpha$ is well accommodated as the sum 2(0.40) + 0.33 = 1.13. Intramolecular chelation of the OH group in tropolone (2-hydroxytropone), on the other hand, is very strong, accordingly $\in \alpha$ is essentially zero (-0.04).

Table 3.
$\in \alpha$ values for hydrocortisone and salicylic derivatives[a]

Molecule	$\in \alpha$	logP(Cl/W)	log P (O/W)	V_x
HOH$_2$C, O / HO / OH **II** hydrocortisone	1.12	0.81	1.61	277.6
O–H / O CH$_3$ **III**	0.20	1.70	1.33	97.5
O–H / O CH$_3$ **IV** / H O	0.29	1.42	1.21	113.1
HO O / H O **V**	0.84	0.56	2.26	99.0
H O / H O **VI**	0.11	2.43	1.81	93.0
O O H / O CH$_3$ **VII**	0.34	1.65	1.59	113.1
HO OH **VIII** diphenylolpropane	1.26	1.16	3.32	186.4

(continued)

Table 3.

∈ α values for hydrocortisone and salicylic derivatives (continued)

Molecule	∈ α	logP(Cl/W)	log P (O/W)	V_x
IX	0.68	0.28	1.19	128.8

[a]logP(Cl/W) and logP(O/W) are logP(chloroform/water) and logP(octanol/water), respectively. V_x is the McGowan-Abraham volume.

The salicylic derivatives in Table 3 give informative ∈ α results. The vanillin (IV), o-OCH_3 phenol (III) and o-hydrogenaldehyde (VI) have ∈ α values (0.29, 0.20, and 0.11, respectively) that are much smaller than that for a free phenolic OH ($\alpha \approx 0.60$) due presumably to significant amounts of intramolecular hydrogen bonding in III, IV and VI. The somewhat greater ∈ α value for IV than III is reasonable in view of the expected electron-withdrawing effects of the p-CHO. For o-hydroxyaldehyde VI, the smaller value of ∈ α (0.11) is reasonable in terms of the more stabilizing chelation ring size. The value of ∈ α (0.84) for salicylic acid (V) is also very significantly less than $\Sigma \alpha_2^H$ (1.2) for isolated OH and CO_2H groups. The chelate structure given in Table 3 for V offers a satisfactory explanation. The ∈ α of 0.34 for VII is also much smaller than expected for an isolated CO_2H (ca. 0.60), indicating also a major degree of intramolecular chelation.

For aspirin (IX) and diphenylolpropane (VIII) the observed ∈ α values (0.68 and 1.26, respectively) are reasonable for the formation of little or no chelate structures. The aspirin ∈ α value is equal to an α_2^H value of unchelated benzoic acid enhanced by the inductive effect of an ortho acetoxy group. The ∈ α value of the diphenylolpropane corresponds to the expected sum of α_2^H values for two phenol HBD functions.

Table 4 examines ∈ α results for amide, carbamate, imide, and urea functions with only one available –CONH–, as well as for amide, carbamate, and urea functions with –$CONH_2$ substructures. The ∈ α values cover the range from .14 to .79, indicating the presence of a number of effects. For similar functions there are characeristic values: N-alkyl nicotinamides (0.45)

Table 4.
Values for $\in \alpha$ for polyfunctional HBD families[a]

Molecule	$\in \alpha$	logP(Cl/W)	logP(O/W)	logP(O/Cl)	V_X
–CONH Compounds:					
p-CNC$_6$H$_4$OCONHCH$_3$	0.14	1.87	0.95	-.92	132.7
C$_6$H$_5$OCONHCH$_3$	0.18	1.78	1.16	-.62	117.2
phthalimide	0.28	1.29	1.15	-0.14	102.1
theobromine	0.30	-4.3	-0.72	-0.29	122.2
C$_6$H$_5$CH$_2$CONHCH$_3$	0.31	1.00	0.86	-0.14	111.4
(CH$_3$)$_2$NCONHC$_6$H$_5$	0.33	1.29	0.98	-0.31	135.4
N-methyl allobarbital	0.35	2.15	1.55	-0.60	171.1
N-methyl acetamide	0.37[b]		-1.05		64.5
N-methyl nicotinamide	0.46	-0.38	0.00	0.38	107.3
N-ethyl nicotinamide	0.44	0.13	0.31	0.18	121.4
N-hydroxyethyl nicotinamide	0.47	-0.32	-0.11	0.21	127.2
N-CH$_3$ cyclobarbital	0.47	2.28	1.84	-0.44	192.7
N-CH$_3$ i-C$_3$H$_7$barbital	0.49	2.48	2.35	-0.23	179.7
acetanilide	0.45	0.85	1.16	0.31	111.4
trifluorocetanilide	0.53	1.69	2.21	0.52	116.7
indole-2,3-dione	0.51	0.23	0.83	0.60	102.1
–CONH$_2$ Compounds:					
CH$_3$CONH$_2$	0.40	-2.00	-1.26	0.74	50.6
C$_2$H$_5$CONH$_2$	0.44	-1.40	-0.66	0.74	64.7
C$_6$H$_5$CO$_2$(CH$_2$)$_4$CONH$_2$	0.45	1.72	1.39	-0.33	175.2
C$_6$H$_5$CONH$_2$	0.47	0.11	0.64	0.53	97.3
CCl$_3$CONH$_2$	0.51	0.31	1.04	0.73	87.3
C$_6$H$_5$OCONH$_2$	0.54	0.40	1.08	0.68	103.2
nicotamide	0.61	-1.37	-0.37	1.00	93.2
carbamazepine	0.71	1.99	2.45	0.46	181.1
C$_6$H$_5$NHCONH$_2$	0.79	-0.063	0.83	1.46	107.3

[a]As in Table 3, Cl, W and O stand for chloroform, water and octanol.
[b]α_2^H value

are increased by the ring azo inductive effect compared to aliphatic or aromatic CONH amides or ureas (0.33); N-alkylbarbitals appear roughly similar (0.443 $\pm$ 0.06) to the N-alkyl nicotinamides; CF$_3$ and CCl$_3$ inductive effects increase $\in \alpha$ for trifluoroacetanilide compared to acetanilide and for CCl$_3$CONH$_2$ compared to CH$_3$CONH$_2$ or C$_2$H$_5$CONH$_2$; the second carbonyl group of indole-2,3-dione increases $\in \alpha$ compared with the value of ca. 0.33 for n-alkyl amides and ureas. The relatively large $\in \alpha$ values for nicotinamide, carbamazepina and phenylurea suggest second (or third) NH site contributions, but the only supporting evidence available is the larger $\in \alpha$ value (0.79) than corresponding α_2^H values (0.60) for phenylurea.

It must be stressed that $\in \alpha$ values act directly to very significantly enhance log P values for wet octanol/alkane partition constants. The molecules indicated to have significant intramolecular chelation in wet octanol (Table 3) may or may not be involved with the same abundance in water. The result will correspondingly affect the contribution to the log P(octanol/water) value.

5. THEORETICAL REPRESENTATIONS OF β_2^H, α_2^H and $\Delta\nu_{OH}$

5.1 General Interaction Properties Function (GIPF) Approach

We have recently developed a general approach which permits the analysis, correlation and prediction of properties that are determined by molecular interactions in fluid media (solute-solvent, solute-solute and/or solvent-solvent), e.g. boiling points [49], critical properties [49], supercritical solubilities [50, 51] and enhancement factors [52], and liquid-liquid partition coefficients [53, 54]. Our methodology involves the use of a General Interaction Properties Function (GIPF) [55, 56], represented by equation 10, and is applicable to properties involving both covalent and

$$\text{Property} = f\left[\text{area}, \bar{I}_{S,min}, V_{S,max}, V_{S,min}, V_{min}, \Pi, \sigma_{tot}^2, \nu\right] \tag{10}$$

noncovalent interactions. Chapter 8 in this book gives a detailed discussion of the methodology and applications of this approach. In this section we will briefly define and discuss the quantities in equation 10; the latter will be used in sections 5.2 and 5.3 in our multivariable correlations with hydrogen-bonding parameters and O–H frequency shifts.

GIPF is a collective term that we apply to a group of relationships involving one or more of the computed quantities within the brackets in equation 10. The molecule's surface area is a measure of its size; $\bar{I}_{S,min}$ is the average local ionization energy on the molecular surface and reflects the tendency for charge transfer [57]; $V_{S,max}$, $V_{S,min}$ and V_{min} are indicators of long range attraction for nucleophiles and electrophiles, the first two being restricted to the molecular surface while the last can be anywhere in three-dimensional space [57, 58]; Π and σ^2_{tot} represent local polarity [59] and the variability of the surface electrostatic potential [49-52], respectively, and ν is an "electrostatic" balance term [49, 52]. It should be emphasized that by means of the GIPF approach, quantities computed for an isolated molecule can be used to correlate and predict solution and liquid phase properties.

We normally evaluate these quantities by *ab initio* self-consistent-field molecular orbital (SCF-MO) procedures, although other approaches, such as density functional methods, could certainly be used as well. We do not include a discussion of computational methodology in this chapter, since excellent reviews are available elsewhere [60, 61]. However one important point that is worth stressing is that obtaining reliable results for molecules containing second-row atoms requires that the basis set include polarization functions [60]; the latter are necessary for properly describing atoms beyond the first row of the periodic table. Our surfaces are defined as the 0.001 au contour of the electronic density [62]; surface areas are obtained by means of grids of equidistant points, converted to units of Å^2 [50-52, 55, 56].

The electrostatic potential $V(\mathbf{r})$ that the nuclei and electrons of a molecule create in the surrounding space is given rigorously by equation 11, where Z_A is the charge on nucleus A, located at R_A, and $\rho(\mathbf{r})$ is the electronic density function of the molecule.

$$V(\mathbf{r}) = \sum_A \frac{Z_A}{|\mathbf{R}_A - \mathbf{r}|} - \int \frac{\rho(\mathbf{r}') \, d\mathbf{r}'}{|\mathbf{r}' - \mathbf{r}|} \tag{11}$$

$V(\mathbf{r})$ has emerged over the course of the past two decades as a very useful analytical tool in the study of molecular reactive behavior [63-66]; it is a real physical property, which can be determined experimentally [64] as well as computationally. The first term on the right side of equation 11 represents the contribution of the nuclei and is positive, while the second term reflects the electronic charge and is negative.

The site-specific quantities derived from the electrostatic potential are the spatial minima, V_{min}, and the surface extrema, $V_{S,min}$ and $V_{S,max}$. Molecular sites reactive toward electrophiles can be identified and ranked by means of either V_{min} or $V_{S,min}$ [57, 58], while the maxima of $V(\mathbf{r})$ on the molecular surface, $V_{S,max}$, serve the analogous purpose for nucleophilic attack [58].

Our global properties defined in terms of V(**r**) are Π, σ^2_{tot} and ν, given by equations 12 - 14. We have introduced these statistically-based quantities in an effort to develop quantitative measures of key aspects of the overall surface potentials of molecules.

$$\Pi = \frac{1}{n}\sum_{i=1}^{n}\left|V(\mathbf{r}_i) - \bar{V}_S\right| \tag{12}$$

$$\sigma^2_{tot} = \sigma^2_+ + \sigma^2_- = \frac{1}{m}\sum_{i=1}^{m}\left[V^+(\mathbf{r}_i) - \bar{V}^+_S\right]^2 + \frac{1}{n}\sum_{j=1}^{n}\left[V^-(\mathbf{r}_j) - \bar{V}^-_S\right]^2 \tag{13}$$

$$\nu = \frac{\sigma^2_+ \sigma^2_-}{\left[\sigma^2_{tot}\right]^2} \tag{14}$$

$V(\mathbf{r}_i)$ is the potential at any point $\mathbf{r}_i$ on the surface, and $\bar{V}_S$ is their average value: $\bar{V}_S = \frac{1}{n}\sum_{i=1}^{n}V(\mathbf{r}_i)$. In a similar fashion, $V^+(\mathbf{r}_i)$ and $V^-(\mathbf{r}_j)$ are the positive and negative values of $V(\mathbf{r})$ on the surface, and $\bar{V}_{S^+}$ and $\bar{V}_{S^-}$ are their averages: $\bar{V}_{S^+} = \frac{1}{m}\sum_{i=1}^{m}V^+(\mathbf{r}_i)$ and $\bar{V}_{S^-} = \frac{1}{n}\sum_{j=1}^{n}V^-(\mathbf{r}_j)$.

Π, as defined by equation 12, represents the average deviation of the electrostatic potential on the molecular surface. We view Π as a measure of the local polarity that is present even in molecules having zero dipole moment, and we have demonstrated its relationship to several empirical polarity-polarizability parameters and also to the dielectric constant [55, 59]. The values of Π for most organic molecules fall in the range from 2 to 15 kcal/mole [55, 56, 59]. An example of one which has a dipole moment of zero but has considerable internal charge separation (local polarity) is p–$C_6H_4(NO_2)_2$, for which Π is 16.5 kcal/mole [59].

σ^2_{tot} (equation 13) is the sum of σ^2_+ and σ^2_-, which are respectively the variances of the positive and negative regions of $V(\mathbf{r})$ on the molecular surface. We have found σ^2_{tot} as well as σ^2_+ and σ^2_- individually to be of key importance in studies of supercritical solubilities [50-52], boiling points and critical properties [49], and partition coefficients [53, 54]. Since σ_{tot}^2, σ_+^2 and σ_-^2 emphasize the positive and negative extrema, they are interpreted as reflecting tendencies for electrostatic interactions.

The "balance" parameter ν (equation 14) helps us to more accurately represent the manner in which σ_{tot}^2 affects electrostatic interactive tendencies. By definition, ν attains its maximum value of 0.250 when σ_+^2 and σ_-^2 are equal; ν is therefore an indicator of the degree of balance between the positive and negative extrema on the molecular surface. For example, benzene has σ_+^2 and σ_-^2 values of 7.1 and 9.3 (kcal/mole)^2, so that $\nu = 0.246$. Benzene is therefore expected to interact to a similar degree through both its positive and negative surface $V(\mathbf{r})$ regions, although weakly, as indicated by the small magnitudes of σ_+^2, σ_-^2 and σ_{tot}^2. We have found that ν is of key importance in representing properties such as boiling points and critical temperatures [49], which depend on how well a molecule interacts electrostatically with others of its own kind.

The quantity $\bar{I}_{S,min}$ in equation 10 is site-specific; it is the lowest value of the average local ionization energy $\bar{I}(\mathbf{r})$ computed on the molecular surface. $\bar{I}(\mathbf{r})$ is defined by equation 15 [67].

$$\bar{I}(\mathbf{r}) = \sum_i \frac{\rho_i(\mathbf{r})|\varepsilon_i|}{\rho(\mathbf{r})} \tag{15}$$

$\rho_i(\mathbf{r})$ is the electronic density of the ith molecular orbital at the point $\mathbf{r}$, ε_i is its orbital energy, and $\rho(\mathbf{r})$ is the total electronic density function. We interpret $\bar{I}(\mathbf{r})$ as the energy required on the average to remove an electron from a point $\mathbf{r}$ in the space of an atom or molecule. In particular, we have found that the positions where $\bar{I}(\mathbf{r})$ has its lowest values on the molecular surface, the $\bar{I}_{S,min}$, are indicative of the sites that are most susceptible to charge transfer with electrophiles [57, 67].

5.2. Correlations with Hydrogen Bond Basicity

We have shown earlier for several families of molecules (azines, alkyl ethers, primary amines and molecules containing double-bonded oxygens), taken separately, that the HF/STO-5G*//HF/STO-3G* V_{min} of the hydrogen-bond-accepting heteroatoms (O or N) correlate well with both the *solvent* hydrogen-bond-acceptor parameter β [58, 68] and the *solute* hydrogen bond acidity term β_2^H [58].

An important advance was that using both $V(\mathbf{r})$ and $\bar{I}(\mathbf{r})$, we were able to correlate the acidities of all of the Group V - VII hydrides with a *single* relationship [53]. This prompted us to explore whether $\bar{I}_{S,min}$ plus an electrostatic potential term (e.g. V_{min} or $V_{S,min}$), and perhaps, others, could be used in a single multi-variable relationship to describe β_2^H without the necessity of grouping molecules into families. We view $\bar{I}_{S,min}$ as reflecting the ease of charge transfer [5], while V_{min} and $V_{S,min}$ indicate the tendency for an electrophilic interaction [5,7]. (Because of the current more frequent use of the *solute* parameters [22] compared to the *solvent* parameters [23], we have chosen to concentrate on relationships with β_2^H instead of β.)

For a group of eighteen molecules we computed HF/6-31G* electrostatic potentials and local ionization energies on 0.001 au molecular surfaces using HF/6-31G* structures. These compounds are listed in Table 5 together with their β_2^H values [22] and our computed $V_{S,min}$, $\bar{I}_{S,min}$ and Π values [21].

We found that β_2^H can be described well by equation 16 [21]; the linear correlation coefficient is 0.985. The most important term in equation 16 is $V_{S,min}$, followed by $\bar{I}_{S,min}$ and Π. The linear correlation coefficient for the

$$\beta_2^H = -0.0275\,V_{S,min} - 0.0365\,\bar{I}_{S,min} - 0.0164\,\Pi + 0.112 \qquad (16)$$

relationship between β_2^H and $V_{S,min}$ alone is 0.920. Addition of either $\bar{I}_{S,min}$ or Π improves the correlation coefficient to 0.954 or 0.957, respectively. The $\bar{I}_{S,min}$ charge transfer term appears to be critical for merging the different families into one relationship. For example, both $(SCH_2CH_3)_2$ and $S(CH_2CH_3)_2$ have very weakly negative $V_{S,min}$ values reflecting the diffuse nature of the sulfur lone pairs and a relatively small propensity for electrophilic intermolecular interactions; at the same time they have the two lowest $\bar{I}_{S,min}$ values in Table 5, indicating relatively facile charge transfer. The net result is that their predicted β_2^H values are relatively close to the experimentally-derived values; predictions based on either $V_{S,min}$ or $\bar{I}_{S,min}$ alone would be underestimated or overestimated, respectively. The case of acetonitrile is opposite; it has a fairly strong $V_{S,min}$ and a high $\bar{I}_{S,min}$, but an

Table 5.
HF/6-31G* calculated properties and experimentally-derived β_2^H and β values.[a]

Molecule	β_2^H	β_2^H (calc) [equation 16]	$V_{S,min}$ (kcal/mole)	$\bar{I}_{S,min}$ (eV)	Π (kcal/mole)
C_6H_6	0.14	0.08	-20.12	11.89	9.06
$(SCH_2CH_3)_2$	0.22	0.23	-23.99	10.88	9.01
$S(CH_2CH_3)_2$	0.29	0.36	-26.89	10.05	7.45
HCO_2CH_3	0.38	0.39	-43.40	15.00	22.22
C_6H_5CHO	0.41	0.45	-40.00	15.15	12.70
CH_3OH	0.41	0.42	-40.03	14.85	14.93
CH_3CN	0.44	0.41	-41.95	13.94	21.09
CH_3CH_2OH	0.44	0.49	-40.16	14.77	11.58
$CH_3CO_2CH_3$	0.45	0.46	-40.54	15.06	12.88
$O(CH_2CH_3)_2$	0.45	0.45	-36.00	14.86	6.37
CH_3COCH_3	0.49	0.47	-41.20	14.92	14.21
pyridine	0.63	0.59	-41.06	12.62	11.65
$(4-CH_3)$pyridine	0.66	0.64	-42.62	12.51	11.10
$(CH_3)_2NCOCH_3$	0.76	0.71	-48.77	14.30	13.73
$(CH_3)_2SO$	0.78	0.84	-57.48	12.83	23.22
$(1-CH_3)$imidazole	0.81	0.77	-50.30	12.41	16.85
pyridine N-oxide	0.82	0.79	-52.65	12.90	18.13
$(CH_3)_3PO$	0.99	1.00	-59.13	12.75	16.58

[a]The calculated properties are from reference [69]; and the β_2^H values are from references [70].

intermediate value of β_2^H. Our predicted value is 0.41, compared to the reported 0.44.

5.3. Correlations with Hydrogen Bond Acidity

We have shown earlier that good correlations exist between the solvatochromic *solvent* hydrogen-bond-donating parameter α and the $V_{S,max}$ associated with the hydrogen to be donated for groups of –OH and –CH acids, taken separately [71]. We also demonstrated that $V_{S,max}$ correlates linearly with the solute hydrogen bond acidity terms α_2^H for groups of aliphatic alcohols and phenols, taken both separately and together [58].

We have now suceeded in eliminating the family dependence of these relationships by means of our GIPF. Table 6 lists our HF/STO-5G*//HF/STO-3G* surface areas, $V_{S,max}$ values and σ_+^2, σ_-^2, σ_{tot}^2, ν and $\nu\sigma_{tot}^2$ values for twenty hydrogen-bond-donating molecules, including –OH, –CH, –NH and –SH acids, along with their α_2^H values.

Equations 17 and 18 give our best two- and three- parameter relationships for α_2^H, drawing from the twenty molecules in Table 6. The linear correlation coefficients are 0.962 and 0.983, respectively.

$$\alpha_2^H = 1.42 \times 10^{-4} \, (\sigma_+^2 \, V_{S,max}) + 5.83 \times 10^{-3} \, (\nu\sigma_{tot}^2) - 0.0289 \tag{17}$$

$$\alpha_2^H = 1.34 \times 10^{-4} \, (\sigma_+^2 \, V_{S,max}) + 6.40 \times 10^{-3} \, (\nu\sigma_{tot}^2) + 8.58 \times 10^6 \, (area)^2 - 0.150 \tag{18}$$

Notice the presence in equations 17 and 18 of $V_{S,max}$, which has earlier been shown to correlate by families with α and α_2^H [58, 71]. In both equations, $V_{S,max}$ is multiplied by σ_+^2 and taken as a single term. In this context it is interesting to note that the best three one-parameter relationships are with $\sigma_+^2 V_{S,max}$, $V_{S,max}$ and σ_+^2; the linear correlation coefficients are 0.936, 0.916 and 0.908, in that order. Apparently the product of σ_+^2 and $V_{S,max}$ is more effective at describing hydrogen bond acidity than is either term alone. We interpret $\sigma_+^2 V_{S,max}$ as a cross term accounting for both global and site-specific nucleophilic intermolecular interactions.

Table 6 gives our predicted α_2^H values using both equations 17 and 18. It can be seen that the inclusion of a term reflecting size is beneficial (with the exception of aniline). It is interesting to note that our predictions differentiate between the three alcohols with the α_2^H value of 0.33; they are ordered by equations 17 and 18 in the same order as their reported solvatochromic α values [11].

5.4 Correlations with O-H Frequency Shifts

Berthelot and Laurence have developed an experimentally-accessible means of quantitating hydrogen-bond-accepting tendencies through the measurement of the IR frequency shift of the methanol O–H peak when

Table 6.
Properties of some hydrogen-bond donors[a]

Molecule	α_2^H	α_2^H (calc) [eq.(17)]	α_2^H (calc) [eq.(18)]	surface area	$V_{S,max}$	σ_+^2	σ_-^2	σ_{tot}^2	ν	$\nu\sigma_{tot}^2$
c-C_6H_{12}	0.0	0.0	0.0	136.8	5.6	2.5	0.7	3.2	0.171	0.5
C_6H_6	0.0	0.0	0.0	115.3	9.0	7.1	9.2	16.3	0.246	4.0
CH_3COCH_3	0.04	0.09	0.06	99.4	14.6	15.9	159.8	175.7	0.082	14.4
CH_3CN	0.09	0.17	0.10	75.9	22.0	23.6	167.8	191.4	0.108	20.7
C_6H_5SH	0.12	0.04	0.09	142.2	13.9	15.8	10.7	26.5	0.241	6.4
CH_3NO_2	0.12	0.24	0.19	81.2	26.8	34.4	81.7	116.0	0.209	24.2
CH_2Cl_2	0.13	0.18	0.13	91.1	22.1	46.3	13.8	60.1	0.177	10.6
$CHCl_3$	0.20	0.20	0.18	107.6	25.7	53.5	7.4	60.9	0.107	6.5
$C_6H_5NH_2$	0.26	0.35	0.38	129.5	25.6	50.4	95.5	145.8	0.226	33.0
$(CH_3)_3COH$	0.33	0.24	0.26	123.5	26.1	31.1	182.7	213.8	0.124	26.5
$CH_3(CH_2)_3OH$	0.33	0.28	0.31	127.9	28.2	35.0	165.9	201.0	0.144	28.9
CH_3CH_2OH	0.33	0.37	0.32	87.1	29.4	45.1	182.4	227.5	0.159	36.2
CH_3OH	0.37	0.41	0.33	64.7	29.8	49.6	181.5	231.0	0.169	39.0
indole	0.44	0.44	0.50	149.1	35.2	76.0	20.7	96.6	0.169	16.3
CF_3CH_2OH	0.57	0.64	0.59	96.2	40.3	85.0	50.2	135.2	0.233	31.5
C_6H_5OH	0.60	0.49	0.52	124.7	34.9	63.8	73.7	137.4	0.249	34.2
2-naphthol	0.61	0.42	0.55	169.8	35.5	56.5	57.4	113.9	0.250	28.5
p-$C_6H_4(Cl)OH$	0.67	0.67	0.71	140.3	41.5	86.2	55.3	141.5	0.238	33.7
p-$C_6H_4(OH)NO_2$	0.82	0.80	0.86	145.9	46.1	83.9	110.0	193.9	0.245	47.5
CF_3CO_2H	0.95	0.98	0.91	95.6	41.3	141.8	38.7	180.5	0.168	30.3

[a]Units are: surface area, in $Å^2$; $V_{S,max}$, in kcal/mole; σ_+^2, σ_-^2, σ_{tot}^2 and $\nu\sigma_{tot}^2$, in $(kcal/mole)^2$; ν is unitless.

methanol's hydroxyl hydrogen is complexed with a hydrogen-bond-accepting molecule [31-37]. This O–H frequency shift is designated Δv_{OH}.

Table 7 lists thirty-two molecules with their experimentally-determined Δv_{OH} values [72] and some calculated HF/6-31G* properties. The molecules in Table 7 represent a variety of families of compounds, including those with second-row hydrogen-bond-accepting atoms.

We have found a good three-parameter correlation, given in equation 19, between Δv_{OH} and the calculated quantities V_{min}, $\bar{I}_{S,min}$ and Π [69].

$$\Delta v_{OH} = -5.508 V_{min} - 25.31 \bar{I}_{S,min} - 7.147 \Pi + 261.1 \qquad (19)$$

The linear correlation coefficient is 0.967 and the standard deviation 27 cm^{-1}. It is interesting to point out the similiarity between the terms in equation 19 and those in the β_2^H relationship, equation 16. In each the electrostatic term is the dominant term, followed by the charge transfer term $\bar{I}_{S,min}$ and then Π. Equation 19 however requires the use of the absolute minimum in the vicinity of the hydrogen-bond accepting atom, while equation 16 includes the surface minimum, $V_{S,min}$. Equations 16 and 19 both indicate that hydrogen-bond accepting abilities as quantitated by β_2^H and Δv_{OH} are determined not only by the electrostatic interaction tendencies but also the charge transfer capabilities of the hydrogen-bond accepting heteroatoms.

6. SUMMARY

The importance of hydrogen bonding interactions in determining macroscopic properties of solutes and/or solvents is the overall theme of this chapter. We have briefly traced the development of the linear solvation energy relationship, beginning with the solvatochromic equation applicable to solvents and developed by Kamlet and Taft, and further extended to solute properties by a number of researchers. Discussions of hydrogen-bonded complexes and Taft's recently-developed $\epsilon \alpha$ parameter for polyfunctional hydrogen-bond-donors have been included with an emphasis on biologically-interesting compounds. Finally, family-independent correlations between hydrogen bond basicity, acidity and O–H frequency shifts and *ab initio* quantum chemical molecular properties have been presented. These suggest that hydrogen-bond-accepting tendencies are determined by both electrostatic

Table 7.

HF/6-31G* calculated properties and experimentally determined $\Delta \nu_{OH}$ values.[a]

Molecule	$\Delta \nu_{OH}$, exp (cm^{-1})	$\Delta \nu_{OH}$, calc (cm^{-1})	V_{min} (kcal/mole)	$\bar{I}_{S,min}$ (eV)	Π (kcal/mole)
Cl$_3$CCN	23.0	40	-39.1	15.20	7.31
ClH$_2$CCN	48.5	41	-48.6	14.50	16.86
HCO$_2$CH$_3$	54.0	44	-58.3	15.00	22.22
C$_6$H$_5$CHO	65.0	94	-55.8	15.16	12.70
(CH$_3$CH$_2$S)$_2$	75.0	95	-31.5	10.88	9.01
CH$_3$CN	75.5	76	-57.8	13.94	21.09
CH$_3$CO$_2$CH$_3$	77.0	100	-56.6	15.06	12.88
CH$_3$OH	116.0	125	-59.9	14.85	12.59
(CH$_3$)$_2$NCN	117.5	149	-64.4	13.33	18.15
CH$_3$CH$_2$OH	120.0	143	-61.4	14.77	11.58
dithiane	121.0	104	-33.1	10.43	10.62
dioxane	126.0	95	-53.7	15.15	11.03
(CH$_3$CH$_2$)$_2$S	146.0	170	-39.3	10.05	7.45
(CH$_3$CH$_2$)$_2$O	150.0	147	-55.8	14.86	6.37
(CH$_3$)$_2$NCHO	150.0	159	-67.6	14.36	15.60
tetramethylene-sulphone	154.0	164	-39.4	10.04	8.40
tetrahydrofuran	158.0	162	-62.0	14.90	8.87
(CH$_3$)$_2$NCOCH$_3$	179.0	180	-68.8	14.30	13.73
(CH$_3$CH$_2$)$_3$P=S	195.0	181	-47.7	9.64	13.85
(2,4–Cl$_2$)pyridine	200.0	169	-55.1	13.45	7.77
(CH$_3$)$_2$S=O	205.0	243	-85.8	12.83	23.22
pyrimidine	213.0	183	-62.8	13.14	12.78
F$_3$CCH$_2$NH$_2$	250.0	223	-66.7	12.36	12.97
(CH$_3$)$_3$P=O	266.0	255	-79.0	12.75	16.58
NH$_3$	275.0	311	-87.9	11.83	18.90
pyridine N-oxide	278.0	234	-77.9	12.90	18.13
pyridine	286.0	249	-70.9	12.62	11.65
(4–CH$_3$)pyridine	304.0	267	-73.0	12.51	11.10
cyclopropylamine	310.0	336	-79.6	11.63	9.67
(1-CH$_3$)imidazole	313.0	279	-82.1	12.41	16.85
CH$_3$NH$_2$	344.0	354	-84.4	11.52	11.25
CH$_3$(CH$_2$)$_3$NH$_2$	354.0	380	-83.8	11.50	7.21

[a]The calculated properties are from reference [69]. The experimentally determined $\Delta \nu_{OH}$ values are from reference [72].

interaction tendencies and the abilities of hydrogen-bond accepting heteroatoms to undergo charge transfer. Hydrogen bond acidity, on the other hand, appears to be primarily electrostatic in nature. It is anticipated that continuing developments in experimental and theoretical research will increase the predictive capabilities afforded by having reliable overall hydrogen-bond-donating and -accepting quantities for both mono- and polyfunctional hydrogen bond acids and bases. Recent developments in this area will be discussed in the next chapter.

ACKNOWLEDGEMENTS

We thank Dr. Peter Politzer for helpful suggestions and greatly appreciate the support provided by ARPA/ONR Contract No. N00014-91-J-1897, administered by ONR. RWT acknowledges support received from the Public Health Service SBIR Phase 2 grant GM46163. Thanks are also due to the coauthors of reference 48, particularly to Dr. A. J. Leo. The present authors are also indebted to Profs. M. Berthelot and C. Laurence for communication of their FTIR results in advance of publication.

REFERENCES

1. L. Pauling, *The Nature of the Chemical Bond*, 3rd. ed. (Cornell University Press, Ithaca, NY, 1960).
2. P. Murray-Rust, W. C. Stallings, C. T. Monti, R. K. Preston and J. P. Glusker, J. Am. Chem. Soc., 105 (1983) 3206.
3. S. Scheiner, in *Reviews in Computational Chemistry*, K. B. Lipkowitz and D. B. Boyd, eds., (VCH Publishers, New York, 1991).
4. P. Politzer and J. S. Murray, in *Supplement D: Chemistry of the Halides*, S. Patai and Z. Rappoport, eds., (Wiley, Chichester, England, in press).
5. M. J. Kamlet and R. W. Taft, J. Chem. Soc. Perkin Trans. II, (1979) 337.
6. M. J. Kamlet, M. E. Jones, J.-L. M. Abboud and R. W. Taft, J. Chem. Soc. Perkin Trans. II, (1979) 342.
7. M. J. Kamlet and R. W. Taft, J. Chem. Soc. Perkin Trans. II, (1979) 349.
8. M. J. Kamlet and R. W. Taft, J. Am. Chem. Soc., 98 (1976) 377.
9. M. J. Kamlet and R. W. Taft, J. Am. Chem. Soc., 98 (1976) 2886.

10. M. J. Kamlet, J.-L. M. Abboud and R. W. Taft, J. Am. Chem. Soc., 99 (1977) 6027.

11. M. J. Kamlet, J.-L. M. Abboud, M. H. Abraham and R. W. Taft, J. Org. Chem., 48 (1983) 2877.

12. P. C. Sadek, P. W. Carr, R. M. Doherty, M. J. Kamlet, R. W. Taft and M. H. Abraham, Anal. Chem., 57 (1985) 2971.

13. J. H. Park, P. W. Carr, M. H. Abraham, R. W. Taft, R. M. Doherty and M. J. Kamlet, Chromatographica, 25 (1988) 373.

14. M. J. Kamlet, M. H. Abraham, P. W. Carr, R. M. Doherty and R. W. Taft, J. Chem. Soc. Perkin Trans. II, (1988) 2087.

15. M. J. Kamlet, R. M. Doherty, M. H. Abraham, Y. Marcus and R. W. Taft, J. Phys. Chem., 92 (1988) 5244.

16. R. W. Taft, D. Gurka, L. Joris, P. v. R. Schleyer and J. W. Rakshys, J. Am. Chem. Soc., 91 (1969) 4801.

17. M. H. Abraham, P. P. Duce, P. L. Grellier, D. V. Prior, J. J. Morris and P. J. Taylor, Tetrahedron Letters, 29 (1988) 1587.

18. M. H. Abraham, P. L. Grellier, D. V. Prior, P. P. Duce, J. J. Morris and P. J. Taylor, J. Chem. Soc. Perkin Trans. II, (1989) 699.

19. M. H. Abraham, P. L. Grellier, D. V. Prior, J. J. Morris, P. J. Taylor, C. Laurence and M. Berthelot, Tetrahedron Letters, 30 (1989) 521.

20. M. H. Abraham, P. L. Grellier, D. V. Prior, J. J. Morris and P. J. Taylor, J. Chem. Soc. Perkin Trans. II, (1990) 521.

21. M. H. Abraham, P. L. Grellier, D. V. Prior, R. W. Taft, J. J. Morris, P. J. Taylor, C. Lauren, M. Berthelot, R. M. Doherty, M. J. Kamlet, J.-L. M. Abboud, K. Sraidi and J. Guiheneuf, J. Am. Chem. Soc., 10 (1988) 8534.

22. M. H. Abraham, P. P. Duce, D. V. Prior, D. G. Barratt, J. J. Morris and P. J. Taylor, J. Chem. Soc. Perkin Trans. II, (1989) 1355.

23. P. C. Maria, J.-F. Gal, J. de Franceshia and E. Fargin, J. Am. Chem. Soc., 109 (1987) 483.

24. C. Tokayama, T. Fujita and M. Nakajima, J. Org. Chem., 44 (1979) 2871.

25. M. H. Abraham, private communication.

26. J.-L. M. Abboud, R. Notario and V. Bottella, private communication, From results of J. Marco, J. M. Orza and J.-L. M. Abboud, Vibr. Spectrosc. 6 (1994) 267.

27. R. W. Taft, unpublished work.

28. W. Saenger, in *Principles of Nucleic Acid Structure*, (Springer-Verlag, New York, 1983), pp. 126-130.

29. M. H. Abraham, J. Phys. Org. Chem., 6 (1993) 684.

30. J.-L. M. Abboud, K. Sraidi, M. H. Abraham and R. W. Taft, J. Org. Chem., 55 (1990) 2230.

31. C. Laurence, M. Berthelot, M. Helbert and K. Sraidi, J. Phys. Chem., 93 (1989) 3799.

32. J.-Y. Le Questel, C. Laurence, A. Lachkar, M. Helbert and M. Berthelot, J. Chem. Soc. Perkin Trans. II, (1992) 2091.

33. F. Besseau, C. Laurence and M. J. Berthelot, J. Chem. Soc. Perkin Trans. II, (1994) 485.

34. C. Laurence, M. Berthelot, M. Lucon and D. G. Morris, J. Chem. Soc. Perkin Trans. II, (1994) 491.

35. C. Laurence, M. Helbert and A. Lachkar, Can. J. Chem., 71 (1993) 254.

36. E. D. Raczyinska, C. Laurence and M. Berthelot, Can. J. Chem., 70 (1992) 2203.

37. M. Berthelot, M. Helbert, C. Laurence and J. Y. LeQuestel, J. Phys. Org. Chem., 6 (1993) 302.

38. C. Hansch, A. Leo and R. W. Taft, Chem. Rev., 91 (1991) 165.

39. J. Y. Le Questel, Université de Nantes, France, Ph.D. Thesis, April 11, 1991.

40. M. J. Kamlet, M. H. Abraham, R. H. Doherty and R. W. Taft, J. Am. Chem. Soc., 106 (1984) 464.

41. R. W. Taft, J.-L. M. Abboud and M. H. Abraham, J. Solvation Chem., 14 (1985) 153.

42. M. H. Abraham, G. S. Whiting, R. Fuchs and E. J. Chambers, J. Chem. Soc. Perkin Trans. II, (1990) 291.

43. M. H. Abraham, Chem. Soc. Rev., 22 (1993) 73.

44. R. W. Taft, plenary lecture at 200th American Chemical Society Meeting, 8/23/93, Chicago, IL.

45. M. H. Abraham and J. C. McGowan, Chromatographia, 23 (1987) 243.

46. M. Frisch and K. B. Wiberg, unpublished results.

47. D. E. Leahy, J. J. Morris, P. J. Taylor and A. R. Wait, J. Chem. Soc. Perkin Trans. II, (1992). The use of μ^2 follows from J. G. Kirkwood, J. Chem. Phys. 2 (1934) 351.

48. R. W. Taft, A. J. Leo, F. Anvia, R. Vasanwala and E. D. Raczynska, Chem. Tech., (submitted).

49. J. S. Murray, P. Lane, T. Brinck, K. Paulsen, M. E. Grice and P. Politzer, J. Phys. Chem., 97 (1993) 9369.

50. P. Politzer, P. Lane, J. S. Murray and T. Brinck, J. Phys. Chem., 96 (1992) 7938.

51. P. Politzer, J. S. Murray, P. Lane and T. Brinck, J. Phys. Chem., 97 (1993) 729.

52. J. S. Murray, P. Lane, T. Brinck and P. Politzer, J. Phys. Chem., 97 (1993) 5144.

53. T. Brinck, J. S. Murray and P. Politzer, J. Org. Chem., 58 (1993) 7070.

54. J. S. Murray, T. Brinck and P. Politzer, J. Phys. Chem., 97 (1994) 13807.

55. J. S. Murray, T. Brinck, P. Lane, K. Paulsen and P. Politzer, J. Mol Struct. (Theochem), 307 (1994) 55.

56. J. S. Murray and P. Politzer, in *Solute/Solvent Interactions*, vol. 1, J. S. Murray and P. Politzer, eds., (Elsevier, Amsterdam, The Netherlands, in press).

57. T. Brinck, J. S. Murray and P. Politzer, Int. J. Quant. Chem., 48 (1993) 73.

58. J. S. Murray and P. Politzer, J. Chem. Res., S (1992) 110.

59. T. Brinck, J. S. Murray and P. Politzer, Mol. Phys., 76 (1992) 609.

60. W. J. Hehre, L. Radom, P. v. R. Schleyer and J. A. Pople, *Ab Initio Molecular Orbital Theory*, (Wiley-Interscience, New York, 1986).

61. J. K. Labanowski and J. W. Andzelm, eds., *Density Functional Methods in Chemistry,* (Springer, Berlin, 1991).

62. R. F. W. Bader, M. T. Carroll, J. R. Cheeseman and C. Chang, J. Am. Chem. Soc., 109 (1987) 7968.

63. E. Scrocco and J. Tomasi, in *Topics in Current Chemistry*, vol. 42, (Springer-Verlag, Berlin, 1973).

64. P. Politzer and D. G. Truhlar, eds., *Chemical Applications of Atomic and Molecular Electrostatic Potentials,* (Plenum Press, New York, 1981).

65. P. Politzer and K. C. Daiker, in *The Force Concept in Chemistry*, B. M. Deb, ed., (Van Nostrand Reinhold Company, New York, 1981), ch. 6.

66. P. Politzer and J. S. Murray, in *Reviews in Computational Chemistry*, vol. 2, K. B. Lipkowitz and D. B. Boyd, eds., (VCH Publishers, New York, 1991), ch 7.

67. P. Sjoberg, J. S. Murray, T. Brinck and P. Politzer, Can. J. Chem., 68 (1990) 1440.

68. J. S. Murray, S. Ranganathan and P. Politzer, J. Org. Chem., 56 (1991) 3734.

69. H. Hagelin, T. Brinck, M. Berthelot, J. S. Murray and P. Politzer, submitted to Can. J. Chem.

70. M. H. Abraham, G. S. Whiting, R. M. Doherty and W. H. Shuely, J. Chem. Soc. Perkin Trans. II, 2 (1990) 1451.

71. J. S. Murray and P. Politzer, J. Org. Chem., 56 (1991) 6715.

72. M. Berthelot, private communication.

P. Politzer and J.S. Murray
Quantitative Treatments of Solute/Solvent Interactions
Theoretical and Computational Chemistry, Vol. 1
© 1994 Elsevier Science B.V. All rights reserved.

New solute descriptors for linear free energy relationships and quantitative structure-activity relationships

Michael H. Abraham

The Chemistry Department, University College London,
20 Gordon Street, London WC1H OAJ, United Kingdom

1. INTRODUCTION

Around the turn of the century, Overton and Meyer independently suggested that the biological activity of a series of compounds could be related to the water-oil partition coefficient of the compounds [1,2]. The activity and the partition coefficient were not expressed in logarithmic form, and the relationship was usually given with the activity as $1/C$ and the partition coefficient as P itself. Here C is, for example, the narcotic concentration of an aqueous solute towards the tadpole, and P the water-olive oil partition coefficient defined as,

$$P = \text{(concentration in organic phase)} / \text{(concentration in water)} \qquad (1)$$

A typical set of results by Meyer [3] is in Table 1. With the exception of propanone, there is an obvious connection between biological activity and partition. Overton [1], in particular, showed with numerous series of compounds that narcotic concentration of aqueous solutes towards the tadpole was related to the solute water-oil partition coefficient, usually that for olive oil. However, it was not until sometime later that the connection was expresed in the quantitative form shown in eqn(2).

$$C \cdot P = \text{constant} \qquad (2)$$

Table 1. Tadpole narcosis and partition [3]

Solute	$1/C$	P (olive oil)
Salicylamide	1300	22.232
Benzamide	500	0.672
Propanone	3	0.146
Monacetin	90	0.099
Chloral hydrate	50	0.053
Ethanol	3	0.026

The earliest application of eqn(2) seems to be that of K.H.Meyer and Gottlieb-Billroth [4], in 1920, who summarised some of the results of Overton and of Meyer. Their analysis is reproduced in Table 2, and shows that the product C·P is reasonably constant. As in Table 1, C is the narcotic concentration of aqueous solutes towards the tadpole, and P is again the water-olive oil partition coefficient.

Table 2. Tadpole narcosis and partition [4]

Solute	C Overton	C Meyer	P Overton	P Meyer	C·P
Ethanol	0.30	0.50	0.03	0.03	0.01-0.02
Methylurethane		0.04		0.04	0.02
Propanone		0.03		0.14	0.04
t-Butyl alcohol	0.13			0.18	0.02
Propan-1-ol	0.11			0.13	0.01-0.02
t-Amyl alcohol	0.057			1.00	0.06
Pentanamide	0.05		0.07		0.004
Ethylurethane	0.04	0.03	0.05	0.14	0.002-0.006
Diethylether	0.024			2.40	0.05
Paraldehyde	0.023		3.00		0.07
Ethylacetoacetate	0.019		4.00		0.08
Acetal	0.012		8.00		0.09
Acetanilide	0.0094		2.00		0.02
Methacetin[a]	0.009		2.00		0.02
Sulfonal[b]	0.009	0.006	4.50	1.10	0.007-0.04
Tetronal[c]	0.0018			4.00	0.007
Trional[d]	0.007	0.0013	16.00	4.50	0.006-0.11
Chloralhydrate	0.006	0.025		0.22	0.001-0.005
Bromalhydrate		0.002		0.70	0.001
Butylchloralhydrate			0.002		1.600.003
Phenol	>0.0053			4.00	0.02
Benzamide	0.003	0.007		0.44	0.003
Phthalide	0.0043		3.30		0.01
Chloroethane	0.004			24.00	0.10
Vanillin	0.0033		3.00		0.01
Phenacitind	0.003		4.00		0.01
2-Methoxyphenol	0.003		30.00		0.09
Bromoethane	0.0023	0.0031		37.00	0.08-0.11
Salicylamide		0.002		14.00	0.03
Piperonal	0.002		100.00		0.20
Chloroform	0.0012			70.00	0.08
4-Methoxyphenol	0.0009		300.00	160.00	0.27-0.14
Chloretone[e]		0.0008		22.80	0.02
Phenylurethane	0.0006		150.00		0.09
Coumarin	0.0006		10.00		0.006
Carbon disulfide	0.0005			50.00	0.03
Menthol	0.0001		250.00		0.03
Thymol	$5.5*10^5$			600.00	0.03
Phenanthrene	$3.7*10^6$			$4.0*104$	0.15
				Mean	0.05

[a] 4-Methoxyacetanilide, [b] $Me_2(SO_2Et)_2$, [c] $Et_2(SO_2Et)_2$, [d] 4-Ethoxyacetanilide, [e] $CCl_3.CMe_2.OH$

Considering the experimental difficulties involved, the product C·P is reasonably constant over a very wide range of concentrations and partition coefficients. Table 2 is given in full from the paper of K.H.Meyer and Gottlieb-Billroth [4] because it represents what seems to be the first quantitative structure-activity relationship, QSAR, ever reported. As will be shown later, eqn.(2) is also a linear free energy relationship, LSER, so that Meyer and Gottlieb-Billroth have predated not only Hammett [5,6] but also Bronsted and Pederson [7] in this respect.

Some years later, Meyer and Hemmi [8] showed that water-oleyl alcohol partition coefficients could be used in eqn(2), again with narcotic concentrations of aqueous solutes towards the tadpole, as shown in Table 3.

Table 3. Tadpole narcosis and partition [8]

Solute	C	P^a	C·P
Ethanol	0.33	0.10	0.033
Propan-1-ol	0.11	0.35	0.038
Butan-1-ol	0.03	0.65	0.020
Pentanamide	0.07	0.30	0.021
Antipyrin	0.07	0.30	0.021
Pyramidon	0.03	1.30	0.039
Benzamide	0.013	2.50	0.033
Dial^b	0.01	2.40	0.024
Salicylamide	0.0033	5.90	0.021
Luminal^c	0.008	5.90	0.048
2-Nitroaniline	0.0025	14.00	0.035
Thymol	0.000047	950.00	0.045
Veronal	0.03	1.38	0.041

^{a} Water-olelyl alcohol.
^{b} Diallylbarbituric acid.
^{c} Ethylphenylbarbituric acid.

It is useful to examine eqn(2) more closely. The chemical potential of a species can be expressed as,

$$\mu = \mu o + RT \ln \mathbf{a} \tag{3}$$

where μo is the chemical potential under a standard set of conditions and $\mathbf{a}$ is the activity. For solutions of solutes, $\mathbf{a}$ is taken as C.f where C is the concentration and f is the activity coefficient, usually defined so that f approaches unity as C approaches zero. Thus through eqn(3), the concentration, C, can be seen to be a free energy (in this context a Gibbs free energy) quantity. A partition coefficient defined as in eqn(1) is simply an equilibrium constant, related to the standard Gibbs free energy change through,

$$\Delta G^\circ = -RT \ln P^\circ \tag{4}$$

Hence eqn(2) is an example of a linear free energy relationship, or LFER, and the work of K.H.Meyer and of Collander in the 1920's and 1930's represents some of the first examples of such a relationship. The term LFER is now usually applied to a relationship involving physicochemical properties; when applied to biological response the term quantitative structure-activity relationship, or QSAR, is employed. Perhaps a general term such as structure-property relationship, SPR, could be used to include LFERs and QSARs [9].

Eqn(2) can be expressed in logarithmic form as eqn(5), or more generally as eqn(6), where SP can be a biological property such as 1/C for tadpole narcosis, or can be a physicochemical property, such as another partition coefficient.

$$\log(1/C) = \log P + \text{constant} \tag{5}$$

$$\log SP = a.\log P + c \tag{6}$$

In an early application of eqn(6), Collander and Barlund [10] examined the logarithmic relationship between the permeability of *Chara Ceratophylla* cells towards aqueous solutes, and water-olive oil and water-ether partition coefficients. Eqn(6) was also applied by Collander[11] to the correlation of different types of partition coefficient,

$$\log P_2 = a.\log P_1 + c \tag{7}$$

Collander was very careful to point out that eqn(7) applied only to partitions in similar solvent systems, otherwise eqn(7) was restricted to homologous series.

Eqn(6) is now referred to as the Overton-Meyer relationship, or sometimes just as the Overton relationship, or rule, and has been applied to numerous series of biological results in aqueous solution, see for example the review of Dearden [12].

The first quantitative analysis of gaseous narcosis was carried out by K.H.Meyer and Gottlieb-Billroth [4], in the same paper in which they discussed aqueous narcosis. They showed that a similar relationship to eqn(2) holds also,

$$C.L = \text{constant} \tag{8}$$

Here, C is the gaseous concentration of the narctic, and L is the Ostwald solubility coefficient of the narcotic in some particular solvent, defined by eqn(9),

$$L = (\text{concentration in solvent})/(\text{concentration in gas phase}) \tag{9}$$

As an example, they related the gaseous narcotic concentration towards mice, to the narcotic L-value in olive oil at 310K, as shown in Table 4, slightly altered from the original table given by Meyer and Gottlieb-Billroth [4]. The product C.L is reasonably constant. This type of analysis is nowadays referred to as the Ferguson rule or relationship, but the work of Meyer and Gottlieb-Billroth [4] predates that of Ferguson [13] by nearly 20 years. It should be pointed out that the Overton rule and the Ferguson rule apply only to 'nonreactive' compounds. Those solutes that interact with the biological system in some specific way are nearly always more potent than calculated - sometimes by orders of magnitude.

Table 4. Gaseous narcotic potency and gaseous solubility [4]

Solute	C(gas, vol%)	L(olive oil)	C.L
Nitrous oxide	100	1.40	140
Dimethyl ether	12	11.6	139
Chloromethane	6.5	14.0	91
Ethylene oxide	5.8	31.0	180
Chloroethane	5.0	40.5	202
Bromomethane	3.5	32.0	112
Pentene	4.0	65.0	260
Diethyl ether	3.4	50.0	170
Methylal	2.8	75.0	210
Bromoethane	1.9	95.0	180
Dimethylacetal	1.9	100.0	190
Diethylformal	1.0	120.0	120
1,2-Dichloroethene	0.95	130.0	124
Chloroform	0.44	265.0	117

mean: 160 ± 49

The results of Meyer and Gottlieb-Billroth [4], given in Table 4, can also be treated through the more usual logarithmic relationship to yield a quite reasonable QSAR,

$$\log C(gas, vol\%) = 2.11 - 0.950 \log L(olive\ oil) \qquad (10)$$

$$n = 14, \quad r = 0.9761, \quad sd = 0.127, \quad F = 242.4$$

Here, and elsewhere, n is the number of data points, r is the overall correlation coefficient, sd is the standard deviation in the dependent variable, and F is the Fisher F-statistic.

K.H.Meyer and Hopff [14] extended this work on gaseous narcotics, and showed that whereas the connection between narcotic concentration and L(olive oil) could be extended to other compounds, there was very little connection between narcotic concentration and L(water). This first attempt to use QSARs to understand the mechanism of gaseous narcosis has received little attention. Only much more recently have QSARs been used to discuss matters such as the site of action of gaseous irritants, as will be shown later in this work.

A rather different type of LFER to that shown in eqn(6) was developed by Hammett [5,6], who put forward the now well known Hammett equation,

$$\log K_X/K_H = \rho \cdot \sigma_X \qquad (11)$$

where K_H and K_X are are the acid dissociation constants of a parent compound and a derivative with substituent X, ρ is a reaction parameter and σ_X is the substituent parameter of X. There have been numerous modifications of the Hammett equation [15], but we deal only with the pioneering work of Hansch [16] who effectively combined previous SPRs, such as eqn(6) and eqn(11), into multiparameter equations,

$$logBR = a.[logP(oct)]^2 + b.logP(oct) + c.\sigma + d.Es + e \qquad (12)$$

$$logBR = b.logP(oct) + c.\sigma + d.Es + e \qquad (13)$$

Here, BR is a set of biological responses towards a series of compounds by a given organism or biological system, logP(oct) is the water-octanol partition coefficient, σ is Hammett's σ_X parameter, representing an electronic effect, and Es is Taft's steric substituent constant [17].

The Hansch equation has been a cornerstone of attempts by pharmaceutical and medicinal chemists to relate biological activity to structure, for over 20 years, and is still widely used today [18]. A recent example, given by Hansch [18], is for the aqueous molar concentrations of some nuclear substituted benzyl nitrosamines that cause mutagenicity in *Salmonella typhimurium*,

$$log(1/C) = 0.92 \, \Delta logP(oct) + 2.08\sigma - 3.26 \qquad (14)$$

$$n = 12, \quad r = 0.981, \quad sd = 0.314, \quad F = 17.4$$

Here, $\Delta logP(oct)$ is just the difference in logP(oct) between the substituted compound and the nuclear unsubstituted one. As shown by Hansch [18], several series of compounds acting in bacterial mutagenicity tests give rise to coefficients in the logP(oct) term of between 0.65–1.14 units.

Although the Hansch equation has been enormously successful in the correlation and prediction of biological response, it suffers from a number of disadvantages. Firstly, it is very difficult to apply to a series of structurally unrelated solutes. For example, consider the set of compounds in Table 2. Neither the Hammett substituent parameter, nor the Taft steric parameter can be used with such a set of disparate structures, and other descriptors must therefore be employed. Secondly, the Hansch equation provides limited information on the mechanism of biological activity. Quite often, the main factor in the equation is the logP(oct) term, but since the factors that influence logP(oct) itself are not known, little can be said about the factors that influence biological activity. Thirdly, the Hansch equation is obviously not appropriate for the analysis of the biological activity of gases and vapors. Some other type of QSAR is needed in such a case. Forthly, it has long been suggested that biological activity, as well as physicochemical partition, depends on a number of solute factors such as polarity, acidity and basicity. Over 60 years ago Wilbrandt [19] and Collander and Barlund [10] proposed that plasma membrane lipoids have varying acidity and basicity depending on the cell, whilst Collander [11] set out in some detail how solvent and solute polarity, acidity, and basicity affect partition. Yet the Hansch equation contains no explicit terms for solute polarity, acidity, or basicity, and hence the influence of these important factors cannot be deduced. It might be useful to point out that Collander [11] was clearly not refering to proton acidity or basicity, because he discussed the partition of undissociated acids in these terms. Collander seems to have been one of the first workers to recognise what we now refer to as 'hydrogen-bond acidity' and 'hydrogen-bond basicity'. If Collander, and many subsequent workers, is correct, then descriptors of hydrogen-bond acidity and basicity could usefully be incorporated into LSERs and QSAR. In the next section is detailed how this may be achieved.

2. 1:1 HYDROGEN-BOND DESCRIPTORS

Athough the early workers in the field [10,11,19] felt that hydrogen-bonding was important in partitioning and in biological activity, it is only recently that any attempt has been made to obtain descriptors of hydrogen-bonding. Thus as recently as 1985, Dearden [12] reviewed QSARs and LFERs, but could refer to no scales of hydrogen-bonding at all. The first attempt to analyse partitioning through SPRs that included hydrogen-bond terms was probably that of Lien et al. [20] who estimated bond strengths for a few hydrogen-bonds. They recognised that this was a preliminary method only, because, for example, all O-H ...O bonds were assigned the same bond strength [20].

As shown in the introduction, most biological responses are expressed as log1/C or logC, both of these terms being related to the Gibbs free energy. Furthermore, all partition coefficients, as logP values, are also Gibbs free energy related, hence the use of LFERs. It seems logical, therefore, to set up scales of hydrogen-bond strengths using Gibbs free energy quantities, rather than, for instance, enthalpic quantities. Taft et al. [21,22] were the first workers to construct a quantitative scale of solute hydrogen-bond strength of bases on these lines, for more than just a few solutes. They defined a quantity pK_{HB} as logK for the complexation of a given base with 4-fluorophenol in tetrachloromethane,

$$B + HOAr = B...HOAr \tag{15}$$

These pK_{HB} values then represent the relative hydrogen-bond basicity of solutes towards the reference acid 4-fluorophenol. Some values are given in Table 5.

Table 5. Values of the pK_{HB} descriptor for solute bases [21]

Solute	pK_{HB}	Solute	pK_{HB}
Benzoyl fluoride	-0.30	Pyridine	1.88
Flurobutane	-0.16	Dimethylformamide	2.06
Anisole	0.02	Trimethylphosphate	2.40
Diethylsulfide	0.11	Tetramethylurea	2.42
Nitrobenzene	0.73	Dimethylsulfoxide	2.53
Methyl acetate	1.00	HMPT	3.56
Propanone	1.18	Et_3PO	3.64

Taft *et al.* [21] showed that logK values for complexation
of bases against various OH reference acids were linearly
related to pK_{HB}. Such a linear relationship did not hold so well
for logK values for bases against 5-fluoroindole [22], except
within families of bases. The major difficulty, however, with
the pK_{HB} scale can be seen by inspection of Table 5. The weakest
base studied by Taft was benzoyl fluoride, with pK_{HB} = -0.30 log
units, but it is not clear what the pK_{HB} value would be for a
nonbasic solute such as an alkane. In other words, the pK_{HB}
scale has no defined origin.

Other workers [23,24] examined various aspects of basicity
scales, but it was not until the perceptive analysis of basicity
dependent properties by Maria and Gal *et al.* [25], that the
problem of family-dependent relationships in hydrogen-bond
basicity was solved [26]. The other problem of a defined origin
had to wait until the analysis of Abraham *et al.* [27] on a scale
of solute hydrogen-bond acidity was completed. These workers
examined complexation constants for acids against reference
bases in dilute solution in tetrachloromethane, so that the
acids and bases all existed as simple monomeric solutes under
the experimental conditions,

$$A-H \; + \; B \; = \; A-H...B \hspace{4cm} (16)$$

Forty five series of acids were studied, and yielded forty five
equations in logK, for the complexation eqn(16),

$$logK(\text{series of acids against base B}) \; = \; L_B.logK^H_A \; + \; D_B \hspace{1cm} (17)$$

These equations could be solved to yield forty five values of
L_B and D_B, characterising the forty five reference bases, and
values of $logK^H_A$ that characterised the acids. Nearly 200 values
of $logK^H_A$ were reported covering a wide range of acids. A very
important finding was that all the equations of type eqn(16)
intersected at a 'magic point' where $logK = logK^H_A = -1.1$ when
K is expressed on the molar concentration scale. Typical
equations against the relatively weak base tetrahydrofuran (THF)
and the strong base dimethylsulfoxide (DMSO) are [27,28],

$$logK(\text{acids against THF}) \; = \; 0.8248 \; logK^H_A \; - \; 0.1970 \hspace{1cm} (18)$$

$$n = 23, \quad r = 0.9960, \quad sd = 0.089, \quad F = 2609$$

$$logK(\text{acids against DMSO}) \; = \; 1.2399 \; logK^H_A \; + \; 0.2656 \hspace{1cm} (19)$$

$$n = 51, \quad r = 0.9947, \quad sd = 0.096, \quad F = 4586$$

The 200 values of $logK^H_A$ now form a quantitative scale of solute
hydrogen-bond acidity. Most importantly, the identification of

the magic point at (−1.1, −1.1), leads to a natural origin for the scale; all solutes with zero hydrogen-bond acidity can be assigned $logK^H_A = -1.1$ units. The origin can be shifted to the more convenient one of zero, and the scale compressed somewhat at the same time, through the definition,

$$\alpha^H_2 = (logK^H_A + 1.1)/4.636 \qquad (20)$$

where the factor 4.636 is now arbitrary. Some values of $logK^H_A$ and the corresponding α^H_2 descriptor are in Table 6. Structural effects on α^H_2 have been discussed at length [27], but it is worth noting that there is little correlation of hydrogen-bond acidity with proton-transfer acidity across families. Thus phenol and acetic acid have almost the same α^H_2 value, whereas their proton-transfer acidities differ by over five powers of ten, in both water and dimethylsulfoxide [29].

Table 6. The hydrogen-bond acidity descriptors $logK^H_A$ and α^H_2

Solute	$logK^H_A$	α^H_2
Hexane	−1.10	0.00
Hept-1-yne	−0.5102	0.13
Chloroform	−0.1848	0.20
Aniline	0.1223	0.26
4-Nitroaniline	0.8534	0.42
N-Methylacetamide	0.6779	0.38
Maleimide	1.2037	0.50
Tetrachloropyrrole	2.2495	0.72
Water	0.5359	0.35
Methanol	0.6027	0.37
TFE	1.5303	0.57
HFIP	2.4737	0.77
Phenol	1.6649	0.60
4-Nitrophenol	2.7184	0.82
Acetic acid	1.5870	0.58
Trifluoroacetic acid	3.3070	0.95

In the same way that logK values for eqn(16) had been used to construct a scale of hydrogen-bond acidity, Abraham et al.[30] set up thirty four series of bases against reference acids. The thirty four equations in logK were now of the form,

$$logK(\text{series of bases against acid A}) = L_A.logK^H_B + D_A \qquad (21)$$

Solution of the thirty four equations yielded values of L_A and D_A for the reference acids, and some 270 values of $logK^H_B$. A typical equation is for bases against the reference acid 4-chlorophenol,

$$logK(\text{bases against 4-chlorophenol}) = 1.065\ logK^H_B + 0.074 \qquad (22)$$

$n = 38, \quad sd = 0.054$

Again, all the set of equations intersected at the magic point, so that the $\log K_B^H$ solute hydrogen-bond basicity scale also has an origin of -1.1 units. The origin can be shifted and the scale compressed again, through the defining equation,

$$\beta_2^H = (\log K_B^H + 1.1)/4.636 \qquad (23)$$

The β_2^H solute hydrogen-bond basicity scale, unlike the pK_{HB} scale, now has a defined origin in that all nonbasic solutes have a zero value. However, the pK_{HB} scale can be incorporated into the β_2^H scale through eqn(24),

$$\log K(\text{bases against 4-fluorophenol}) = 1.000 \log K_B^H + 0.000 \qquad (24)$$

n = 74, sd = 0.089

It is no coincidence that the slope and intercept in eqn(24) are unity and zero, respectively; 4-fluorophenol is the most common standard base in use today, and eqn(24) simply connects logK, ie pK_{HB}, with the $\log K_B^H$ and hence the β_2^H scale. Values of $\log K_B^H$ and β_2^H are given in Table 7 for some representative bases [30].

Table 7. The hydrogen-bond basicity descriptors $\log K_B^H$ and β_2^H

Solute	$\log K_B^H$	β_2^H
Hexane	-1.10	0.00
1-Chlorobutane	-0.608	0.11
Benzene	-0.422	0.15
Benzoyl fluoride	-0.300	0.17
Diethylsulfide	0.220	0.28
Nitrobenzene	0.482	0.34
Diethyl ether	0.988	0.45
Acetophenone	1.268	0.51
Pyridine	1.797	0.62
Dimethylformamide	1.973	0.66
Triethylamine	2.001	0.67
N-Methylacetamide	2.217	0.72
Dimethylsulfoxide	2.492	0.77
HMPT	3.536	1.00
Triethylphosphine oxide	3.617	1.02

Inspection of Tables 5 and 7 illustrates the value of a scale that incorporates a zero point. The pK_{HB} value of -0.30 for benzoyl fluoride gives no indication of how near the origin, or zero point, is this compound, whereas a β_2^H value of 0.17 allows the exact position of benzoyl fluoride on a hydrogen-bond scale to be determined. Structural effects on β_2^H have been discussed before [30], and will not be repeated, except to note that across families of solutes there is little connection with proton- transfer basicity.

The α_2^H and β_2^H scales can be combined into one general equation for logK values for 1:1 complexation of solutes in tetrachloromethane at 298K,

$$logK = 7.354\ \alpha_2^H \cdot \beta_2^H - 1.094 \qquad (25)$$

$$n = 1312, \quad r = 0.9956, \quad sd = 0.093, \quad F = 147882$$

Knowing α_2^H for some 150 solutes, and β_2^H for over 500, it is now possible to estimate further logK values for some 75,000 acid –base combinations in tetrachloromethane [31]. Eqn(25) is also useful in the evaluation of further values of either α_2^H or β_2^H, knowing one or the other, and, of course, logK. In this way, α_2^H values for a number of imines were obtained [32]. More recently, Berthelot and Laurence et al. [33–38] have considerably extended the data base of α_2^H and, especially, β_2^H values through further measurements of logK values in tetrachloromethane.
    All the above has referred to 1:1 equilibria in the solvent tetrachloromethane, but Abboud et al.[39] have shown that 1:1 equilibria in cyclohexane and in the gas phase can be treated similarly. Thus for the complexation of hydrogen–bond acids with pyridine–N–oxide (PyO) in cyclohexane they find,

$$\alpha_2^H = 0.185\ logK\text{(acids against PyO)} - 0.069 \qquad (26)$$

$$n = 22, \quad r = 0.9930, \quad sd = 0.03, \quad F = 1408$$

For hydrogen–bond acidity against the iodide ion in the gas phase, as logKp with Kp in atm^{-1}, Abboud et al.[39] found,

$$\alpha_2^H = 0.0534\ logKp\text{(acids against I}^-\text{)} + 0.111 \qquad (27)$$

$$n = 18, \quad r = 0.9803, \quad sd = 0.032, \quad F = 394$$

Although the α_2^H and β_2^H methodology is the most general system for the construction of scales of hydrogen–bonding using 1:1 complexation constants, it is not the only one available. Raevsky et al. [40] have devised a quite similar system, again using 1:1 complexation constants in tetrachloromethane, that they summarise in the equation,

$$\Delta G° = 2.43\ C_A \cdot C_B + 5.70 \qquad (28)$$

$$n = 936, \quad r = 0.9840, \quad sd = 1.11, \quad F = 28556$$

Here, $\Delta G°$ is the Gibbs free energy change for complexation in kJ mol^{-1}, and C_A and C_B are the solute hydrogen–bond acidity and basicity respectively. Note that the nomenclature in eqn(28) is not quite the same as that of Raevsky et al.[40]. Eqn(28) is not quite as good as eqn(25), the standard deviation of 1.11 in $\Delta G°$ corresponding to one of 0.194 in logK; compare the sd of 0.093 in eqn(25). Values of C_A and C_B are listed [40] for a wide range of acids and bases, but no values are given for solutes that are nonacidic or nonbasic. Eqn(28), like the similar eqn(25), is cast in terms of Gibbs free energy, but Raevsky et al.[40] also set out the only general equation yet developed in terms of enthalpy,

$$\Delta H^\circ = 4.96 \ E_A.E_B \qquad\qquad (29)$$

$$n = 936, \quad r = 0.9540, \quad sd = 2.70, \quad F = 9553$$

Now ΔH° is the standard enthalpy change for 1:1 complexation in tetrachloromethane in kJ mol^{-1}, and E_A and E_B are the solute hydrogen-bond acidity and basicity. Again, the nomenclature of eqn(29) is not quite the same as Raevsky's. The goodness of fit of eqn(29) is appreciably less than that of eqn(28), but this is certainly due to the larger experimental error in ΔH°, many values being obtained by the temperature variation of logK. E_A and E_B are given [40] for a range of acids and bases, but, once more, no values for nonacidic or nonbasic compounds are listed. However, the Gibbs free energy parameters of Raevsky et al.[40] have been used with considerable success in the interpretation of complexation of crown ethers and cryptands [41].

Taylor et al. [42] have measured a considerable number of 1:1 complexation constants in the solvent 1,1,1-trichloroethane (TCE). They define a hydrogen-bond acidity descriptor, logKα, as logK for complexation of a series of acids with the standard base N-methylpyrrolidinone, and a hydrogen-bond basicity descriptor, logKβ, as logK for complexation of a series of bases with the standard acid, 4-nitrophenol. Although logKα and logKβ have been used successfully in LFERs for partitions in several solvents [43], the system again suffers from lack of a clearly defined origin, Taylor et al. [43] estimating the zero point as around -0.60 in TCE. It would be useful to combine the TCE and the tetrachloromethane scales, but at the moment there is not enough information to do so.

3. OVERALL HYDROGEN-BOND DESCRIPTORS

All the hydrogen-bond descriptors mentioned above have been derived from 1:1 complexation constants, and it is by no means certain that these descriptors are appropriate for situations in which the solute molecule is surrounded by an excess of solvent molecules. Symons and Eaton have suggested that propanone is largely mono-hydrogen-bonded in methanol but di-hydrogen-bonded in water [44], and that triphenylphosphine oxide is mono- and di- hydrogen-bonded in methanol but tri-hydrogen-bonded in water [45]. Much more recently, Nagy et al. [46] concluded that in water, the phenolic group in 4-hydroxybenzoic acid is strongly hydrogen-bonded to one water molecule via the hydrogen atom, and weakly bonded to two water molecules via the oxygen atom. The carboxylate group is also tri-hydrogen-bonded, with one strong bond from water to the hydrogen atom, one weaker bond to the carbonyl oxygen, and another weaker bond to the oxygen of the OH group [46]. Under such circumstances, descriptors tied to 1:1 complexation may not be useful. In any case, such descriptors cannot deal with the problem of multiple functionality that is invariably encountered in pharmaceutical

and medicinal chemistry. For these reasons, descriptors of
hydrogen-bonding that refer to the 'overall' or 'effective'
hydrogen-bond acidity and basicity have to be developed.

There have been only two realistic attempts to obtain these
overall descriptors through experiment, at least for any
extended series of solutes. Abraham and Carr have separately
worked on this problem, and their investigations are outlined
in turn.

3.1 The method of Abraham

The method of Abraham [28] starts with the definition [47]
of a new solute descriptor, $logL^{16}$, as the gas-hexadecane
partition coefficient or the Ostwald solubility coefficient,

$$logL^{16} = (conc\ in\ hexadecane)/(conc\ in\ gas\ phase) \qquad (30)$$

Both the symbols L and K are used for the partition coefficient
as defined by corresponding equations to eqn(30). Values of
$logL^{16}$ can be obtained through gas-chromatographic measurements
using a hexadecane stationary phase, provided that the solute
is not too involatile [47]. Abraham et al. [48] then defined
another descriptor, R_2, as the excess molar refraction. The
molar refraction itself was determined from the equation,

$$MR = (\eta^2 -1)/(\eta^2 + 2).V_x \qquad (31)$$

where η is the refractive index for the sodium D-line at 298K,
and V_x is McGowan's characteristic volume that can be
calculated simply from molecular structure [49]. R_2 is then MR
for the solute less MR for an alkane of the same characteristic
volume. For solutes that were too involatile to study at 298K,
values of $logL^{16}$ could be obtained through gas-chromatographic
(GLC) measurements on nonpolar phases such as squalane or
apiezon at elevated temperature. GLC retention data for a
series of solutes were fitted [50,51] to an equation of the
form,

$$logSP = c + r.R_2 + 1.logL^{16} \qquad (32)$$

Here, SP can be retention volume, or relative retention time,
or logSP can be the retention index,I. Thus for the large
series of solutes studied by Dutoit [52] on a hydrocarbon phase
at 383K,

$$I/10 = 6.669 + 8.918 R_2 + 20.002 logL^{16} \qquad (33)$$

n = 138, r = 0.9995, sd = 0.449, F = 67950
so that further values of $logL^{16}$ can be found for any solute for
which I, or I/10, is available. In this way, some 1500 $logL^{16}$
values have been obtained [50,51].
Now suppose that a polar stationary phase is used instead
of a nonpolar one. A good LFER can only be constructed if a new

solute descriptor is included in the equation. The new parameter is a measure of solute dipolarity/polarisability, it being not possible to separate the two effects. The stationary phase is chosen so that it is nonacidic, therefore the solute basicity will not be relevant. If the series of solutes contains no acidic compounds, retention data can be fitted to an equation,

$$\log SP = c + r.R_2 + s.\pi_2^H + l.\log L^{16} \tag{34}$$

where π_2^H is the new solute descriptor. If some of the solutes are hydrogen-bond acids, then a descriptor derived from α_2^H has to be included. Now, in general, cf Nagy et al. [46], a hydrogen- bond acidic group such as a phenol or carboxylic acid will be bound to only one solvent base, so that the α_2^H descriptor might still be applicable to the case of a solute surrounded by solvent molecules. Indeed, it turns out that for a mono-acid the simple α_2^H descriptor can be taken as equivalent to an overall desriptor that is designated as $\Sigma\alpha_2^H$. There are a few minor differences, but the α_2^H scale can rather easily be used to set up the required $\Sigma\alpha_2^H$ descriptor for monofunctional solutes. In such cases, the new π_2^H descriptor can be obtained by a modification of eqn(34), as follows,

$$\log SP = c + r.R_2 + s.\pi_2^H + a.\Sigma\alpha_2^H + l.\log L^{16} \tag{35}$$

An example [51] of such an equation that was used to determine values of π_2^H for a few extra phenols is,

$$I/10 = 2.991 - 22.683\ R_2 + 52.865\ \pi_2^H + 21.692\ \Sigma\alpha_2^H + 20.579\ \log L^{16} \tag{36}$$

$$n = 30, \quad r = 0.9987, \quad sd = 0.949, \quad F = 2399$$

Here, I is the retention index on OV-1701 at 423K, from Engewald et al. [53]. In this way, not only were values of the new π_2^H descriptor obtained, but a start was made on the overall acidity scale, or descriptor, $\Sigma\alpha_2^H$.

In principle, the GLC method could be used to set up an overall hydrogen-bond basicity scale, denoted as $\Sigma\beta_2^H$, using acidic stationary phases. This proved difficult in practice, partly because equations of the necessary goodness of fit could not be obtained. Another process that depends markedly on solute hydrogen-bond basicity is that of water-solvent partition. Furthermore, numerous partition coefficients, P, are available for a wide variety of water-solvent systems. The method adopted [54] was to set up LFERs for a series of logP values in a given water-solvent system, using the β_2^H descriptor for simple mono-functional solutes. Values of $\Sigma\beta_2^H$ could then be obtained by back calculation, and could then be used to adjust the LFERs, that took the general form,

$$\log P = c + r.R_2 + s.\pi_2^H + a.\Sigma\alpha_2^H + b.\Sigma\beta_2^H + v.V_x \tag{37}$$

The process was repeated in a round-robin procedure, until the

$\Sigma\beta_2^H$ values had converged to a constant set, and until the LFER coefficients had also converged. During the round-robin process, the $\Sigma\alpha_2^H$ descriptor was occasionally adjusted as well. Sixteen water-solvent systems were used in the round-robin process, including water-gas as a special case. The final set of LFER coefficients for the sixteen systems are in Table 8, together with a brief summary of the statistical analysis [54]. As can be seen, the sixteen LFERs are all of excellent goodness of fit, with sd values ranging from 0.10 to 0.19 log unit. This is very important if the LFERs are to be used to calculate $\Sigma\beta_2^H$ values. In general, the error in any back-calculated parameter cannot be less than the LFER standard deviation divided by the coefficient of the parameter. For $\Sigma\beta_2^H$ an average error is thus likely to be around 0.14/4.8 or about 0.03 units.

The form of the LFER eqn(38) was deduced from a number of preliminary studies in which it became clear [28] that for processes of the type gas-condensed phase, a general SPR was,

$$logSP = c + r.R_2 + s.\pi_2^H + a.\Sigma\alpha_2^H + b.\Sigma\beta_2^H + 1.logL^{16} \qquad (38)$$

but for processes within condensed phases, a better SPR was usually, cf eqn(37),

$$logSP = c + r.R_2 + s.\pi_2^H + a.\Sigma\alpha_2^H + b.\Sigma\beta_2^H + v.V_x \qquad (39)$$

In both of these equations, SP is a solute property such as L, the gas-liquid partition coefficient in eqn(38) or P, the water-solvent partition coefficient in eqn(39). The V_x descriptor is McGowan's characteristic volume in units of (ml mol^{-1})/100, and the other descriptors have been outlined before. The reason why earlier work on the π_2^H descriptor had to be carried out, is that no simple descriptor of solute dipolarity was found to be very useful in equations such as eqn(38) and eqn(39).

Analysis of the system of sixteen equations leads not only to the LFER coefficients in Table 8, but also to average values of $\Sigma\beta_2^H$ across the sixteen equations. In this way, some 500 $\Sigma\beta_2^H$ values were obtained, not counting zero values for nonbasic solutes, with an average standard deviation of around 0.03 units [54]. It turns out that $\Sigma\beta_2^H$ values for homologous series are constant, except for the first one or two members, so that the list of 500 can be considerably extended. A full list of the 500 values is available [54], and so just a small selection is given in Table 9, together with β_2^H for comparison.

As might be expected, there is reasonable agreement between the two sets of basicities for monofunctional solutes, with one or two exceptions. However, there are now available $\Sigma\beta_2^H$ values for polyfuctional bases, and in such cases there will be litle agreement with a β_2^H value obtained from a 1:1 complexation constant. To show how constant is the $\Sigma\beta_2^H$ value, when averaged from back-calculations using partition coefficient in different systems, some results [54] are collected in Table 10. Again, a standard deviation of about 0.03 units is obtained.

Table 8. The LFER coefficients for water-solvent partitions

Solvent	No	c	r	s	a	b	v	n	ρ	sd
Isobutanol	1	0.249	0.480	-0.639	-0.050	-2.284	2.758	35	0.9903	0.119
Octanol	2	0.081	0.585	-1.090	0.033	-3.401	3.810	584	0.9965	0.133
n-Butyl acetate	3	-0.468	0.712	-0.397	0.010	-3.743	3.865	47	0.9892	0.152
Diethyl ether	4	0.462	0.571	-1.035	-0.024	-5.508	4.346	84	0.9897	0.195
PGDP	5	0.287	0.338	-0.638	-0.908	-5.038	4.093	56	0.9909	0.175
Nitrobenzene	6	-0.181	0.576	0.003	-2.356	-4.420	4.263	84	0.9913	0.172
Chlorobenzene	7	0.046	0.259	-0.466	-3.047	-4.819	4.660	93	0.9973	0.115
Benzene	8	0.017	0.490	-0.604	-3.013	-4.628	4.587	112	0.9953	0.119
Toluene	9	0.015	0.594	-0.781	-2.918	-4.571	4.533	87	0.9957	0.114
Olive Oil	10	-0.086	0.575	-0.861	-1.447	-4.945	4.295	111	0.9972	0.129
Chloroform	11	0.125	0.118	-0.372	-3.390	-3.467	4.521	112	0.9972	0.112
Tetrachloromethane	12	0.223	0.564	-1.151	-3.510	-4.536	4.501	111	0.9980	0.102
Cyclohexane	13	0.124	0.844	-1.800	-3.727	-4.923	4.692	152	0.9955	0.155
Alkane	14	0.281	0.647	-1.687	-3.520	-4.848	4.326	173	0.9982	0.120
Hexadecane	15	0.103	0.686	-1.624	-3.566	-4.880	4.444	366	0.9983	0.123
Gas phase	16	1.003	-0.630	-2.496	-3.901	-4.843	0.852	353	0.9967	0.160

Table 9. Some values of the overall hydrogen-bond basicity descriptor, $\Sigma\beta_2^H$ [54]

Solute	$\Sigma\beta_2^H$	β_2^H
Hexane	0.00	0.00
1-Chlorobutane	0.10	0.11
Benzene	0.14	0.15
Diethylsulfide	0.32	0.28
Nitrobenzene	0.28	0.34
Acetonitrile	0.32	0.44
Diethyl ether	0.45	0.45
Acetophenone	0.48	0.51
Propanone	0.49	0.50
Dimethylformamide	0.74	0.66
Triethylamine	0.79	0.67
N-Methylacetamide	0.72	0.72
Benzenesulfonamide	0.80	
Dimethylphthalate	0.88	
Trimethylphosphate	1.00	0.76
HMPT	1.60	1.00
Phenol	0.30	0.22
Benzoic acid	0.40	
Acetic acid	0.44	
Ethanol	0.48	0.44
4-Methoxyphenol	0.48	
4-Aminophenol	0.83	

Table 10. Some calculations of $\Sigma\beta_2^H$ from partition coefficients

System	Propanone	Ethanol	Acetic acid	Phenol	Acetanilide
Isobutanol		0.577	0.496		
Octanol	0.519	0.526	0.437	0.321	0.641
Bu acetate			0.421	0.313	
Diethylether	0.440	0.490	0.406	0.311	0.685
PGDP				0.289	0.673
Nitrobenzene	0.485	0.487		0.289	
Benzene	0.458	0.477	0.464	0.272	
Toluene		0.500	0.414	0.293	
Chloroform	0.507	0.464	0.423	0.283	0.666
CCl$_4$	0.485	0.472	0.447	0.331	
Cyclohexane	0.486	0.446	0.434	0.285	0.611
Alkane	0.488	0.508	0.474	0.304	0.700
Hexadecane	0.506	0.500	0.469	0.323	
Gas phase	0.463	0.498	0.442	0.298	0.690
Average:	0.484	0.495	0.444	0.301	0.667
sd:	0.024	0.033	0.027	0.018	0.031

One type of solute for which no hydrogen-bond acidities or basicities were hitherto availble is that possessing internal hydrogen-bonds. Some values for 2-, 3-, and 4-

substituted phenols obtained by the partition coefficient method are in Table 11.

Table 11. Values of solvation parameters for some phenols [28,54]

Phenol	R_2	π_2^H	$\Sigma\alpha_2^H$	$\Sigma\beta_2^H$	V_x
2-Cl	0.853	0.88	0.32	0.31	0.8795
3-Cl	0.909	1.06	0.69	0.15	0.8795
4-Cl	0.915	1.08	0.67	0.20	0.8795
2-OMe	0.837	0.91	0.22	0.52	0.9747
3-OMe	0.879	1.17	0.59	0.39	0.9747
4-OMe	0.900	1.17	0.57	0.48	0.9747
2-NO$_2$	1.015	1.05	0.05	0.37	0.9402
3-NO$_2$	1.050	1.57	0.79	0.23	0.9402
4-NO$_2$	1.070	1.72	0.82	0.26	0.9402

With all three sets of phenols, the 2-substituted solute has a much lower acidity than 3- or 4- isomers, through intra-molecular hydrogen-bonding. Indeed, the acidity of 2-nitrophenol is very nearly zero. But internal hydrogen-bonding of the hydrogen on the phenolic group not only reduces the acidity, and reduces the overall dipolarity of the molecule, but also increases the basicity of the phenolic oxygen. The total effect of internal hydrogen-bonding on partition coefficients is thus not as clear cut as might be thought. A comparison of the effect for 2- and 4-nitrophenol is given in Table 12 for the water- octanol and water-cyclohexane systems [54], with a breakdown of eqn(39) into its constituent terms. In the case of the water- cyclohexane partition, the lack of acidity of the 2-isomer is the main factor that leads to a much increased logP value, with a decreased dipolarity also contributing. The increased basicity of the 2-isomer counteracts these effects to some extent, but still there is a very large difference in the logP values. This is not so in the water-octanol system, where the logP values are almost the same for the 2-and 4-nitrophenols. This might at first sight be taken as evidence that 2-nitrophenol is not intramolecularly hydrogen-bonded in this system. Inspection of Table 12, however, shows the reasons for the similar logP values. In the water- octanol system, solute acidity plays no part, see Table 8, and so the only factors left are dipolarity and basicity. These, as has been seen, counteract each other, the net result being that the two logP values are very close.

Table 12. The effect of internal hydrogen-bonding in nitrophenols on partition coefficients

Isomer	c	r.R_2	s.π_2^H	a.$\Sigma\alpha_2^H$	b.$\Sigma\beta_2^H$	v.V_x	calc[a]	obs[a]
Water-cyclohexane								
2-	0.12	0.87	-1.89	-0.19	-1.82	4.45	1.54	1.45
4-	0.12	0.90	-3.10	-3.05	-1.27	4.45	-1.95	-2.05

Table 12 (cont). The effect of internal hydrogen-bonding in nitrophenols on partition coefficients

Isomer	c	$r.R_2$	$s.\pi_2^H$	$a.\Sigma\alpha_2^H$	$b.\Sigma\beta_2^H$	$v.V_x$	calc[a]	obs[a]
Water-octanol								
2-	0.08	0.59	-1.14	0.00	-1.26	3.62	1.89	1.85
4-	0.08	0.63	-1.87	0.03	-0.88	3.62	1.61	1.91

[a] These are the calculated and observed values of logP.

In their work on LFERs and partitioning in four water-solvent systems, Taylor et al. [43] observed that for a number of bases, the relative basicity, as back-calculated from the LFERs, did not remain constant. The particular bases identified by Taylor et al. [43] were a number of P:O and S:O bases, including Ph$_3$PO and sulfoxides but not sulfones or sulfonamides. Abraham [54] confirmed this in an analysis of the sixteen systems in Table 8, but also showed that phosphate esters had a constant $\Sigma\beta_2^H$ value, and that alkylanilines and alkylpyridines showed a variable relative basicity. Note that Taylor et al.[43] examined no pyridines, and so the only solutes that seem anomalous as between the two studies are the anilines. In Table 13 are given details of the back-calculated $\Sigma\beta_2$ values for some anilines and pyridine [54].

Table 13. Calculated $\Sigma\beta_2$ values for anilines and pyridine

System[a]	Aniline	p-Toluidine	N,N-Dimethylaniline	Pyridine
1				0.44
2	0.53	0.55	0.47	0.43
3	0.46	0.50		0.48
4	0.49			0.51
5	0.43		0.41	0.52
8	0.40	0.47		0.54
9	0.41	0.47		0.55
11	0.41	0.43	0.39	0.47
12	0.40	0.44	0.45	0.53
13	0.41	0.46	0.43	0.53
14	0.40	0.46	0.40	0.52
15		0.45	0.42	0.53
16		0.43	0.37	0.52
Av 1-3:	0.50	0.52	0.47	0.45
Av 5-16:	0.41	0.45	0.41	0.52
sd 5-16:	0.01	0.02	0.03	0.02

[a] Solvent systems are numbered as in Table 8.

A comparison of results in Table 13 with those in Table 8 shows that in the 'wet' solvents 1-3, and possibly 4 as well, the $\Sigma\beta_2$ values for the anilines and pyridine are out-of-line by about +0.9 and -0.7 units respectively. This behaviour is shown by anilines generally, other than the nitroanilines, and by pyridine and some substituted pyridines. It was suggested [54] that the $\Sigma\beta_2$ values obtained from systems 5-16 are the 'normal' $\Sigma\beta_2{}^H$ values, because they are applicable to a number of other systems as well. The $\Sigma\beta_2$ values from systems 1-4 are denoted as $\Sigma\beta_2{}^0$ values, since they must be used in the important water- octanol system. The reasons for the variable relative basicity observed by Taylor and by Abraham are not clear; Abraham [54] notes that the observations could be explained if the anilines are preferentially solvated by the organic component in the organic layer, and if the pyridines are peferentially solvated by the water component in the organic layer, but there is no evidence to hand on this matter. A full list of the solutes that exhibit variable relative basicity has been given [54], and a selection is in Table 14. It should be stressed that the LFER equations summarised in Table 8 do not include results from solutes of varying relative basicity.

Table 14. Some solutes of variable relative basicity [54]

Solute	$\Sigma\beta_2{}^0$	$\Sigma\beta_2{}^H$
Dimethylsulfoxide	0.75	0.88
Methylphenylsulfoxide	0.75	0.91
Aniline	0.50	0.41
p-Toluidine	0.52	0.45
N-Methylaniline	0.48	0.43
N,N-Dimethylaniline	0.47	0.41
4-Chloroaniline	0.35	0.31
4-Methoxyaniline	0.72	0.65
Pyridine	0.47	0.52
4-Methylpyridine	0.43	0.55
4-Acetylpyridine	0.75	0.84
Quinoline	0.51	0.54
Isoquinoline	0.47	0.54

The solvation parameters required for use in eqn (38) and eqn (39), including the hydrogen-bond parameters discussed above, have been listed for numerous solutes [28] and constitute a self consistent set of descriptors. The application of eqn (38) and eqn (39) to various processes will be considered later, but first the method of Carr will be outlined.

3.2 The method of Carr

This method starts off in a very similar way to that of Abraham, as described in Section 3.1. Retention data for various series of solutes on a number of GLC stationary phases are analysed through an LFER equation that is analogous [55] to eqn (35),

$$\log SP = c + d.d_2 + s.\pi_2{}^c + a.\alpha_2{}^c + b.\beta_2{}^H + 1.\log L^{16} \qquad (40)$$

Here, logSP is a set of retention data. The descriptors are: d_2 Kamlet's [56] polarisability correction parameter, taken as 1.0

for aromatic solutes, 0.5 for polychloroaliphatic solutes, and zero for the rest, π_2^C a dipolarity/polarisability parameter, α_2^C the hydrogen-bond acidity, β_2^H the 1:1 hydrogen-bond basicity parameter of Abraham [30] and $\log L^{16}$ as above. Data on nineteen GLC phases were analysed using a round-robin procedure, until the regression coefficients and the π_2^C and α_2^C descriptors had converged. The other descriptors were taken as fixed values. Over 200 solutes were examined by Carr et al. [55] and in Table 15 are given some of their results, together with those of Abraham [28] for comparison. Note that the zero point for the π_2^C scale is that for cyclohexane, whereas Abraham takes alkanes as the zero point for the π_2^H scale.

Bearing in mind the different origin for the π_2 scales, there is reasonable agreement between the two sets of descriptors as pointed out by Carr et al. [55]. The correlations for the training set of sixteen phases studied by Carr are all excellent, and so the only way to differentiate between the two sets of descriptors is through application to test sets of data that have not been used to set up the descriptor scales. It should be noted that the descriptors obtained by Carr are 'overall' descriptors, since the solute is surrounded by an excess of solvent molecules.

Table 15. Some values of the π_2^C and α_2^C descriptors [55]

Solute	π_2^C	α_2^C	π_2^H	$\Sigma\alpha_2^H$
Butane	-0.17	0.00	0.00	0.00
Hexane	-0.16	0.00	0.00	0.00
Octane	-0.12	0.00	0.00	0.00
Decane	-0.11	0.00	0.00	0.00
Dodecane	-0.09	0.00	0.00	0.00
Tetradecane	-0.07	0.00	0.00	0.00
Cyclohexane	0.00	0.00	0.10	0.00
Hex-1-ene	-0.04	0.00	0.08	0.00
Oct-1-yne	0.16	0.04	0.23	0.12
Methanol	0.35	0.35	0.44	0.43
Ethanol	0.29	0.29	0.42	0.37
Propan-2-ol	0.21	0.29	0.36	0.33
TFE	0.37	0.66	0.60	0.57
HFIP	0.47	1.11	0.55	0.77
tert-Butanol	0.19	0.25	0.30	0.31
Propanone	0.38	0.01	0.70	0.04
Butanone	0.39	0.00	0.70	0.00
Dodecan-2-one	0.46	0.00	0.68	0.00
Acetonitrile	0.62	0.05	0.90	0.07
Dichloromethane	0.34	0.06	0.57	0.10
Trichloromethane	0.27	0.16	0.49	0.15
Acetic acid	0.50	0.72	0.65	0.61
Octanoic acid	0.68	0.41	0.60	0.60
Methyl acetate	0.30	0.00	0.64	0.00
Benzene	0.29	0.00	0.52	0.00
Butylbenzene	0.30	0.00	0.51	0.00
Phenol	0.77	0.69	0.89	0.60
Benzyl alcohol	0.71	0.43	0.87	0.33
Aniline	0.76	0.20	0.96	0.26

Following their development of the π_2^C and α_2^C descriptors, Carr et al. [57] then proceeded to set up a scale of solute hydrogen-bond basicity that they denoted as β_2^C, although the scale is actually an 'overall' basicity scale. Their method was to synthesise two GLC phases, 4-dodecyl-α, α-bis(trifluoromethyl) benzyl alcohol and the corresponding methyl ether, denoted here as COH and COMe. Relative retention data , as K', were obtained for a set of 84 solutes on each phase, and the ratio logK'(COH)/K'(COMe) was shown to have no dependence on any parameter except the Kamlet d_2 and the Abraham β_2^H descriptors,

$$\log K'(COH)/K'(COMe) = -0.089 + 2.15 \beta_2^H - 0.23 d_2 \qquad (41)$$

$$n = 84, \quad r = 0.967, \quad sd = 0.22, \quad F = 583$$

Eqn(41) can then be used to back-calculate further values of the solute hydrogen-bond basicity, defined as,

$$\beta_2^C = \{ \log K'(COH)/K'(COMe) + 0.089 + 0.23 d_2 \}/2.15 \qquad (42)$$

It might be thought that such a procedure would yield further values that were close to β_2^H or to $\Sigma\beta_2^H$, but this proved not to be the case. Some of the further 59 solutes studied are given in Table 16, with their various β_2 values. There is a rough connection between β_2^C and either β_2^H or $\Sigma\beta_2^H$, but it is clear that for the strong hydrogen-bond bases, β_2^C is invariably much larger. Although the defining equation, eqn(41) looks to be statistically reasonable, it conceals the same effect that occurs within the 84 solute set. If the total 142 solutes (these are the 147 solutes given, less methanol, cyclohexanol, and three carboxylic acids for which other values were suggested by Carr) are taken together, the plot of β_2^C against $\Sigma\beta_2^H$ is clearly not a straight line, Figure 1. This is not due to the $\Sigma\beta_2^O/\Sigma\beta_2^H$ behaviour, mentioned above, because a similar plot with β_2^C against $\Sigma\beta_2^O$ also shows the curved behaviour, Figure 2.

Table 16. Some values of β_2^C calculated [57] via eqn(42)

Solute	β_2^C	β_2^H	$\Sigma\beta_2^H$
Ethylcyclohexane	0.00	0.00	0.00
Oct-1-ene	0.02	0.07	0.07
1-Chloropentane	0.08	0.11	0.10
Hexanal	0.38	0.39	0.45
Ethyl propanoate	0.45	0.45	0.45
Butylamine	1.00	0.71	0.61
Hexylamine	0.88	0.69	0.61
Diethylamine	0.93	0.70	0.69
Dibutylamine	0.87	0.71	0.69
3-Methylbutanol	0.52	0.45	0.48
Styrene	0.11	0.18	0.16
Mesitylene	0.13	0.20	0.19
Naphthalene	0.10	0.21	0.20

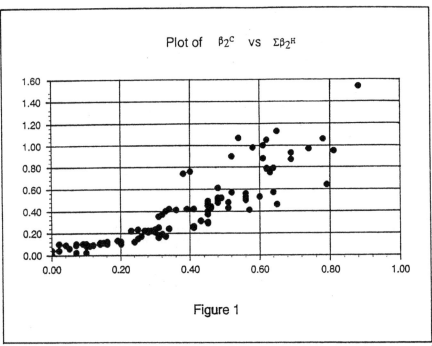

Figure 1

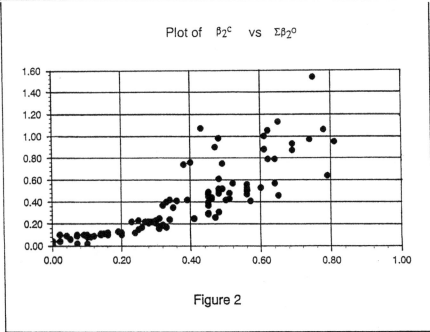

Figure 2

Table 16 (cont.). Some values of β_2^C calculated [57] via eqn(42)

Solute	β_2^C	β_2^H	$\Sigma\beta_2^H$
Phenyl ethyl ether	0.19	0.26	0.32
Propriophenone	0.43	0.51	0.51
p-Toluonitrile	0.42	0.42	0.34
4-Ethylaniline	0.47	0.42	0.45
4-Chloroaniline	0.35	0.34	0.31
4-Methylphenol	0.25	0.24	0.31
4-Fluorophenol	0.22	0.21	0.23
2-Methylpyridine	0.98	0.63	0.58
4-Methylpyridine	1.07	0.66	0.54
3-Chloropyridine	0.76	0.49	0.40

It must be concluded that the β_2^C descriptor of Carr et al. [57] represents another hydrogen-bond basicity than the $\Sigma\beta_2^H$ descriptor of Abraham [54]. Which descriptor is the most general can only be ascertained through LFER regressions of test data sets, ie data sets that have not been used to set up the descriptors.

3.3 Other hydrogen-bond descriptors

The only other set of experimentally determined overall hydrogen-bond descriptors is the basicity descriptor obtained by Taylor et al. [43] as the result of an LFER analysis of four sets of partition coefficients. It is difficult to compare Taylor's descriptor with those of Abraham or of Carr, because the former is calculated as two parts – $\Sigma\beta$ and $n\beta$, where n is the number of lone pairs in the solute base, after the first.

An interesting new development is that of the theoretical calculation of descriptors. Murray and Politzer [58] have shown that there is a reasonable relationship between the calculated surface maxima, $V_{s,max}$, of the electrostatic potential associated with hydrogen atoms in the solute, and the descriptor α_2^H. Their results are summarised in Table 17.

Table 17. The calculated electrostatic potential, $V_{s,max}$ [58]

Solute	$V_{s,max}$	α_2^H	$\Sigma\alpha_2^H$
Perfluoro-t-butanol	48.7	0.86	0.86[a]
Hexafluoropropan-2-ol	43.1	0.77	0.77
Trifluoroethanol	40.3	0.57	0.57
Methanol	29.8	0.37	0.43
Ethanol	29.4	0.33	0.37
Propan-1-ol	27.8	0.33	0.37
Butan-1-ol	28.2	0.33	0.37
Propan-2-ol	27.8	0.32	0.33
2-Methylpropan-2-ol	26.1	0.32	0.31
4-Nitrophenol	46.1	0.82	0.82
4-Chlorophenol	41.5	0.67	0.67
4-Fluorophenol	37.3	0.63	0.63

Table 17 (cont.). The calculated electrostatic potential,
$V_{s,max}$ [58]

Solute	$V_{s,max}$	α_2^H	$\Sigma\alpha_2^H$
Phenol	34.9	0.60	0.60
4-Methoxyphenol	34.1	0.57	0.57
4-Methylphenol	34.0	0.57	0.57

[a] Estimated by comparison to α_2^H.

As Murray and Politzer [58] pointed out, there is a good correlation between α_2^H and $V_{s,max}$,

$$\alpha_2^H = -0.371 + 0.0257 \, V_{s,max} \qquad (43)$$

$n = 15, \quad r = 0.9685, \quad sd = 0.05, \quad F = 199.6$

There is just as good a correlation if $\Sigma\alpha_2^H$ is used instead,

$$\Sigma\alpha_2^H = -0.316 + 0.0246 \, V_{s,max} \qquad (44)$$

$n = 15, \quad r = 0.9731, \quad sd = 0.04, \quad F = 222.4$

Since the usual error in hydrogen-bond parameters is around 0.03 units, it is possible that further values of $\Sigma\alpha_2^H$ could be calculated, at least for monofunctional oxygen acids.
Murray and Politzer [58] also calculated the electrostatic potential minimimum, V_{min}, for a series of oxygen bases and a series of nitrogen bases. Results are in Table 18. The oxygen and nitrogen bases clearly have to be taken separately. For the sixteen oxygen bases there is a reasonable correlation between V_{min} and β_2^H,

$$\beta_2^H = -0.228 - 0.0134 \, V_{min} \qquad (45)$$

$n = 16, \quad r = 0.9554, \quad sd = 0.065, \quad F = 146.4$

With the exception of the point for methyl acetate, both $\Sigma\beta_2^H$ and β_2^C are also well correlated with V_{min}, although the spread of data is much smaller.

The seven nitrogen bases lead [58] to the correlation,

$$\beta_2^H = -0.657 - 0.0141 \, V_{min} \qquad (46)$$

$n = 7, \quad r = 0.9758, \quad sd = 0.027, \quad F = 99.5$

so that possibly eqn(46) could be used to estimate further values of β_2^H.

However, it is interesting that the overall basicity descriptors, $\Sigma\beta_2^H$ and $\Sigma\beta_2^O$, do not relate well to V_{min}. The two-site acceptors, pyrimidine and pyrazine, being out of line. Note that N-methylimidazole is a single site acceptor. It remains to be seen if the calculated electrostatic potential in general refers to one basic site in a polybasic molecule.

Table 18. The calculated electrostatic potential, V_{min} [58]

Solute	V_{min}	β_2^H	$\Sigma\beta_2^{Ha}$	β_2^C
Trimethylphosphine oxide	−94.5	0.98		
Pyridine−N−oxide	−73.4	0.81		
Tetramethylurea	−68.0	0.74		
Dimethylsulfoxide	−69.5	0.77	0.88	1.54
Dimethylacetamide	−65.2	0.73	0.78	1.06
Dimethylformamide	−60.5	0.66	0.74	0.97
Acetophenone	−56.4	0.51	0.48	0.49
Propanone	−56.7	0.50	0.49	0.52
Dimethyltrifluoroacetamide	−48.2	0.46		
Benzaldehyde	−53.5	0.42	0.39	0.42
Methyl acetate	−46.6	0.40	0.45	0.47
Propionaldehyde	−54.0	0.39	0.45	0.37
Methyl formate	−52.5	0.38	0.38	
1,3-Dichloropropanone	−45.0	0.35		
4-Nitrobenzaldehyde	−44.1	0.36		
Hexafluoropropanone	−27.2	0.20		
N-Methylimidazole	−102.3	0.81	0.83(0.60)	
2,6-Dimethylpyridine	−93.5	0.64	0.63(0.49)	
Methylpyridine	−92.6	0.62	0.54(0.44)	
4-Methylpyridine	−93.1	0.66	0.54(0.43)	
3-Bromopyridine	−84.9	0.51	0.38(0.38)	
Pyrimidine	−82.2	0.53	0.65(0.65)	
Pyrazine	−79.9	0.48	0.62(0.62)	

[a] Values in parentheses are $\Sigma\beta_2^O$.

Famini *et al.* [59-63] have recently computed theoretical descriptors for use in SARs, including two hydrogen-bond acidity and two hydrogen-bond basicity descriptors. The hydrogen-bond acidity is represented by (i) a covalent term, e_a, the magnitude of the difference between the lowest unoccupied molecular orbital (LUMO) of the solute and the highest occupied molecular orbital (HOMO) of water, and (ii) an electrostatic term, $q+$, being the most positive formal charge on a hydrogen atom in the solute. In a similar way, the hydrogen-bond basicity is defined by (i) a covalent term, e_b, the difference between the HOMO of the solute and the LUMO of water, and (ii) an electrostatic term, $q-$, the most negative formal charge in the solute molecule. Other descriptors needed in order to set up a Theoretical Linear Free Energy Relationship, or more generally a Theoretical Structure Activity Relationship, TSAR, are a van der Waals volume, V_{mc}, calculated using the method of Hopfinger [64], and a dipolarity/ polarisability parameter, π_i. The latter is calculated by dividing the polarisation volume by V_{mc} so that π_i is not well correlated with volume. The TSAR then takes the form,

$$logSP = c + A.V_{mc} + B.\pi_i + C.e_b + D.q- + E.e_a + F.q+ \qquad (47)$$

There is a marked resemblance between eqn.(47) and eqn.(39), both incorporating descriptors of solute dipolarity/polarisability, hydrogen-bond acidity, hydrogen-bond basicity, and volume. As might be expected, there are reasonable correlations between q+ and q- and the Abraham descriptors $\Sigma\alpha_2^H$ and $\Sigma\beta_2^H$ [65], at least for the monofunctional solutes [60].

4. APPLICATIONS OF HYDROGEN-BOND DESCRIPTORS TO GAS-CONDENSED PHASE PROCESSES

The descriptors used by Carr (Section 3.2), have only been applied to GLC processes, and those of Murray and Politzer and by Famini (Section 3.3) are dealt with in other Chapters. Hence only the descriptors of Abraham (Section 3.1) will be considered. It should be noted that equations used to define the descriptors, that is so-called training equations, cannot then be used as examples of the application of the descriptors. What are required as examples are either sets of solutes that considerably extend those used in the training examples, or, preferably, completely different examples altogether. In the latter case, the examples can then be regarded as test equations. In nearly all the applications that follow, the systems studied are different to those of the training equations. Thus any SPR set up will be a test equation that represents how well the descriptors can account for processes other than those of the training equations.
It is of interest to note that all the descriptors in the two general SPRs, eqn(38) and eqn(39), are free energy related, as ,indeed, is required for a linear free energy relationship. The π_2^H, $\Sigma\alpha_2^H$, $\Sigma\beta_2^H$ and logL^{16} descriptors are all experimentally determined from equilibrium constants, as logK values, and hence are directly related to Gibbs free energy. A volume can also be regarded as a free energy related property: since PV= RT, it follows that at constant pressure V is proportional to RT, the latter being a free energy term. It is expected, therefore, that the two general equations should be applicable to the correlation of free energy related processes, and this has invariably been the case. There is no particular reason why eqn(38) or eqn(39) should be applicable to enthalpy related properties, unless there is some proportional entropic term, and it remains to be seen how these equations deal with enthalpy related properties.

4.1 The general solvation equation (38); GLC data.

$$logSP = c + r.R_2 + s.\pi_2^H + a.\Sigma\alpha_2^H + b.\Sigma\beta_2^H + 1.logL^{16} \qquad (38)$$

This equation, as suggested above, is useful in the examination of gas-condensed phase processes. The first application was to various sets of gas chromatographic retention data. Poole et al.[66] had very carefully determined gas-liquid partition coefficients, denoted as K or L, corrected for any possible effects due to adsorption, for series of solutes on a variety of GLC stationary phases. The gas-liquid partition coefficient is related to the specific retention volume of a solute, V_G at the column temperature, through K = V_G.d where d is the density of the liquid stationary phase, again at the column temperature. Equation (38) was applied to each phase, but for all 24 phases, the $\Sigma\beta_2^H$ term was found to be statistically not significant, thus showing that the phases were all nonacidic [67]. The reduced eqn(48) was then applied to the 24 phases; in Table 19 are listed the phases, and in Table 20 is given a summary of the regression equations obtained.

$$logK(orL) = c + r.R_2 + s.\pi_2^H + a.\Sigma\alpha_2^H + 1.logL^{16} \qquad (48)$$

Table 19. The 24 stationary phases of Poole et al. [66][a]

Code	Stationary phase	
A	Squalane	
B	SE-30	
C	OV-3	
D	OV-7	
E	OV-11	
F	OV-17	
G	OV-22	
H	OV-25	
I	OV-105	poly(cyanopropylmethyldimethylsiloxane)
J	OV-225	poly(cyanopropylmethylphenylsiloxane)
K	OV-275	poly(dicyanoallylsiloxane)
L	OV-330	poly(dimethylsiloxane)/carbowax copolymer
M	QF-1	poly(trifluoropropylmethylsiloxane)
O	Carbowax 20M	
P	DEGS	poly(diethylene glycol succinate)
Q	TCEP	1,2,3-tris(2-cyanoethoxy)propane
R	PPE-5	poly(phenyl ether), five rings
S	Tetraethylammonium 4-toluene sulfonate	
T	Tributylammonium 4-toluene sulfonate	
U	Tetrabutylammonium 4-toluene sulfonate	
V	Tetrabutylammonium picrate	
W	Tetrabutylammonium methanesulfonate	
X	Tetrabutylammonium N-(2-acetamido)-2-aminoethanesulfonate	
Y	Tetrabutylammonium 3-[tris(hydroxymethyl)methylamino-2-hydroxy-1-propanesulfonate	
O2[b]	Carbowax	
P2[b]	Diethylene glycol succinate	
R2[b]	Poly(phenyl ether), six rings	
TC[b]	Tricyanoethoxypropane	
ZE7	Zonyl E-7	
AJ[c]	Apiezon	
O3[c]	Carbowax 1540	
P3[c]	Diethylene glycol succinate	
R3[c]	Poly(phenyl ether), five rings	
PL[c]	Pluronic L72	
ZE7	Zonyl E-7	

[a]These are phases A-Y, [b]The Laffort phases [69]: note that O2, P2 and R2 are the same phases as O, P and R in the Poole set. [c]McReynolds phases [70]: note that O3, P3 and R3 are same phases as O, P and R.

Table 20. A summary of regression equations for GLC phases[a]

Phase	t/°C	c	r	s	a	l	n	r	sd
A	121	-0.20	0.12	0.02	-0.10	0.581	39	0.9985	0.033
B	121	-0.19	0.02	0.19	0.12	0.498	39	0.9989	0.022
C	121	-0.18	0.03	0.33	0.15	0.503	39	0.9992	0.021
D	121	-0.23	0.06	0.43	0.16	0.510	39	0.9989	0.025
E	121	-0.30	0.10	0.54	0.17	0.516	39	0.9985	0.029
F	121	-0.33	0.13	0.61	0.15	0.509	38	0.9978	0.036
G	121	-0.33	0.20	0.66	0.19	0.489	38	0.9979	0.034
H	121	-0.27	0.28	0.64	0.18	0.472	39	0.9973	0.042
I	121	-0.21	-0.04	0.40	0.37	0.499	39	0.9987	0.026
J	121	-0.51	0.02	1.21	0.96	0.462	39	0.9979	0.035
K	121	-0.64	0.39	1.90	1.64	0.241	32	0.9935	0.080
L	121	-0.43	0.10	1.06	1.42	0.481	36	0.9954	0.051
M	121	-0.25	-0.36	1.10	0.05	0.416	39	0.9853	0.077
O	121	-0.56	0.29	1.29	1.80	0.450	39	0.9957	0.059
P	121	-0.50	0.35	1.68	1.72	0.311	38	0.9899	0.096
Q	121	-0.49	0.28	1.91	1.68	0.290	40	0.9972	0.056
R	121	-0.39	0.23	0.83	0.34	0.527	39	0.9972	0.044
S	121	-1.01	0.36	2.06	3.61	0.340	29	0.9941	0.076
T	121	-0.72	0.11	1.55	2.92	0.466	30	0.9922	0.069
U	121	-0.62	0.01	1.66	3.36	0.440	34	0.9885	0.106
V	121	-0.54	0.10	1.56	1.42	0.445	36	0.9935	0.061
W	121	-0.63	0.09	1.60	3.41	0.437	34	0.9895	0.097
X	121	-0.67	0.28	1.81	3.42	0.329	34	0.9902	0.100
Y	121	-0.69	0.28	1.82	2.86	0.305	29	0.9932	0.080
O2	120	-2.01	0.25	1.26	2.07	0.429	199	0.997	0.07
P2	120	-1.77	0.35	1.58	1.84	0.383	199	0.997	0.07
R2	120	-2.51	0.14	0.89	0.67	0.547	199	0.997	0.06
TC	120	-1.69	0.26	1.93	1.88	0.365	199	0.998	0.06
ZE7	120	-1.99	-0.41	1.46	0.77	0.432	199	0.995	0.07
AJ	120	-0.48	0.24	0.15	0.13	0.596	165	0.999	0.02
O3	120	-0.75	0.22	1.37	1.92	0.456	169	0.998	0.04
P3	120	-0.97	0.26	1.76	1.80	0.375	158	0.995	0.05
R3	120	-0.69	0.14	0.92	0.61	0.560	168	0.999	0.02
PL	120	-0.54	0.09	0.93	1.42	0.529	163	0.998	0.03
ZE7	120	-0.76	-0.42	1.55	0.78	0.448	170	0.991	0.07

[a]Note that the c-constants for the last two sets are not comparable with those for the Poole set, because the retention data used are logVG for the McReynolds set and logK(rel) for the Laffort set; this does not affect the regressions in any other way, however.

In general, the regression equations summarised in Table 20 are all good, with r values between 0.9992 and 0.9885 so that the test sets in Tables 19 and 20 confirm the usefulness of the general eqn(38) or the reduced eqn(48). It should be noted that over all the 24 phases, only one point was dropped as an outlier - that of 2,6-dimethylaniline on phase V.

The regression coefficients for the phases in eqn(48) serve to characterise the phases in terms of chemical properties. The r-constant shows the tendency of the phase to interact with solutes through π- and n- electron pairs. Usually the r-constant is positive, but for phases that contain fluorine atoms, the r-constant can be negative. The s-constant gives the tendency of the phase to interact with dipolar/ polarizable solutes, the a-constant denotes the hydrogen-bond basicity of the phase (because acidic solutes will interact with a basic phase), and the b-constant is a measure of the hydrogen-bond acidity of the phase (because basic solutes will interact with an acidic phase). It is important to note that for gas-phase processes, the s-, a-, and b-constants must always be positive (or zero), because interaction between the phase and a solute will increase the solubility of a gaseous solute. The r-constant is an exception, because it is tied to hydrocarbons as a zero; hence fluoro compounds as phases may give rise to a negative r-constant. The l-constant is a combination of exoergic dispersion forces that make a positive contribution to the l-constant, and an endoergic cavity term that makes a negative contribution. In the event, the dispersion interaction nearly always dominates, so that the l-constant is positive. In GLC or GSC work, the l-constant is very important, since it is a measure of the ability of the phase to separate homologues in any homologous series [68].

These regression coefficients, or characteristic constants, can therefore be used as a system for characterisation of GLC stationary phases, in a much more rigorous and chemically interpretable way than the old system based on McReynolds probes. GLC phases can be selected for the separation of homologues, for example the relatively nonpolar phases A-F all have large l-constants. If a phase is required to separate solutes mainly on the basis of their dipolarity/polarizability, then phase M, which has a large s-constant but only a very small a-constant can be chosen. There are many basic phases in the A-Y group, most of which have large s-constants as well. But phases L and O have larger a-constants than s-constants, and also have reasonably large l-constants, the latter a definite advantage in GLC work. Of the molten salts, many are very highly basic due to the anionic group. Of these, phases T, U and W have reasonably large l-values, and perhaps could be useful in specialized separations.

Prior to the formulation of eqn(38), a similar equation was used [48,68] to characterise the five GLC phases of Laffort et al. [69] and the 77 GLC phases studied by McReynolds [70]. The new eqn(38) yields regression equations that are very close to those of the earlier equation. Details of the equations for the Laffort phases, and for six of the McReynolds phases are in Tables 19 and 20. For the remaining 71 McReynolds phases, the constants given before [68] can be taken as valid for eqn(38). Eqn(38) gives a good account of the Laffort and McReynolds phases, and provides a new method for the characterization of GLC stationary phases.

It should be noted that the regression equation remains the same, except for the c-constant, no matter whether the dependent variable is logK(L), or logV$_G$ or even logτ, where τ is the relative retention time. However, the retention index, I, cannot be used to characterize stationary phases through eqn(38), although it is useful in other contexts.

Carr et al. [71] have also examined a large number of GLC phases, using an equation slightly different to eqn(38). They obtain results very comparable to those in Table 20, and, again, suggest that the LFER approach can be used to character-ise GLC phases in a chemically more useful way than hitherto.

All the phases listed in Tables 19 and 20 are nonacidic GLC stationary phases, so that the b.$\Sigma\beta_2^H$ term is redundant. A number of acidic phases were synthesised by Abraham et al.[72] for use as candidate coatings for piezoelectric sorption detectors. Since such detectors are used at ambient temperature, the phases had all to be either non-volatile liquids or amorphous polymers above the glass transition point. Values of the gas-liquid partition coefficient, K or L, were obtained at 25°C for a range of solutes, using the liquids or polymers as GLC stationary phases.It had been established previously [73,74] that the sorption of vapors by materials made up as GLC stationary phases on conventional packed col-umns, matched the response of surface acoustic wave sensors to the vapors when the sensor was coated with the same materials. Of the thirteen materials studied at 25°C, the phenolic compound 2,2-bis(4-hydroxy-3-propylphenyl)-hexafluoro-propane was the most acidic [72], giving rise to the regression equation,

$$logL = -1.21 -0.38 R_2 + 1.38 \pi_2^H + 0.71 \Sigma\alpha_2^H + 5.31 \Sigma\beta_2^H$$
$$+ 0.984 logL^{16} \qquad (49)$$

n = 28,   r = 0.9887,   sd = 0.18

It must be noted that the various coefficients in eqn(49) cannot be compared to those given in Table 20, because there is a large temperature effect. In general the s-, a-, b- and l-constants all decrease with increase in temperature. The constants in eqn(49) show that not only is the bis-phenol a very strong hydrogen-bond acid, with a b-constant no less than 5.31 units, but it is a rather weak hydrogen-bond base (a=0.71). Hence the bis-phenol can discriminate strongly between solutes that are hydrogen-bond acids and those that are hydrogen-bond bases. Grate and Abraham [75] were able to set out an analysis on these lines for the design of chemically selective coatings for chemical sensors - an interesting application of the general eqn(38). More recently, Abraham and Grate et al.[76] have characterized a new set of fourteen candidate coatings in the same way, including phases that are strongly basic as well as strongly acidic.

One of the phases synthesised before [72] was a solid at room temperature, and hence was not useful as a possible chemical sensor coating. However, as it was strongly acidic, it seemed of interest to examine it as a GLC phase above its melting point. The phase, bis(3-allyl-4-hydroxyphenyl)sulfone, was made up as a GLC stationary phase in a packed column and logK values were obtained for a rather large series of solutes at 121 and 176°C, fully corrected for interfacial adsorption [77]. The obtained regression equations were,

$$logK(121°C) = -0.568 - 0.051 R_2 + 1.323 \pi_2^H + 1.266 \Sigma\alpha_2^H$$
$$+ 1.457 \Sigma\beta_2^H + 0.418 logL^{16} \quad (50)$$

$$n = 58, \quad r = 0.9940, \quad sd = 0.069, \quad F = 856$$

$$logK(176°C) = -0.749 + 0.165 R_2 + 1.160 \pi_2^H + 0.808 \Sigma\alpha_2^H$$
$$+ 1.290 \Sigma\beta_2^H + 0.332 logL^{16} \quad (51)$$

$$n = 54, \quad r = 0.9738, \quad sd = 0.134, \quad F = 176$$

Although the equation at 176°C is not at all good, both eqn(49) and eqn(50) show that the bis-phenol sulfone is certainly acidic. The variation of the characteristic constants with temperature is quite marked, and this is why neither eqn(50) nor eqn(51) can be compared with eqn(49), run at 25°C.

The examination [77] of the bis-phenol sulfone was useful in another context. Poole and Abraham et al. [78] had shown that the free energy of solvation model used by Poole to analyse GLC retention data was quite compatible with the LFER model set out, above. However, none of the GLC phases investigated by Poole and Abraham et al.[78] were acidic, and so the retention data for the bis-phenol sulfone were used also to confirm the essential equivalence of the two models [77].

Another acidic GLC stationary phase was synthesized by Carr et al.[57] in order to define a solute scale of hydrogen - bond basicity, as described in Section 3.2, above. They obtained relative K' values at 80°C on the acidic stationary phase 4-dodecyl-α,α-bis(trifluoromethyl)benzyl alcohol, COH, and on the corresponding methyl ether, COMe. Because the solute basicities found by Carr et al.[57] define another scale than the $\Sigma\beta_2^H$ descriptor, there is no reason why eqn(38) should well correlate the logK'(COH) values. This was, indeed, found [77],

$$logK'(COH) = -1.598 - 0.223 R_2 + 0.477 \pi_2^H + 2.686 \Sigma\beta_2^H$$
$$+ 0.678 logL^{16} \quad (52)$$

$$n = 143, \quad r = 0.9598, \quad sd = 0.265, \quad F = 402.9$$

Eqn(52) confirms the high acidity of the bis(trifluoromethyl) benzyl alcohol, and shows also that the phase has zero basici - ty. It will therefore be highly selective towards solutes that are hydrogen-bond bases. The rather poor correlation coefficient and standard deviation are almost certainly because the alcohol defines a different hydrogen-bond basicity than the $\Sigma\beta_2^H$ scale.

The regression equation for the corresponding ether, eqn (53) is very good, and shows that the ether has zero basicity (a = 0), a result that might be expected in view of the bis(trifluoromethyl) groups. The slight acidity (b=0.166) of the ether, however, is unexpected.

$$\log K'(\text{COMe}) = -1.440 + 0.411\ \pi_2^H + 0.166\ \Sigma\beta_2^H$$
$$+ 0.691\ \log L^{16} \qquad (53)$$

$$n = 140, \quad r = 0.9977, \quad sd = 0.053, \quad F = 9761.6$$

As part of their investigation into solvation models, above, Poole and Abraham *et al.* [78] had synthesised a number of new phases that might be acidic in nature. These are listed in Table 21, and the characteristic constants are given in Table 22. In the event, none of the phases showed any hydrogen-bond acidity, and so the reduced eqn(48) was applied.

Table 21. Some GLC stationary phases examined [78]

EGAD	Poly(ethylene glycol adipate)
THPED	N,N,N',N'-Tetrakis(2-hydroxypropyl)ethylenediamine
U50HB	Poly(ethylene glycol) (Ucon 50 HB 660)
DEHPA	Di(2-ethylhexyl)phosphoric acid
QBES	Tetrabutylammonium-N,N-(bis-2-hydroxyethyl)-2-aminoethane sulfonate

Table 22. A summary of the regressions for the phases in Table 21

Phase	t/°C	c	r	s	a	l	n	r	sd
EGAD	121	-0.544	0.088	1.491	1.720	0.410	54	0.9917	0.088
THPED	121	-0.399	0.093	1.203	2.014	0.466	54	0.9930	0.084
U50HB	121	-0.229	0.107	0.836	1.320	0.491	56	0.9889	0.106
DEHPA	121	-0.251	0.021	0.565	1.528	0.556	50	0.9946	0.087
QBES	121	-0.613	0.220	1.533	3.189	0.358	42	0.9635	0.167

Comparison of the results in Table 22 with those in Table 20 shows that the phases in Table 22 are all quite basic, with QBES being very basic indeed. It might be surprising that phases such as tributylammonium 4-toluene sulphonate (T in Table 20) or di(2-ethylhexyl)phosphoric acid (DEHPA in Table 22) do not act as hydrogen-bond acids. The reason is no doubt due to intramolecular or internal hydrogen-bonding between the active hydrogen atom, and some basic part of the molecule. Thus in order to design a GLC stationary phase that is a good hydrogen-bond acid, it is necessary to incorporate only weak basic sites, or basic sites that are spatially incapable of internal hydrogen-bonding with the acidic group(s) in the molecule.

The temperature variation of the characteristic constants in eqn(38) has been noted, above. Recently, Poole and Kollie [79] have applied eqn(38) to a study of the effect of temperature on retention data obtained for some 60 solutes on ten stationary phases. In all cases, the data were corrected for interfacial adsorption and were used to determine the gas-liquid partition coefficients, as logK values. The stationary phases studied were squalane, OV-17, OV-105, OV-225, QF-1, Carbowax 20M, THPED, TCEP, DEGS and tetrabutylammonium 4-sulfonate. As found before, these phases had either no or little hydrogen-bond acidity, so that the $\Sigma\beta_2^H$ term was statistically small or insignificant. As an example, the regression coefficients for Carbowax 20M are given in Table 23, as a function of temperature.

Table 23. Regression coefficients for Carbowax 20M [79]

t/°C	c	r	s	a	l	n	r	sd
61.2	-0.437	0.318	1.692	2.746	0.635	60	0.9992	0.042
81.2	-0.497	0.309	1.486	2.231	0.538	60	0.9996	0.027
101.2	-0.498	0.314	1.333	2.019	0.482	60	0.9994	0.028
121.2	-0.560	0.317	1.256	1.883	0.447	60	0.9992	0.032
141.2	-0.554	0.298	1.208	1.564	0.412	60	0.9996	0.021

The regressions obtained by Poole and Kollie [79] are excellent, thus showing not only what can be achieved by very careful experimentation, but also that the descriptors used must have been reasonably good, in terms of GLC regressions. Poole and Kollie [79] found that for the phases studied, the s-, a- and l-constant invariably decreased with increase in temperature.However, plots of the constants against 1/T were usually curved, rather than being straight lines, as might have been expected. Hence the ranking of phases using a given constant may alter with temperature. Thus QF-1 is more dipolar/polarizable than OV-225 at 61°C (s= 1.638 and 1.565) but less so at 141°C (s= 1.087 and 1.231). This highlights the necessity of using a common temperature when comparing the properties of two or more phases.

Eqn(38) is useful in a number of ways other than the characterization of phases or materials in terms of their constants. For example, two commercial samples of a well-known GLC stationary phase, OV-25, were studied at 25°C in order to ascertain the effect of phenyl groups in a siloxane [80]. The phase is represented as a poly(methylphenylsiloxane) with 75% phenyl substitution, and so might be expected to give rise to a large s-constant, and, of course, a negligible b-constant. The phases were made up as GLC packed columns in the usual way, and values of logL obtained for a wide range of solutes. Application of eqn(38) led to the regression equations,

$$\log L(OV\text{-}25A) = -0.48 + 0.22\ R_2 + 0.95\ \pi_2^H + 0.70\ \Sigma\alpha_2^H$$
$$+ 1.22\ \Sigma\beta_2^H + 0.815\ \log L^{16} \qquad (54)$$

n = 78,   r = 0.9928,   sd = 0.12

$$\log L(OV\text{-}25B) = -0.79 + 0.24\ R_2 + 1.23\ \pi_2^H + 0.63\ \Sigma\alpha_2^H$$
$$+ 0.49\ \Sigma\beta_2^H + 0.866\ \log L^{16} \qquad (55)$$

n = 42,   r = 0.9932,   sd = 0.10

Certainly, both samples of OV-25 have an enhanced s-constant, although it must be remembered that this refers to a temperature of 25°C only, and both samples are hydrogen-bond bases, as expected from their structure (a = 0.70 and 0.63 respectively). But what is unexpected is the hydrogen-bond acidity shown by the phases (b= 1.22 and 0.49). This is not an artifact through adsorption on the support, for example, but relates to the actual chemical constitution of the liquid phases themselves. The infra-red spectra of both samples showed the characteristic absorption of the OH stretch at around 3500 cm^{-1}, with that for sample OV-25A being the stronger [80]. This agrees with results from the LFER analysis, and shows how such an analysis can identify the presence of unsuspected groups.

4.2 The general solvation equation (38); polymers.

The analysis of polymer-probe interactions is an important application of eqn(38). Munk *et al.*[81] had determined gas-polymer partition coefficients for 42 probes on nine hydrocarbon polymers at 100°C, and this data, as logL values, was later [82] analysed through eqn(38). Two examples are given, those for poly (isobutene) and for poly(1,2-butadiene), denoted as H and I respectively,

$$logL(H) = -0.448 + 0.226\ R_2 + 0.110\ \pi_2^H + 0.643\ logL^{16} \qquad (56)$$

$$n = 43, \quad r = 0.9982, \quad sd = 0.037$$

$$logL(I) = -0.285 + 0.131\ R_2 + 0.361\ \pi_2^H + 0.130\ \Sigma\alpha_2^H + 0.626\ logL^{16} \qquad (57)$$

$$n = 43, \quad r = 0.9988, \quad sd = 0.029$$

Since both poymers are hydrocarbons, neither has much dipolarity/polarizability (s= 0.110 and 0.361), and neither is at all acidic (b = 0). The saturated material poly(isobutene) is nonbasic, but poly(1,2-butadiene) that contains alkene bonds is slightly basic, with an a-constant of 0.130, so that the general eqn(38) can identify the hydrogen-bond basic character of very weak bases.

A key quantity in polymer chemistry is the Flory-Huggins interaction parameter, X, that contains information about the strength of the interactions between a polymer and a bulk liquid. Possibly the most convenient method of determining X - values is to use the polymer as a stationary phase on a conventional packed GLC column, and then to determine V_G for a given solute probe. The polymer-bulk liquid probe X-value can then be calculated from V_G through well-known equations [83,84], as can also the weight fraction activity coefficient of the bulk liquid probe in the polymer at infinite dilution, Ω. In this way, Romdhane and Danner [84] determined values of X and $log\Omega$ for 25 probes at 80-100°C on a sample of poly(butadiene). Now since the definition of X and $log\Omega$ includes not only V_G, but the saturated vapor pressure of the bulk liquid probe, it is extremely difficult to predict X and $log\Omega$, simply because they involve properties of both the probe as a solute and the probe as a bulk liquid.

Abraham and Sakellariou *et al.*[85] overcame this difficulty through an indirect prediction. Firstly, they applied eqn(38) to the $logV_G$ values that Romdhane and Danner [84] had determined, using the poly(butadiene) as a GLC stationary phase. Of the regressions using data at 80, 90, and 100°C, that at 90°C is shown as an example,

$$logV_G(90) = -0.102 + 0.296 \ R_2 \ + 0.327 \ \pi_2^H + 0.387 \ \Sigma\alpha_2^H$$
$$+ 0.610 \ logL^{16} \qquad (58)$$

$$n = 24, \quad r = 0.9981, \quad sd = 0.041$$

Equations (57) and (58) are reasonably close, considering that the poly(butadiene) sample of Romdhane and Danner was not the same as that of Munk et al.[81], and contained 52% of trans-1,4-, 40% of cis-1,4-, and 8% of vinyl-1,2. Abraham and Sakellariou et al.[85] then showed that eqn(58) could be used to predict further values of $logV_G$ with an average deviation of 0.039 and a standard deviation of 0.064 log units. Then from the known [83,84] equations connecting V_G with X and $log\Omega$, it was possible to predict X with an average deviation of 0.090 and a standard deviation of 0.131 units, and to predict $log\Omega$ with an average deviation of 0.039 and a standard deviation of 0.057 log units. This seems to be the first example of the prediction of X and $log\Omega$ using the GLC plus LFER method.

To see how the predictive method would work with a polymer that was functionally substituted, the commercially available polymer OV-202, a poly(trifluoropropylmethylsilox-ane), was then chosen [85]. Values of $logV_G$ were known [86] for 50 varied solutes at 25°C, and application of eqn(38) led to,

$$logV_G(OV-202) = -0.489 - 0.480 \ R_2 \ + 1.298 \ \pi_2^H + 0.441 \ \Sigma\alpha_2^H$$
$$+ 0.705 \ \Sigma\beta_2^H + 0.626 \ logL^{16} \qquad (59)$$

$$n = 50, \quad r = 0.9967, \quad sd = 0.071$$

The regression eqn(59) is chemically reasonable. The r-constant is negative because fluorinated compounds are often less polarizable than hydrocarbons, the s-constant shows that the polymer is dipolar/polarizable, the a- and b-constants indicate that the polymer is slightly basic and acidic, whilst the l-constant shows that the polymer has a quite marked ability to separate homologues. Once an equation has been established that can be used to predict further values of $logV_G$, then, again, values of X and $log\Omega$ can be indirectly predicted through the $logV_G$ values. Some examples are given in Table 24.

Of the probes in Table 24(i), tetrahydrofuran is the most compatible with OV-202, as shown by the Flory-Huggins X - parameter, but cyclohexanone and dimethylformamide are both predicted to be more compatible with the polymer than is tetrahydrofuran [85].

A related study to the above has been carried out by Abraham and Whiting [86] on soybean oil. Although not tech-nically a polymer, the same information is required, notably on the weight fraction activity coefficient. King and List [87] had determined specific retention volumes for 22 solutes on soybean oil as a GLC phase, from 58.7 to 123.4°C, and Abraham and Whiting obtained regression equations for $logV_G$, through eqn (38) in the usual way. A summary of the equations is in Table 25; only 21 solutes were studied, because decane was an outlier.

A reduced form of eqn(38) was applied, because the r-constant and the b-constant were zero throughout. This is

Table 24. Values of X and logΩ for probes on OV-202 at 25°C [85]

Probe	logV_G	X	logΩ
(i) From observed values of logV_G			
Propan-2-ol	1.822	3.22	2.03
Butan-1-ol	2.582	3.14	1.98
Cyclohexane	1.862	1.98	1.50
n-Butylbenzene	3.772	1.74	1.35
Dichloromethane	1.632	1.55	1.08
Trichloromethane	1.972	1.33	0.93
Benzene	2.272	1.25	1.13
Toluene	2.732	1.22	1.12
Tetrahydrofuran	2.362	0.60	0.84
(ii) From predicted values of logV_G			
Benzyl alcohol	4.201	3.67	2.08
Aniline	3.877	2.50	1.60
p-Xylene	3.103	1.39	1.19
Propanone	2.065	1.04	1.08
Ethyl acetate	2.450	0.75	0.90
Cyclohexanone	3.889	0.41	0.72
Dimethylformamide	4.118	0.37	0.71

Table 25. Regression coefficients for logV_G on soybean oil[a]

t/°C	c	s	a	l	n	r	sd
58.7	-0.415	0.815	1.602	0.820	21	0.988	0.09
79.0	-0.421	0.735	1.322	0.744	21	0.990	0.08
100.9	-0.414	0.649	1.089	0.671	21	0.990	0.07
123.4	-0.427	0.584	0.901	0.611	21	0.991	0.06

[a] The r- and b-constants were zero [86]

Table 26. Comparison of characteristic constants at 120°C

Phase	c	r	s	a	l
Soybean oil	-0.42	–	0.59	0.92	0.618
Di-2-ethylhexyl adipate[a]	-0.36	0.13	0.55	0.87	0.590
Dioctyl sebacate[a]	-0.35	0.12	0.49	0.79	0.594
Diethyleneglycol succinate[b]	-0.50	0.35	1.68	1.72	0.311
Ethyleneglycol adipate[c]	-0.54	0.09	1.49	1.72	0.410

[a] Ref[86].
[b] From Table 22, phase P.
[c] From Table 23, phase EGAD.

reasonable in view of the chemical constitution of soybean oil as a triglyceride of linoleic acid (9,12-octadecanodienoic acid) with a molecular weight of around 880. The s- and a-constant at 120°C, estimated from those at other temperatures in Table 25, can be compared with other phases, especially esters, see Table 26. It can be seen that the characteristic constants for soybean oil are very close to those for di-2-ethylhexyl adipate or dioctyl sebacate.

King and List [87] used their V_G values to calculate weight fraction activity coefficients for the bulk liquid probes on soybean oil, as an indication of the mutual solubility of the liquid probe with soybean oil. However, it is possible also to use the regression equations summarised in Table 26 to predict other $\log V_G$ values, and thence to predict other $\log \Omega$ or Ω values exactly as for the polymers, above. In Table 27 are some values of $\log V_G$ and of Ω, the weight fraction activity coefficient of the liquid probe in soybean oil, obtained in these ways [86].

Table 27. Values of $\log V_G$ and Ω for probes on soybean oil at 79°C [86]

Probe	$\log V_G$	Ω
(i) From observed values of $\log V_G$		
Butan-1-ol	2.17	12.9
Propanone	1.42	9.7
n-Hexane	1.65	5.8
Ethylbenzene	2.08	3.3
Dichloromethane	1.65	2.4
Trichloroethene	2.22	1.8
Tetrachloromethane	2.02	1.7
(ii) From predicted values of $\log V_G$		
Acetonitrile	1.65	17.9
Ethyl acetate	1.76	5.5
Benzyl alcohol	3.92	5.4
Dimethylformamide	2.92	5.1
Cyclohexanone	3.03	3.2
1,2-Dichloroethane	2.10	2.7
Trichloromethane	1.98	1.5

Since descriptors for eqn(38) are known for hundreds of compounds, it is very easy to survey possible solvents for soybean oil, once the necessary regression equation for $\log V_G$ has been set up. Thus dimethylformamide is predicted to be not a very good solvent for soybean oil, but trichloromethane should be a much better solvent.

Application of eqn(38) to polymers and related materials is therefore useful in a number of ways. The polymer can be characterized in chemical terms through the characteristic constants in the equation, and the constants can be compared (at the same temperature) to those for various liquid phases. Specific polymer-probe interactions can be identified and quantified. Important properties such as the Flory-Huggins interaction parameter and the weight fraction activity coefficient can be predicted indirectly through the prediction of $\log V_G$ values. There seems to be considerable scope here for further work in this important area.

4.3 The general solvation equation (38); solvents.

Application of eqn(38) to solvents is no different in principle to the above applications to GLC phases and to poly - mers. A set of gas-solvent partition coefficients is required for a range of solutes, and the logL(or K) values are used as the dependent variable in eqn(38). The logL values can be obtained directly, and can also be obtained by combining infinite dilution activity cofficients with vapor pressures. In this way, sets of logL values were obtained for solutes in a number of solvents at 25°C, as a first step towards a character - ization scheme for solvents [88]. A summary of the regression equations in logL, on applying eqn(38), are in Table 28.

Table 28. Regression equations for logL on solvents at 25°C [88]

Solvent	c	r	s	a	b	l	n	r	sd
CH_2I_2	-0.74	0.32	1.34	0.38	1.19	0.866	37	0.9979	0.09
$CHCl_3$	0.10	-0.35	1.26	0.60	1.18	0.994	35	0.9969	0.15
CCl_4	0.23	-0.20	0.35	0.07	0.27	1.041	89	0.9993	0.07
DCE[a]	-0.01	-0.28	1.72	0.73	0.59	0.926	40	0.9977	0.10

[a] 1,2-Dichloroethane

The equations summarized in Table 28 are all reasonably good, and the constants are generally in line with general chemical intuition. Thus the s-constant is very small for tetra - chloromethane, and somewhat larger for the other three more dipolar solvents, the a-constants are all quite small as required for solvents that are weak bases, and the b-constant is very small for tetrachloromethane (it would actually be expected to be zero) but larger for dichloromethane and trichloromethane, both of which have active hydrogen atoms. The l-constants are all very large (note that by definition the l - constant is unity for solvent hexadecane at 25°C) and show that all four phases are very hydrocarbon-like in their dispersion/ cavity effects. The only anomaly is in the r-constant that might be expected to be small and positive for the three chlorinated solvents and larger and positve for diiodomethane. It is possible that the small negative constants arise from lone pair-lone pair repulsions.

The characteristic constants for the four halogenated solvents can be compared with those for a variety of others, in order to set out a first preliminary characterization of solvents, see Table 29.

Although there are only twelve solvents in Table 29, they span a quite wide range of properties. The r-constant seems not easy to interpret if lone pair-lone pair repulsions play a part, but the other constants fall into a sensible pattern. The s-constant, reflecting dipolarity/polarizability interactions, is very large for NFM and for water, and it is likely that most s-constants will be in the range 0.00-3.00 for solvents at 25°C. The solvent hydrogen-bond basicity, as the a-constant, covers a range of 0.00-4.32, at present, with water and the amide

Table 29. The characterization of solvents at 25°C via eqn(38)

Solvent	c	r	s	a	b	l
Hexadecane	0.00	0.00	0.00	0.00	0.00	1.000
CH_2I_2	-0.74	0.32	1.34	0.38	1.19	0.866
$CHCl_3$	0.10	-0.35	1.26	0.60	1.18	0.994
CCl_4	0.23	-0.20	0.35	0.07	0.27	1.041
DCE[a]	-0.01	-0.28	1.72	0.73	0.59	0.926
OV-25[b]	-0.79	0.24	1.23	0.63	0.49	0.866
4-Butylphenol[c]	-0.06	-0.56	0.91	1.40	2.76	0.846
3-Ethylphenol[c]	-1.08	-0.20	0.87	1.80	3.42	0.899
Bis-phenol[d]	-1.21	-0.38	1.38	0.71	5.31	0.984
NFM[e]	-0.53	-	2.57	4.32	-	0.730
2-EHP[f]	-0.07	-0.26	0.91	3.74	-	0.955
Water[g]	-1.28	0.87	2.70	4.01	4.80	-0.210

[a] 1,2-Dichloroethane
[b] Eqn(55)
[c] From ref [72]
[d] 2,2-Bis(4-hydroxy-3-propylphenyl)-hexafluoropropane, eqn(49)
[e] N-Formylmorpholine, from ref [28]
[f] 2-Ethylhexylphosphate, from ref [28]
[g] From ref [28]

N-formylmorpholine having the largest basicity. This is totally at variance with results of the Kamlet-Taft solvatochromic comparison method, that show water as having a much smaller solvent basicity than amides. Thus β_1 for water is either 0.18 [89] or 0.50 [90], whereas β_1 for amides ranges from 0.70-0.78 units [89]. The b-constant, a measure of the solvent hydrogen-bond acidity, is large for water and even larger for the bis-phenol. The l-constant is very interesting in that it is actually negative for water - the only pure solvent studied to date with a negative l-constant. Hence the solubility of gases in water decreases along any homologous series [91], in contrast to the solubility of a homologous series of liquids. The negative l-constant seems quite clearly to be a manifestation of the "hydrophobic effect" and sets water apart from other pure solvents. In these cases, the general dispersion interaction that favors solubility of solutes is larger than the cavity effect that opposes the solubility of solutes, but for water the converse applies.

Once the solubility logL values have been correlated through the LFER equation (38), the various contributing factors to the overall solubility can be identified term-by-term [28,88]. Some representative calculations on these lines are in Table 30.

For the nonaqueous solvents, the $l.logL^{16}$ term is often larger than any other. Now since this term is a combination of a rather large negative cavity effect that is outweighed by an even larger positive effect due to general dispersion inter-actions, it follows that the general dispersion interaction

Table 30. Contribution of terms in eqn(38) to logL values at 25°C

	$r.R_2$	$s.\pi_2^H$	$a.\Sigma\alpha_2^H$	$b.\Sigma\beta_2^H$	$l.logL^{16}$
Solvent CH_2I_2					
Hexane	0.00	0.00	0.00	0.00	2.31
Benzene	0.20	0.70	0.00	0.17	2.41
Propanone	0.06	0.94	0.03	0.58	1.47
Ethanol	0.08	0.56	0.31	0.57	1.29
Solvent $CHCl_3$					
Hexane	0.00	0.00	0.00	0.00	2.65
Benzene	-0.21	0.66	0.00	0.17	2.77
Propanone	-0.06	0.88	0.02	0.58	1.69
Ethanol	-0.09	0.53	0.22	0.57	1.48
Solvent bis-phenol					
Hexane	0.00	0.00	0.00	0.00	2.63
Benzene	-0.23	0.72	0.00	0.74	2.74
Propanone	-0.07	0.97	0.03	2.60	1.67
Ethanol	-0.09	0.58	0.26	2.55	1.46
Solvent N-formylmorpholine					
Hexane	0.00	0.00	0.00	0.00	1.91
Benzene	0.00	1.34	0.00	0.00	2.03
Propanone	0.00	1.80	0.17	0.00	1.24
Ethanol	0.00	1.08	1.60	0.00	1.08
Solvent water					
Hexane	0.00	0.00	0.00	0.00	-0.56
Benzene	0.53	1.40	0.00	0.67	-0.59
Propanone	0.16	1.89	0.16	2.35	-0.36
Ethanol	0.21	1.13	1.48	2.30	-0.31

itself will be much larger than any dipolarity/ polarizability term or hydrogen-bonding term. Of course, differences in the solubility of gases will depend on such terms: this is why benzene is more soluble than hexane in all solvents, and is why ethanol is very soluble in acidic and in basic solvents.

Similar analyses to the above can be carried out for any combination of solute and solvent for which the solute descriptors and solvent characteristic constants are known.

4.4 The general solvation equation (38); adsorbents.

A very useful method of obtaining data on gas-solid partitions is that of gas-solid chromatography, GSC, in which a solid adsorbent is used as the stationary phase. Pankow [92] used the GSC method to obtain $logV_G$ values at 20°C for 38 solutes on Carbotrap a graphitized carbon material. Since this material is used to sample gaseous organic compounds, it would be helpful to be able to predict $logV_G$ values. Pankow [92] suggested that either the solute boiling point, T_b/K, or vapor pressure at 20°C, as logP/torr, might be a suitable descriptor, but found that only by dividing the 38 solutes into various

subsets could any reasonable correlations be obtained.   For all
38 compounds, rather poor equations are found [93],

$$logV_G = -12.64 + 0.040 \ T_b/K \qquad (60)$$

n = 38,   r = 0.8262,   sd = 2.14,   F = 77

$$logV_G = 5.14 - 2.00 \ logP/torr \qquad (61)$$

n = 38,   r = 0.7869,   sd =2.34,   F = 59

Neither eqn(60) nor eqn(61) are very suitable for the
estimation of further $logV_G$ values, and so Abraham and Walsh
[93] applied eqn(38) to the set of 38 values. Most of the descriptors were not significant, leaving two possible equations,

$$logV_G = \ -4.73 -2.27 \ R_2 + 2.65 \ logL^{16} \qquad (62)$$

n = 38,   r = 0.9737,   sd = 0.88,   F = 318

$$logV_G = \ -4.82 + 2.41 \ logL^{16} \qquad (63)$$

n = 38,   r = 0.9570,   sd = 1.10,   F = 392

Either of the above equations could be used to estimate $logV_G$ for
a host of compounds for which the relevant descriptors are known.
       Abraham and Walsh [93] also showed that eqn(38) could be
applied successfully to the $logV_G$ values obtained by Cao [94]
for solutes on three microporous carbons, again with standard
deviations low enough for the regression equations to be useful
for the estimation of further $logV_G$ values.   It is very
interesting that one of the first applications [95] of the
$logL^{16}$ descriptor was not to Gibbs energies (or logK values),
but to enthalpies of adsorption of gaseous solutes on a variety
of carbonaceous materials. Arnett et al.[95] determined $\Delta H°$
values for a range of (nonacidic) gaseous solutes on graphite,
anthracite coal, Ambersorb XE-348, Carbopack B, and Carbopack
F. The most extensive series was on Carbopack F, for which 31
$\Delta H°$ values were obtained. Various correlations of $\Delta H°$ with
electronic polarizability, orientation polarizability, and
linear combinations of the two were carried out [95], but the
best correlation was simply against $logL^{16}$ itself,

$$-\Delta H° = 1.715 + 2.592 \ logL^{16} \qquad (64)$$

n = 31,   r = 0.959

Arnett et al.[95] suggested that adsorption on Carbopack F was
the result of nonspecific intermolecular forces. Whether or not
the Gibbs free energy descriptor, $logL^{16}$, is generally useful in
the correlation of enthalpies of reaction remains to be seen.
       The properties of fullerenes have been the subject of
intense investigations over the last few years [96], but their
adsorbent properties have only just been studied; Abraham et
al. [97] used the GSC method to determine gas -solid partition
coefficients for 22 gaseous solutes at 25°C on a sample of

fullerene that was 83.9% C_{60} and 16.1% C_{70}. The partition coefficient was defined as,

$$K_c = (C_s/C_g); \quad C_g - 0 \tag{65}$$

where C_s and C_g are the concentration of solute in the solid (g/g) and in the gas phase (g/dm^3). Application of eqn(38) to the logK_c values resulted in the regression equation,

$$\log K_c = -1.58 - 0.24 \, R_2 + 0.72 \, \pi_2^H + 1.04 \, \Sigma\alpha_2^H + 0.48 \, \log L^{16} \tag{66}$$

$$n = 22, \quad r = 0.951, \quad sd = 0.12, \quad F = 40$$

This equation shows that the fullerene will separate gases and vapors mainly through interactions with solutes that are dipolar/ polarizable (s = 0.72) and are hydrogen-bond acids (a = 1.04), as well as through general dispersion interactions (l = 0.48). The characteristic constants in eqn(66) show that the fullerene is only weakly polarizable, is a hydrogen-bond base, and is only partially hydrocarbon-like in its general dispersion interaction/ cavity effect. The weak polarizability of the fullerene is in line with the suggestion that fullerenes do not behave as highly aromatic molecules, but rather as giant closed-cage alkenes [96].

4.5 The general solvation equation (38); QSARs.

Some years ago, Abraham *et al.* [98] applied a preliminary version of eqn(38) to the upper respiratory tract irritation of male Swiss OF_1 mice by nonreactive airborne chemicals. A re-analysis using eqn(38) yields a very similar result [28],

$$-\log FRD'_{50} = 0.96 + 0.81 \, \pi_2^H + 2.55 \, \Sigma\alpha_2^H + 0.722 \, \log L^{16} \tag{66}$$

$$n = 39, \quad r = 0.987, \quad sd = 0.12$$

In eqn(66), FRD'_{50} is the concentration of airborne chemical, in mol dm^{-1}, required to elicit a 50% decrease in respiratory rate. This equation is just good enough to be used to estimate further values of FRD'_{50} for nonreactive compounds. In addition, eqn(66) yields valuable information as to the chemical properties of the receptor site or area, through a comparison of the characteristic constants with those for gaseous solubility in various solvent phases. Inspection of Table 30 shows that the receptor area can not be at all aqueous-like, because the characteristic constants of solvent water do not resemble those in eqn(66) at all. It seems as though the receptor area is somewhat dipolar/polarizable (s = 0.81), is quite strongly basic (a = 2.55) but is nonacidic (b = 0.00). This is an interesting application of eqn(38) that has so far only been touched upon.

5. APPLICATIONS OF HYDROGEN-BOND DESCRIPTORS TO PROCESSES WITHIN CONDENSED PHASES

As pointed out, above, the training sets that have been used to construct scales of solute descriptors cannot then be used as test sets to demonstrate how `good' the descriptors are. In order to examine how well equations that incorporate the descriptors can be used to explain and interpret processes, either a test set for a given process must considerably be expanded, or, preferably, new processes altogether must be studied. In setting up the solute descriptor scales for eqn(38) and eqn(39), considerable use was made of condensed phase processes, notably water-solvent partitions. Thus eqn(39) can be tested on these processes only by considerably increasing the solute set used, or by examining new water-solvent partitions. For gas-condensed phase processes, the coeffi - cients in the SAR are characteristic of the condensed phase under study. But in the application of equations to processes within condensed phases, the coefficients will characterise the difference between the phases concerned. Eqn(39) has proved the more useful for processes in condensed phases [28], and various applications will now be considered.

5.1 The general solvation equation (39); partition.

$$\log SP = c + r.R_2 + s.\pi_2^H + a.\Sigma\alpha_2^H + b.\Sigma\beta_2^H + v.V_x \qquad (39)$$

In a few cases, the water-solvent systems listed in Table 8 have been re-examined using larger and more varied data sets [99] and some examples are given below. Here, P(oct) refers to the water-octanol system, P(alk) to water-alkane partitions, P(16) specifically to water-hexadecane partitions, and P(cyc) to the water-cyclohexane system. Note that the solutes in the water - octanol regression do not include any with a variable basicity.

$$\log P(oct) = 0.088 + 0.562\ R_2 - 1.054\ \pi_2^H + 0.034\ \Sigma\alpha_2^H$$
$$- 3.460\ \Sigma\beta_2^H + 3.841\ V_x \qquad (67)$$
$$n = 613, \quad r = 0.9974, \quad sd = 0.116, \quad F = 23162$$

$$\log P(alk) = 0.287 + 0.649\ R_2 - 1.657\ \pi_2^H - 3.516\ \Sigma\alpha_2^H$$
$$- 4.818\ \Sigma\beta_2^H + 4.282\ V_x \qquad (68)$$
$$n = 200, \quad r = 0.9978, \quad sd = 0.121, \quad F = 8902$$

$$\log P(16) = 0.087 + 0.667\ R_2 - 1.617\ \pi_2^H - 3.587\ \Sigma\alpha_2^H$$
$$- 4.869\ \Sigma\beta_2^H + 4.433\ V_x \qquad (69)$$
$$n = 370, \quad r = 0.9982, \quad sd = 0.124, \quad F = 20236$$

$$\log P(cyc) = 0.127 + 0.816\ R_2 - 1.731\ \pi_2^H - 3.778\ \Sigma\alpha_2^H$$
$$- 4.905\ \Sigma\beta_2^H + 4.646\ V_x \qquad (70)$$
$$n = 170, \quad r = 0.9968, \quad sd = 0.121, \quad F = 5123$$

The above equations contain a few more solutes than those listed in Table 8, and so can hardly be regarded as test sets. However, they are more definitive than the earlier equations, and the coefficients can be taken as differences in the characteristic constants for the two phases in each system. Because the alkanes all have a zero acidity and zero basicity, the a-constant and the b-constant should be the same in eqn(68) - (70), and this is approximately the case. Hence on this `water - solvent' scale, water has a hydrogen-bond acidity of around 4.86 units, and a hydrogen-bond basicity of about 3.62 units. In like vein, the dipolarity/polarizability of water, by comparison to the alkane solvents can be taken as 1.67 units. The r-constant is interesting as it now seems to represent more of a general dispersion interaction term - the alkanes being more prone to interact in this way than is water. The v-constant can be regarded as representing a lipophilic or alkane - like scale. It would not be expected that v-constants larger than about 4.7 units would be observed in any usual water - solvent system, compare Table 8 again.

In the light of this analysis, the constants in eqn(67) are of considerable interest. The magnitude of the r-constant and the s-constant is a little less than those in the water - alkane systems, as expected if octanol (or wet octanol) is somewhat nearer water as regards general dispersion interactions and in dipolarity/polarizability. Likewise, the lower magnitude of the b-constant and the v-constant shows that wet octanol has some acidity (ie it is nearer water than the alkanes are), and is lipophilic, but less so than the alkanes. All this is comprehendable in the light of comparative properties of water, wet octanol, and alkanes. What is not so obvious is the almost zero a-constant that can only arise if water and wet octanol have the same hydrogen-bond basicity. Yet the Kamlet-Taft solvent basicity of water, 0.18 [89] or 0.50 [90] units, as discussed in Section 4.3, is much less than that of alcohols, eg 0.88 for pentan-1-ol [90]. Dallas and Carr [100] have shown that the solvent hydrogen-bond basicity of wet octanol (0.79 or 0.95) is of the same order as that of dry octanol (0.86 or 0.96) so there is a clear conflict between results from the characteristic constants and from solvatochromic parameters. This is in the same direction as found from the constants for gas-water partitions, see Table 30, which also show water to be more basic than the solvatochromic method indicates. Testa et al.[101] have also used LFERs to analyse water-solvent partition coefficients, using a set of descriptors that are comparable to those in eqn(39). They also find that the hydrogen-bond basicity of water and wet octanol to be the same. The LFER equation of Taylor et al.[43] uses a set of descriptors that are not the same as those in eqn(39) at all, yet again the hydrogen-bond basicity of water and wet octanol turn out to be almost identical. The different results from LFER analyses of partition coefficients, and from the solvatochromic method are not therefore due to the choice of some particular solute des - criptors. Further work here is needed to resolve the problem.

The ΔlogP parameter of Seiler [102] has found considerable use in the correlation of biochemical properties, as a possible alternative to logP(oct), see for example ref [103]. Seiler's parameter is defined as,

$$\Delta logP = logP(oct) - logP(cyc) \tag{71}$$

although water-alkane partition coefficients are sometimes used instead of the water-cyclohexane coefficients in eqn(71). It is possible to analyse ΔlogP just as the difference in constants between eqn(67) and eqn(68)-eqn(70), but a better statistical method is to calculate ΔlogP for a set of solutes, and then to regress the values against descriptors using eqn(39). In this way, Abraham et al.[99] obtained the following equation for ΔlogP as defined using the water-hexadecane partitions instead of water-cyclohexane.

$$\Delta logP(16) = - 0.072 - 0.093 \; R_2 + 0.528 \; \pi_2^H + 3.655 \; \Sigma\alpha_2^H$$
$$+ 1.396 \; \Sigma\beta_2^H - 0.521 \; V_x \tag{72}$$
$$n = 288, \quad r = 0.9833, \quad sd = 0.173, \quad F = 1646$$

The factors that influence ΔlogP(16), and this is also the case for ΔlogP(alk) and ΔlogP(cyc), are mainly solute hydrogen-bond acidity with solute dipolarity/polarizability, hydrogen-bond basicity and volume also playing a part. Testa et al.[101] also analysed the ΔlogP parameter and concluded that it was mainly a measure of solute hydrogen-bond acidity. However, eqn(72) rather collapses if only $\Sigma\alpha_2^H$ is retained as a descriptor,

$$\Delta logP(16) = 0.038 + 4.495 \; \Sigma\alpha_2^H \tag{73}$$
$$n = 288, \quad r = 0.9072, \quad sd = 0.396, \quad F = 1329$$

and the corresponding equations using water-cyclohexane or water - alkane partitions are no better [99]. Thus although the a.$\Sigma\alpha_2^H$ term is the main one in the ΔlogP regression equation, the other terms cannot be overlooked.

There is one set of partition coefficients that has been used as a test set. Abraham and Leo [104] have applied eqn(39) to water-decanol partition coefficients, as logP(dec), not used in setting up the descriptor scales,

$$logP(dec) = 0.008 + 0.485 \; R_2 - 0.974 \; \pi_2^H + 0.015 \; \Sigma\alpha_2^H$$
$$- 3.798 \; \Sigma\beta_2^H + 3.945 \; V_x \tag{74}$$
$$n = 51, \quad r = 0.9929, \quad sd = 0.124, \quad F = 630$$

Eqn(39) thus gives a good account of water-decanol partition coefficients, with constants that are reasonable by comparison with those for the water-octanol partition. Interestingly, on eqn(74), water and wet decanol must have the same hydrogen-bond basicity.

5.2 The general solvation equation (39); HPLC.

Larrivee and Poole [105] have used eqn(39) to characterize the retention properties of particle-loaded membranes in solid-phase extraction, and Miller and Poole [106] have applied eqn (39) to HPLC data in connection with the characterization of a sorbent for solid-phase extraction by HPLC. The solid sorbent was an octadecyl packing, and Miller and Poole [106] obtained capacity factors, k', for 19 varied solutes over a range of water - methanol compositions of the mobile phase. A summary of the regression equations obtained for logk' values is in Table 31.

Table 31. Application of eqn(39) to HPLC capacity factors [106][a]

% MeOH	c	r	s	a	b	v	r	sd	F
100	-0.98	0.01	-0.19	-0.03	-0.62	1.22	0.932	0.08	17
90	-0.70	-0.59	-0.18	0.03	-1.10	1.29	0.978	0.07	58
80	-0.75	0.10	-0.49	-0.32	-1.11	1.79	0.983	0.08	74
70	-0.71	0.16	-0.44	-0.32	-1.59	2.12	0.990	0.08	124
60	-0.54	0.18	-0.62	-0.31	-1.79	2.49	0.990	0.09	123
50	-0.41	0.34	-0.74	-0.31	-2.00	2.75	0.988	0.11	103
40	-0.57	0.28	-0.67	-0.33	-2.45	3.43	0.992	0.10	165
30	-0.32	0.44	-0.58	-0.43	-2.52	3.34	0.995	0.07	241

[a]The number of data points is 19 in all cases. The % methanol in the water-methanol phase is given as v/v.

The regressions summarized in Table 31 are quite good, with correlation coefficients ranging from 0.932 to 0.995, and the constants make general chemical sense. As Miller and Poole [106] point out, the two main solute factors that influence logk' values in a given system are solute hydrogen-bond basicity that favors partition into the mobile phase, and solute volume that favors partition into the stationary phase. The former is due to the very high hydrogen-bond acidity of bulk aqueous methanol, and the latter to the more hydrocarbon - like (or lipophilic) character of the stationary phase. It must be noted, however, that the octadecyl-bonded stationary phase will be saturated with the mobile phase. The variation of the b - constant and the v-constant with solvent composition is as expected. As the solvent becomes more aqueous, the magnitude of both increases, because (i) the mobile phase becomes even more acidic, and (ii) the difference in lipophilicity between the mobile and stationary phase increases.

5.3 The general solvation equation (39); aqueous anesthesia.

A preliminary version of eqn(39) has been applied [107] to the effect of aqueous solutes on the inhibition of activity of the firefly luciferase enzyme. A re-analysis [28] using eqn(39) leads to a very similar equation,

$$-\log EC_{50} = 0.58 + 0.72\ R_2 - 3.44\ \Sigma\beta_2^H + 3.77\ V_x \qquad (75)$$
$$n = 42, \quad r = 0.989, \quad sd = 0.33, \quad F = 566$$

As was originally pointed out [107], the crucial factors that determine the potency of aqueous compounds are solute hydrogen-bond basicity that decreases potency, and solute volume that increases potency. The characteristic constants in eqn(75) show that the anesthetic-binding site on the enzyme has about the same dipolarity and hydrogen-bond basicity as water (s = 0 and a = 0), but but is much less acidic than water (b = -3.44) and is much more lipophilic (v = 3.77). A similar re-analysis of the general anesthetic potency of aqueous solutes towards the tadpole leads to the equation,

$$-\log EC_{50} = 0.69 + 0.66\ R_2 - 4.44\ \Sigma\beta_2^H + 4.15\ V_x \qquad (76)$$
$$n = 28, \quad r = 0.9900, \quad sd = 0.22, \quad F = 392$$

The interpretation of eqn(76) is almost exactly the same as that of eqn(75). Solute hydrogen-bond basicity decreases potency, and solute size or volume increases potency. The anesthetic target site must be similar in nature to that in the firefly luciferase enzyme, as regards dipolarity/polariza-bility, hydrogen-bond acidity and basicity, and lipophilicity.

These particular properties of the target sites can be put on a quantitative basis through a comparison [107] of the characteristic constants in eqn(75) and (76) with those for various water-solvent partitions. Although the constants have changed slightly from those found in the original analysis, the conclusions are almost quantitatively the same. Taking water as 100 and hexadecane as 0, the hydrogen-bond acidity of the target site is 29 for the luciferase enzyme and 10 in general anesthesia whilst if the lipophilicity of water is taken as 0 and that of hexadecane as 100, the lipophilicity of the luciferase enzyme is 85 and that in general anesthesia is 94. Of course, the dipolarity/polarisability and hydrogen-bond basicity of the two target sites will be approximately the same as that of water.

6. CONCLUSIONS

It has been possible to devise a set of solute descriptors, particularly those of effective or overall hydrogen-bond acidity and basicity that can be used to construct LFERs and QSARs. Two general equations have been formulated, one that is the more useful for gas-condensed phase processes, and one that is better for processes within condensed phases. Notably, the descriptors in the two equations are the same, except for the `size' parameters $\log L^{16}$ and V_x. The two equations work reasonably well as LFERs and QSARs when applied to a wide variety of solubility-related processes. There seems no reason why descriptors for a very wide range and number of additional solutes cannot be determined by the methods given above, and thus lead to further applications to physicochemical and biochemical processes.

However, there is no descriptor in the two general equations that is related to solute `shape'. This must be an important factor in the binding of solutes to specific sites, or to specific lock-and-key interactions, so that there is much further work to be done to improve the two general equations and to increase their applicability as LFERs and QSARs.

ACKNOWLEDGEMENTS.

A great deal of the work that has been summarized briefly in this review has been carried out at University College London, and it is a pleasure to acknowledge the contributions made by members of our group over the past few years. They include Dr Gary S. Whiting, Dr David P. Walsh, Dr Jenik Andonian-Haftvan, Harpreet S. Chadha, Chau My Du, and Juliet P. Osei-Owusu.

I am grateful to Dr Colin F. Poole, Dr George R. Famini, and Dr Peter W. Carr who kindly supplied manuscripts of papers prior to publication.

REFERENCES.

1.   E. Overton, *Studien über die Narkose*, Fischer, Jena, 1901.
2.   H. Meyer, *Arch. exp. Path. Pharmak. (Naunyn-Schmiedebergs)* 1899, **42**, 109-118.
3.   H.Meyer, *Arch. exp. Path. Pharmak. (Naunyn-Schmiedebergs)* 1901, **46**, 338-346.
4.   K.H.Meyer and H.Gottlieb-Billroth, *Zeit. Physiol. Chem.*, 1920, **112**, 55-79.
5.   L.P.Hammett, *Chem. Rev.*, 1935, **17,** 125-136.
6.   L.P.Hammett, *Physical Organic Chemistry,* McGraw-Hill, New York, 1940.
7.   J.N.Bronsted and K.J.Pedersen, *Zeit. Physikal. Chem.*, 1924, **108**, 185-195.
8.   K.H.Meyer and H.Hemmi, *Biochem. Zeit.*, 1935, **277**, 39-71.
9.   H. van de Waterbeemd, *UK QSAR Discussion Group,* Oct. 1993.
10. R.Collander and H.Barland, *Acta Botanica Fennica II,* 1933, 79-96.
11. R.Collander, *Acta Physiol. Scand.*, 1947, **13**, 363-381.
12. J.C.Dearden, *Envir. Health Persp.*, 1985, **61**, 203-228.
13. J.Ferguson, *Proc.Roy.Soc.(London)*, 1939, **B127**, 387-404.
14. K.H.Meyer and H.Hopff, *Zeit. Physiol. Chem.*, 1923, **126**, 281-298.
15. H.Maskill, *The Physical Basis of Organic Chemistry*, Oxford University Press, Oxford, 1985.
16. C.Hansch, *Acc. Chem. Res.*, 1969, **2**, 232-239.
17. R.W.Taft,Jr., in *Steric Effects in Organic Chemistry*, ed. M.Newman, John Wiley, New York, 1956.
18. C.Hansch, *Acc. Chem. Res.*, 1993, **26**, 147-153.
19. W.Wilbrandt, *Pflugers Arch.*, 1931, **229**, 86-99.

20. G.Yang, E.Lien and Z.Guo, *Quant Struct-Act Relat*, 1986, **5**, 12-18.
21. R.W.Taft, D Gurka, L.Joris, P.von R.Schleyer and J.W. Rakshys, *J. Am. Chem. Soc.*, 1969, **91**, 4801-4808.
22. J.Mitsky, L.Joris and R.W.Taft, *J. Am. Chem. Soc.*, 1972, **94**, 3442-3445.
23. B.Panchenko, N.Oleinik, Yu.Sadovsky, V.Dadali and L.Litvinenko, *Organic Reactivity (USSR)*, 1980, **17**, 65-87.
24. J.-L.M.Abboud and L.Bellon, *Ann. Chim.*, 1970, **5**, 63-74.
25. P.-C.Maria, J.-F.Gal, J.de Franceschi and E.Fargin, *J. Am. Chem. Soc.*, 1987, **109**, 483-492.
26. M.H.Abraham, P.L.Grellier, D.V.Prior, J.J.Morris, P.J.Taylor, P.-C.Maria and J.-F.Gal, *J.Phys.Org.Chem.*, 1989, **2**, 243-254.
27. M.H.Abraham, P.L.Grellier, D.V.Prior, P.P.Duce, J.J.Morris and P.J.Taylor, *J.Chem.Soc.,Perkin Trans.2*, 1989, 699-711.
28. M.H.Abraham, *Chem. Soc. Revs.*, 1993, **22**, 73-83.
29. R.W.Taft and F.G.Bordwell, *Acc. Chem. Res.*, 1988, **21**, 463-469.
30. M.H.Abraham, P.L.Grellier, D.V.Prior, J.J.Morris and P.J.Taylor, *J.Chem.Soc.,Perkin Trans.2*, 1990, 521-529.
31. M.H.Abraham, P.L.Grellier, D.V.Prior, R.W.Taft, J.J.Morris, P.J.Taylor, C.Laurence, M.Berthelot, R.M.Doherty, M.J.Kamlet, J.-L.M.Abboud, K.Sraidi and G.Guiheneuf, *J. Am. Chem. Soc.*, 1988, **110**, 8534-8536.
32. M.H.Abraham, P.L.Grellier, D.V.Prior, J.J.Morris, P.J. Taylor and R.M.Doherty, *J. Org. Chem.*, 1990, **55**, 2227-2229.
33. E.D.Raczynska and C.Laurence, *J.Chem.Res.(S)*, 1989, 148-149.
34. C.Laurence, M.Berthelot, E.Raczynska, J.-Y. Le Questel, E.Duguay and P.Hudhomme, *J.Chem.Res.(S)*, 1990, 250-251.
35. C.Laurence and R.Queignec, *J. Chem. Soc.,Perkin Trans.2*, 1992, 1915-1917.
36. J.-Y. Le Questel, C.Laurence, A.Lachkar, M.Helbert and M.Berthelot, *J.Chem.Soc.,Perkin Trans.2*, 1992, 2091-2094.
37. M.Berthelot, M.Helbert, C.Laurence, J.-Y. Le Questel, F.Anvia and R.W.Taft, *J.Chem.Soc.,Perkin Trans.2*, 1993, 625-627.
38. M.Berthelot, M.Helbert, C.Laurence and J.-Y. Le Questel, *J.Phys.Org Chem.*, 1993, **6**, 302-306.
39. J.-M.L.Abboud, K.Sraidi, M.H.Abraham and R.W.Taft, *J. Org.Chem.*, 1990, **55**, 2230-2232.
40. O.A.Raevsky, V.Yu.Grigor'ev, D.B.Kireev and N.S.Zefirov, *Quant. Struct.-Act. Relat.*, 1992, **11**, 49-63.
41. H.-J. Schneider, V.Rudiger and O.A.Raevsky, *J. Org. Chem.*, 1993, **58**, 3648-3653.
42. M.H.Abraham, P.P.Duce, D.V.Prior, D.G.Barrett, J.J.Morris and P.J. Taylor, *J.Chem.Soc.,Perkin Trans.2*, 1989, 1355-1375.
43. D.E.Leahy, J.J.Morris, P.J.Taylor and A.R.Wait, *J. Chem. Soc.*, Perkin Trans.2, 1992, 705-722.
44. M.C.R.Symons and G.Eaton, *J.Chem.Soc., Faraday Trans.1*, 1985, **81**, 1963-1977.
45. M.C.R.Symons and G.Eaton, *J.Chem.Soc., Faraday Trans.1*, 1982, **78**, 3033-3044.
46. P.I.Nagy, W.J.Dunn III, G.Alagona and C.Ghio, *J.Phys.Chem.*, 1993, **97**, 4628-4642.

47. M.H.Abraham, P.L.Grellier and R.A.McGill, *J.Chem.Soc., Perkin Trans.2*, 1987, 797-803.
48. M.H.Abraham, G.S.Whiting, R.M.Doherty and W.J.Shuely, *J.Chem. Soc., Perkin Trans.2*, 1990, 1451-1460.
49. M.H.Abraham and J.C.McGowan, *Chromatographia*, 1987, **23**, 243-246.
50. M.H.Abraham, G.S.Whiting, R.M.Doherty and W.J.Shuely, *J. Chromatogr.*, 1991, **587**, 213-228.
51. M.H.Abraham, *J.Chromatogr.*, 1993, **644,** 95-139.
52. J.-C. Dutoit, *J.Chromatogr.*, 1991, **555**, 191-204.
53. W. Engewald, U. Billing, I. Topalova and N. Petsev, *J. Chromatogr.*, 1988, **446**, 71-77.
54. M.H.Abraham, *J.Phys.Org Chem.*, 1993, **12** , 660-684.
55. J.Li, Y.Zhang, A.J.Dallas and P.W.Carr, *J.Chromatogr.*, 1991, **550**, 101-134.
56. M.J.Kamlet, R.M.Doherty, J.-L.M.Abboud, M.H.Abraham and R.W.Taft, *CHEMTECH*, 1986, **16**, 566-576.
57. J.Li, Y.Zhang, H.Ouyang and P.W.Carr, *J. Am. Chem. Soc.*, 1992, **114**, 9813-9828.
58. J.S.Murray and P.Politzer, *J.Chem.Res.(S)*, 1992, 110-111.
59. G.R.Famini, R.J.Kassel, J.W.King and L.Y.Wilson, *Quant. Struct. -Act. Relat.*, 1991, **10**, 344-349.
60. L.Y.Wilson and G.R.Famini, *J.Med.Chem.*, 1991, **34**, 1668-1674.
61. G.R.Famini, W.P.Ashman, A.P.Mickiewicz and L.Y.Wilson, *Quant. Struct.-Act. Relat.*, 1992, **11**, 162-170.
62. G.R.Famini, L.Y.Wilson and C.A.Penski, *J.Phys.Org Chem.*, 1992, **5**, 395-408.
63. C.J.Cramer, G.R.Famini and A.H.Lowrey, *Acc.Chem.Res.*, in press.
64. A.J.Hopfinger, *J. Am. Chem. Soc.*, 1980, **102**, 7196-7206.
65. M.H.Abraham and G.R.Famini, unpublished work.
66. B.R.Kersten, S.K.Poole and C.F.Poole, *J.Chromatogr.*, 1989, **411**, 43-59.
67. M.H.Abraham and G.S.Whiting, *J.Chromatogr.*, 1992, **594**, 229-241.
68. M.H.Abraham, G.S.Whiting, R.M.Doherty and W.J.Shuely, *J. Chromatogr.*, 1990, **518**, 329-348.
69. F.Patte, M.Etcheto and P.Laffort, *Anal. Chem.*, 1982, **54**, 2239-2247.
70. W.O.McReynolds, *Gas Chromatographic Retention Data*, Preston Technical Abstracts, Evanston, IL, 1966.
71. J.Li, Y.Zhang and P.W.Carr, *Anal. Chem.*, 1992, **64**, 210-218.
72. M.H.Abraham, I.Hamerton, J.B.Rose and J.W.Grate, *J.Chem.Soc., Perkin Trans.2*, 1991, 1417-1423.
73. J.W.Grate, A.Snow, D.S.Ballantine,jr., H.Wohltjen, M.H.Abraham, R.A.McGill and P.Sasson, *Anal. Chem.*, 1988, **60**, 869-875.
74. J.W.Grate, M.Klusty, R.A.McGill, M.H.Abraham, G.Whiting and J. Andonian-Haftvan, *Anal. Chem.*, 1992, **64**, 610-624.
75. J.W.Grate and M.H.Abraham, *Sensors and Actuators B*, 1991, **3,** 85-111.
76. M.H.Abraham, J.Andonian-Haftvan, C.M.Du, V.Diart, G.S. Whiting, J.W.Grate and R.A.McGill, submitted.

134

77. M.H.Abraham, J.Andonian-Haftvan, I.Hamerton, C.F.Poole and T.O.Kollie, *J. Chromatogr.*, 1993, **646,** 351-360.
78. T.O.Kollie, C.F.Poole, M.H.Abraham and G.S.Whiting, *Anal. Chim. Acta*, 1992, **259**, 1-13.
79. C.F.Poole and T.O.Kollie, *Anal.Chim.Acta*, 1993, **282**, 1-17.
80. M.H.Abraham, G.S.Whiting, J.Andonian-Haftvan, J.W.Steed and J.W.Grate, *J. Chromatogr.*, 1991, **588**, 361-364.
81. Q.Du, P.Hattam and P.Munk, *J.Chem.Eng.Data*, 1990, **35**, 367-371.
82. M.H.Abraham, G.S.Whiting, W.J.Shuely and P.Sakellariou, *Polymer*, 1992, **33**, 2162-2167.
83. M.Galin, *Polymer*, 1983, **24**, 865-870.
84. I.H.Romdhane and R.P.Danner, *J.Chem.Eng.Data*, 1991, **36**, 15-20.
85. M.H.Abraham, C.M.Du, J.P.Osei-Owusu, P.Sakellariou and W.J. Shuely, *Eur. Polymer J.*, in press.
86. M.H.Abraham and G.S.Whiting, *J.Am.Oil Chem.Soc.*, 1990, **69**, 1236-1238.
87. J.W.King and G.List, *J.Am.Oil Chem.Soc.*, 1990, **67**, 424-430.
88. M.H.Abraham, J.Andonian-Haftvan, J.P.Osei-Owusu, P.Sakellariou, J.S.Urieta, M.C.Lopez and R.Fuchs, *J.Chem.Soc, Perkin Trans.2*, 1993, 299-304.
89. M.J.Kamlet, J.-L.M.Abboud, M.H.Abraham and R.W.Taft, *J.Org. Chem.*, 1983, **48**, 2877-2887.
90. R.M.C.Goncalves, A.M.N.Simoes, L.M.P.C.Albuquerque, M.Roses C.Rafols and E.Bosch, *J.Chem.Res.(S)*, 1993, 214-215.
91. M.H.Abraham, *J.Chem.Soc.Faraday Trans.1*, 1984, **80**, 153-181.
92. J.F.Pankow, *J.Chromatogr.*, 1991, **547**, 488-493.
93. M.H.Abraham and D.P.Walsh, *J.Chromatogr.*, 1992, **627**, 294-299.
94. X.-L.Cao, *J.Chromatogr.*, 1991, **586**, 161-165.
95. E.M.Arnett, B.J.Hutchinson and M.H.Healy, *J.Am.Chem.Soc.*, 1988, **110**, 5255-5260.
96. R.Taylor and D.R.M.Walton, *Nature,* 1993, **363**, 685-693.
97. M.H.Abraham, C.M.Du, J.W.Grate, R.A.McGill and W.J.Shuely, *Chem. Comm.*, 1993, 1863-1864.
98. M.H.Abraham, G.S.Whiting, Y.Alarie, J.J.Morris, P.J.Taylor, R.M.Doherty, R.W.Taft and G.D.Nielsen, Quant. Struct.-Act. Relat., 1990, **9**, 6-10.
99. M.H.Abraham, H.S.Chadha, G.S.Whiting and R.C.Mitchell, *J. Pharm. Sci.*, in press.
100. A.J.Dallas and P.W.Carr, *J.Chem.Soc., Perkin Trans.2*, 1992, 2155-2161.
101. N.El Tayer, R.-S.Tsai, B.Testa, P.-A.Carrupt and A.Leo, *J.Pharm.Sci.*, 1991, **80**, 590-598.
102. P.Seiler, *Eur.J.Med.Chem.*, 1974, **9**, 656-671.
103. N.El Tayer, R.-S.Tsai, B.Testa, P.-A.Carrupt, C.Hansch and A.Leo, *J.Pharm.Sci.*, 1991, **80**, 744-749.
104. M.H.Abraham and A.Leo, unpublished work.
105. M.L.Larrivee and C.F.Poole, *Anal. Chem.*, 1994, **66** 139-146.
106. K.G.Miller and C.F.Poole, *J.High Res. Chromatogr.*, in press.
107. M.H.Abraham, W.R.Lieb and N.F.Franks, *J.Pharm.Sci.*, 1991, **80,** 719-724.

P. Politzer and J.S. Murray
Quantitative Treatments of Solute/Solvent Interactions
Theoretical and Computational Chemistry, Vol. 1
© 1994 Elsevier Science B.V. All rights reserved.

Hydrogen Bonding in the Gas Phase and in Solution. New Experimental Developments*.

J.-L.M. Abboud[a], R. Notario[a] and V. Botella[b]

[a]Instituto de Química Física "Rocasolano", C.S.I.C., c/Serrano 119, E-28006 Madrid, Spain.

[b]Instituto de Estructura de la Materia, C.S.I.C., c/Serrano 119, E-28006 Madrid, Spain.

1. INTRODUCTION

An experimentalist's contribution to the "cross-pollination" between theory and experiment is based on "raw data" endowed with a precise physical meaning. In the case of molecular interactions, these data include, for example:
1. Changes in thermodynamic state functions pertaining to the formation of a complex A···B starting from species A and B in the gas phase, reaction 1, or in solution in a solvent S, reaction 2:

$$A(g) + B(g) \longrightarrow A···B(g) \tag{1}$$

$$A(S) + B(S) \longrightarrow A···B(S) \tag{2}$$

2. Structural information such as bond lengths, bond angles and force constants in "free" A and B and in the complex.
Very often, the experimentalist goes one step further and uses correlation analysis techniques in order to establish empirical relationships between structural properties of the isolated species, say A(g/S) and B(g/S) and the thermodynamic stability and/or structural properties of A···B(g/S). This approach is by no means limited to 1:1 interactions, for it is frequently extended to the treatment of "bulk solvent (or medium) effects" on the reactivity (thermodynamic or kinetic) and/or physical (mostly spectroscopic) properties of solutes.

(*) This work is supported by grants PL87-0357 and PB 90-0228-C02-02 from the Spanish D.G.I.C.Y.T., and is dedicated to Prof. Louis Bellon.

We feel that one of the most important challenges facing both theoreticians and experimentalists is the establishment of quantitative and conceptual relationships between empirical models encoding experimental data and physical magnitudes of the interacting species as determined, whenever possible, by means of quantum - mechanical methods.

In keeping with this spirit, we shall examine thermodynamic (and sometimes structural) data for the formation of 1:1 complexes between hydrogen bond donors (HBD) A-H and hydrogen bond acceptors (HBA) B in the gas phase, reaction 1a, and in solution, reaction 2a:

$$\text{A-H(g)} + \text{B(g)} \rightleftharpoons \text{A-H}\cdots\text{B(g)} \tag{1a}$$

$$\text{A-H(S)} + \text{B(S)} \rightleftharpoons \text{A-H}\cdots\text{B(S)} \tag{2a}$$

2. EXPERIMENTAL METHODS.

The concept of hydrogen bonding (HB) emerged from the study of self-associated liquids (notably water) [1] and its importance in complex biological systems is paramount [2]. This notwithstanding, the study of HB interactions in the gas phase is relevant because it provides direct information on the structure and stability of these complexes in the absence of the perturbations induced by the solvent.

Different experimental methods are used in these studies depending on whether the interactions involve neutral (reaction 1a) or charged species (reactions 1b and 1c):

$$\text{(A-H)}^+\text{(g)} + \text{B(g)} \longrightarrow \text{(A-H}\cdots\text{B)}^+\text{(g)} \tag{1b}$$

$$\text{A-H(g)} + \text{B}^-\text{(g)} \longrightarrow \text{(A-H}\cdots\text{B)}^-\text{(g)} \tag{1c}$$

2.1. Neutral species
Methods leading to the determination of the geometry, dissociation energy and force constants, the most important quantities characterizing the pairwise potential between the HB-donor (such as A-H) and the HB acceptor (such as B) are those based on: (i) microwave absorption spectroscopy [3] (including Stark-modulated microwave absorption spectroscopy), (ii) molecular beam electric resonance spectroscopy (MBERS) [4-5] and, (iii) pulsed-nozzle molecular beam, Fourier Transform Microwave Spectroscopy [6]. For recent reviews on microwave techniques, see, e.g., references (7) and (8). In the case of chelated species, gas phase electron diffraction [9] provides molecular geometries. Both the microwave techniques and MBERS yield the rotational constants for the isolated HB complexes. From these, the moments of inertia are determined. Assuming that monomer geometries remain essentially unaffected by association, the geometry of the complex can be determined.

Furthermore, absolute intensity measurements of vibrational satellites in microwave absorption spectra lead to vibrational spacings [3] and contribute to force constants determinations [10].

Under conditions of thermodynamic equilibrium, absolute intensity measurements in microwave absorption spectra allow the determination of the dissociation energies D_0 and D_e of the HB complex A-H···B [11], Figure 1.

From the knowledge of the moments of inertia, D_0, vibrational frequencies and spacing, the partition functions [12] for A-H, B and A-H···B can be calculated that lead to the equilibrium constant, K_p, for equilibria 1 or 1a as well as to the changes in the standard thermodynamic state functions [13] pertaining to these reactions. (In principle, this information can not be obtained directly from non-equilibrium techniques such as MBERS and pulsed-nozzle FT microwave spectrometry [13]). Representative results are discussed in the next section.

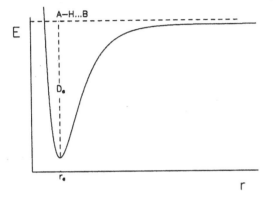

Figure 1a. Pertaining to the definition of D_e

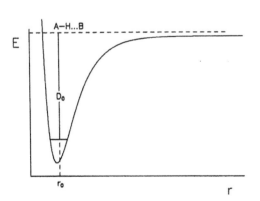

Figure 1b. Pertaining to the definition of D_0

The combination of experimental spectroscopic data with high-level theoretical calculations [14] provides an interesting alternative means for the determination of changes in the thermodynamic state functions for reactions 1 and 1a.

Vibrational spectroscopy is a time-honored technique [15-16] that has played and still plays a key role in the detection and even the operational definition [15a] of HB. Besides providing the necessary vibrational frequencies for the construction of the partition functions, this technique allows:

1. The detection of very weak HB interactions, as achieved at very low temperatures. This increases the concentration of complexes and simplifies the spectra. These

conditions prevail, for example, in solid rare-gas matrices. Interesting examples are those involving HF and CCl_4 in solid argon [17] and HF and saturated hydrocarbons in neon and argon matrices [18]. These studies conclusively show that, albeit weak, the HB complexes F-H···Cl-CCl_3 and F-H···C-C bonds to indeed exist. Interestingly, CCl_4 and saturated hydrocarbons are the "inert solvents" par excellence in solution work.

2. The determination of equilibrium constants for reactions 1a and 2a. Vibrational spectroscopy is possibly the most widely used experimental technique for this purpose [15 - 19]. In recent years, the development of the Fourier Transform Infrared Spectroscopy (FTIR) [20] has vastly increased its accuracy and reliability. Given the inherent difficulties of the microwave methods, FTIR should be the choice technique for the determination of the position of the equilibrium 1a. Unfortunately, the extension of the experimental protocol used for solution studies to the treatment of gaseous systems is by no means straightforward, as considerable difficulties arise from condensation and/or adsorption phenomena [21-22]. Recently, however, methods have been developed by Orza and Marco that allow to circumvent these obstacles [21-22]. These methods have provided a small but highly consistent set of equilibrium constants that shall be used in the next section.

Non-spectroscopic techniques are also available for the determination of equilibrium constants for reaction 1a. The most important of them are based on the analysis of deviations from the ideal gas behavior and on thermal conductivity data for gaseous systems. These methods (as well as the spectroscopic ones) have been recently reviewed by Curtiss and Blander [23] and full details can be found in their excellent study. Suffice to say here that the constants based on PVT methods are model-dependent and that equilibrium constants obtained from thermal conductivity results involve an indirect method. For all these shortcomings, these results are quite valuable and shall be used in the next section.

2.2. Charged species
The cases of reactions 1b and 1c, namely:

$$(A-H)^+(g) + B(g) \rightleftarrows (A-H\cdots B)^+(g) \tag{1b}$$

$$A-H(g) + B^-(g) \rightleftarrows (A-H\cdots B)^-(g) \tag{1c}$$

require specific techniques and this, for three main reasons: (i) charged species must be selectively separated from ions of the opposite sign, (ii) they must be stored long enough for them to reach a state of thermodynamic equilibrium with their surroundings while avoiding collisions with the walls of the reaction chamber and, (iii) because of their very low concentrations, they must be monitored by highly sensitive methods.

Several books and reviews are available that deal in depth with the experimental techniques suitable for the study of these reactions. They have been briefly summarized by Keesee and Castleman [24] in a mayor review of thermodynamic data for ion-molecule clustering reactions in the gas phase.

The main experimental techniques used to build the database presented in this study are as follows:

A. Ion cyclotron resonance spectroscopy (ICR). One of the most relevant features of this technique is the low range of total gas pressures: typically between 10^{-7} and 10^{-4} Torr [25]. In the drift-cell instruments, the time for ion-molecule reaction and thermalization of the various species is in the order of milliseconds, with pressures of the order of 10^{-4} Torr. In the trapped-ion cell instruments (the most widely used nowaways) reaction times of up to 10^2 seconds can be used while keeping the total pressures below 10^{-5} Torr. This "trapping" involves the simultaneous use of electric and magnetic fields. In general, both methods are able to ensure enough ion-molecule reactions so that equilibrium conditions can be established.

Operation of the ICR spectrometer under equilibrium conditions allows the determination of the equilibrium constants K_p or K_c for reactions 3 and 4:

$$(A_1\text{-H}\cdots B)^-(g) + A_2\text{-H} \rightleftarrows (A_2\text{-H}\cdots B)^-(g) + A_1\text{-H} \tag{3}$$

$$(A\text{-H}\cdots B_1)^+(g) + B_2(g) \rightleftarrows (A\text{-H}\cdots B_2)^+(g) + B_1(g) \tag{4}$$

Reactions involving higher clusters can not be studied by ICR because of the low pressures of the reagents. Notice that this method provides accurate (±0.1-0.2 kcal mol^{-1}) *differences* in complexation Gibbs energies. Absolute binding Gibbs energies require combining these results with at least one "absolute" value obtained by other techniques.

The development of the Fourier Transform ICR (FTICR) has increased both the sensitivity and the precision of this technique [26].

Double resonance experiments, a most useful feature of ICR, allow to ascertain the existence of a state of thermodynamic equilibrium.

B. High Pressure Mass Spectrometry (mostly pulsed - high pressure mass spectrometry) (HPMS or PHPMS) [27]

In this technique, the ionization is carried out by a short pulse of high-energy electrons in a field-free chamber. The partial pressures of the various neutral reactants reach several mTorr and large amounts (~10 Torr) of a neutral gas are added for thermalization purposes. Some of the ions thus generated diffuse through a slit into a low-pressure region, where they are accelerated and mass analyzed. This sampling method might involve in some cases some undesirable side-effects, such as collision-induced decomposition (CID) of ions on their way to the detector. Also, at the relatively high pressures used in the reaction chamber, clustering of several neutral molecules in a given ion often takes place. The importance of clustering can be reduced by operating at high temperatures. On the other hand, the possibility of cluster control is a wonderful feature of this technique: it allows the study of the stepwise binding of neutral molecules to ions and the buildup of quite large clusters [28].

The high dynamic range of this technique is also a valuable asset, for it permits the direct study of systems like 1b and 1c often characterized by very large equilibrium constants.

With respect to ICR methods, PHPMS techniques use much larger pressures and thus enable a much better control of the temperature. In this way, the changes in enthalpy and entropy pertaining to reactions 1b and 1c can be obtained by determining the equilibrium constants at various temperatures.

C. Flowing afterglow [29]. In a manner similar to that of HPMS, ions are generated in a field-free zone, the reaction tube. The carrier gas - frequently hydrogen or helium - and the reagents enter the flow tube upstream, through separate leak values. A flowing plasma is generated by ionizing electrons with energies in the 35-70 eV range. The ions are thermalized by collisions with the reagents and the carrier gas (at pressures near 0.5 Torr). The partial pressures of the reagents are typically 2×10^{-3} - 5×10^{-3} Torr. The ions present in the gaseous mixture are sampled through a small hole mounted at the tip of a nose come situated at the end of the reaction zone and mass analyzed with a quadrupole mass filter.

This method is well suited for the study of stepwise clustering processes in the gas phase at temperatures near 298 K. Also, it allows the direct determination of the reaction rate constants, k_f and k_r for to the formation and decomposition of HB ion-molecule complexes in the gas phase, for example:

$$A\text{-}H(g) + B^-(g) \xrightarrow{k_f} (A\text{-}H\cdots B)^-(g) \tag{1d}$$

$$(A\text{-}H\cdots B)^-(g) \xrightarrow{k_r} A\text{-}H(g) + B^-(g) \tag{1e}$$

The equilibrium constant for the binding process is given by the ratio k_f / k_r.

3. TREATMENT OF THE EXPERIMENTAL DATA: A GENERALIZED FORMALISM.

3.1. Taft's and Abboud-Bellon's formalisms.

In the late fifties and early sixties, the development of commercial spectrometers (IR, UV-Visible and ^{1}H NMR) and convenient experimental methods allowed the systematic determination of reliable equilibrium constants for reaction 2a in solution. The books by Pimentel and McClellan [15a] and Joesten and Schaad [19] illustrate the "long march" towards these values: In the mid-sixties it became clear that, given the huge number of possible hydrogen bonded systems, some sort of conceptual and quantitative framework was necessary in order to guide the research in this field.

Fortunately, it was already known that equilibrium constants for 1:1 HB complexation in solution show a very clear pattern of structural dependence: As an example, we present in Table 1 a set of unpublished equilibrium constants K_c (in l mol^{-1}) for reaction 5, the association between 4-substituted thiophenols and DMSO in CCl$_4$ at

32±1 °C as determined by ^{1}H NMR [30].

$$\text{4-X-C}_6\text{H}_4\text{SH} + \text{DMSO} \overset{\text{CCl}_4}{\rightleftarrows} \text{4-X-C}_6\text{H}_4\text{SH}\cdots\text{DMSO} \qquad K_c \qquad\qquad (5)$$

Table 1
Experimental equilibrium constants K_c (l mol^{-1}) for reaction 5 in CCl$_4$ solution at 32 ± 1°C [30].

X	K_c
OCH$_3$	0.313
(CH$_3$)$_3$	0.344
CH$_3$	0.360
H	0.516
F	0.898
Cl	0.937
CF$_3$	1.43
CN	2.01
NO$_2$	3.44

Data sets as this were well analyzed in terms of Hammett [31] or Taft's [32] parameters. Indeed, if we use the more recent Taft-Topsom formalism [33], we find that log K_c can be nicely described in terms of field and resonance effects, as measured respectively by σ_F and σ_R, equation 6:

$$\log K_c = -0.349 \,(\pm\, 0.040) + 1.216 \,(\pm\, 0.087) \,\sigma_F + 1.094 \,(\pm\, 0.040) \,\sigma_R \qquad (6)$$

with $r^2 = 0.989$, n = 9, sd = 0.06 log units

These facts were taken to indicate the possibility of systematically *ranking and predicting* the HB "acidity" and "basicity" of homologous series of molecules. This was attempted simultaneously and independently by Taft and coworkers [34] and by Abboud and Bellon [35].

Taft defined a thermodynamic scale of HB basicity through the study in CCl$_4$ at 25.0 °C of the 1:1 associations between 4-fluorophenol and a wide variety of HB acceptors, characterized by a pK$_{HB}$ value [36], defined as:

$$\text{pK}_{HB} = \log K_c \qquad\qquad (7)$$

where K_c is the equilibrium constant (in l mol^{-1}) for reaction 8:

$$4\text{-F-C}_6\text{H}_4\text{-OH} + \text{B} \underset{}{\overset{\text{CCl}_4}{\rightleftarrows}} 4\text{-F-C}_6\text{H}_4\text{-OH}\cdots\text{B} \tag{8}$$

The choice of this reference HB donor was brilliant, as its associations can be studied both by IR and by ^{19}F NMR spectroscopies.

Abboud and Bellon [35] were guided in their approach by Taft's work on structure-reactivity relationships [36], Loffler and Grunwald's remarkable treatment of "extrathermodynamic" relationships [37] and by Arnett's review on very weak bases [38]. The main question they addressed in their work was very much in the "structuralist" mood of the time [39]:

Given two sets of molecules, A_i-H (HB donor, HBD) and B_j (HB acceptor, HBA) in solution in an "inert solvent", is it possible to find two sets of descriptors, α_i and β_j respectively ascribed to A_i-H and B_j and such that log $K_c(A_iB_j)$ for reaction 9 could be represented by a simple function of α_i and β_j?[*)]

$$A_i\text{-H} + B_j \rightleftharpoons A_i\text{-H}\cdots\text{B}_j \tag{9}$$

[In what follows we shall use log K_{ij} instead of log $K_c(A_iB_j)$].

An experimental database was built by determining K_{ij} values for a set of fifteen 3- and 4-substituted phenols and seventeen aliphatic and alicyclic ethers in cyclohexane solution at 20.0 °C. *This is one of the largest homogeneous data sets ever constructed.*

It was found that within the limited range of HB "acidities" and "basicities" spanned by these compounds, log K_{ij} can be expressed as in equation 10:

$$\log K_{ij} = \text{constant} + \alpha_i + \beta_j \tag{10}$$

where $\alpha_i = \log(K_{i0}/K_{00})$ and $\beta_j = \log(K_{0j}/K_{00})$.

K_{00} is the equilibrium constant for A_i-H = phenol and B_j = diisopropyl ether, compounds taken as references (notice that constant = log K_{00}).

Plots of log K_{ij} *vs.* α_i or β_j lead to sets of *parallel* lines. The general experimental patterns are represented in Figures 2 and 3.

These workers also considered the case wherein both HBD's and HBA's would span wider structural differences. Then, the expression of the dependence of log K_{ij} on α_i (or β_j) would lead to sets of *intersecting* (instead of parallel lines). They also stated [40] that these lines should likely have a single intersection point, namely the *envelope* (in the mathematical sense) of the full set of lines.

These cases are shown in Figures 4 and 5.

(*) For the sake of simplicity, these notations are slightly different from those used in the original paper.

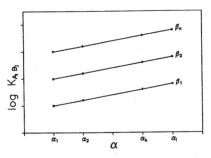

Figure 2. Role of α in equation 10

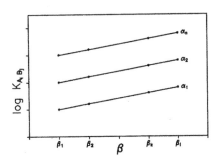

Figure 3. Role of β in equation 10

Under these conditions, the general expression was predicted to be of the form of equation 11:

$$\log K_{ij} = k_1 \cdot \alpha_i \cdot \beta_j + \alpha_i + \beta_j + k_2 \qquad (11)$$

where k_1 and k_2 are constants.

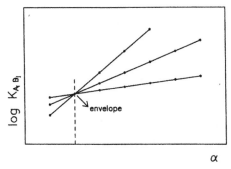

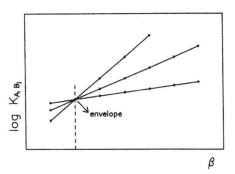

Figure 4. Role of α in equation 11   Figure 5. Role of β in equation 11

Figures 4 and 5 clearly show the symmetric role played by α_i and β_j in the full bilinear equation 11. It follows from the symmetric role of A_i-H and B_j in reactions 2a or 9. Equation 11 rests on the hypothesis that HBD and HBA abilities can be accurately described by a single parameter. From the experimental data available at that time, Abboud and Bellon envisaged the case wherein the sets of lines would not have an envelope. Then, a "vectorial" representation of log K_{ij} was set forth [40] that takes this possibility into consideration. This possibility has been confirmed by an important principal component analysis [41] of acid-base reactions in general. This study showed that a full quantitative description of these reactions requires the consideration of two components for the acidity and two components for the basicity. Fortunately, there are extensive sets of HBAs and HBDs for which the ratio of the weights of these contributions is seen to vary within narrow limits [42]. This explains the wide applicability of expressions such as 11.

To our knowledge, equations 10 and 11 represent the first description of changes in Gibbs energy for reaction 2a as a bilinear form. *For unknown reasons, this work was ignored for nearly twenty years.* Finally, the validity of equation 11 was confirmed in 1988 by Taft and Abboud and independently by Abraham and coworkers [43].

3.2. The Abboud-Bellon-Taft-Abraham approach.
Nowadays, a "condensed" form of equation 11, equation 12, is widely used [43]:

$$\log K_c = 7.354 \, \alpha^H_2 \cdot \beta^H_2 - 1.094 \tag{12}$$

Mathematically, equations 11 and 12 differ only in the choice of the origin. α^H_2 and β^H_2 are descriptors of HB "acidity" and HB "basicity" conceptually identical to α and β in the previous section. Large sets of α^H_2 and β^H_2 values have become available through Abraham's remarkable statistical analyses of log K_c values for reaction 2a in CCl_4 at 25.0 °C. Given values of K_c for the association between families of A_i-H and B_j in CCl_4 at 25.0 °C, Abraham [44-45] defined α^H_2 and β^H_2 as follows:
a) For the complexation of a series of B_j's, against a number of reference acids, equation 13 holds:

$$\log K_c^{(i)} = L_A \cdot \log K_B^{H(j)} + D_A \tag{13}$$

b) Solving the multiple systems of equations thus generated, values of L_A and D_A are obtained that characterize the HBDs and log $K_B^{H(j)}$ that characterize B_j. In order to take advantage of the fact that the *envelope* of the set of lines defined by equation 13 is located at log $K_c = -1.1$, Abraham defined $\beta_2^{H(j)}$ as:

$$\beta_2^{H(j)} = (\log K_B^{H(j)} + 1.1) \, / \, 4.636 \tag{14}$$

where 4.636 is a convenient scaling factor [44].
Abraham has carried out a formally identical treatment of log K_c for series of acids against sets of reference bases [45] that leads to the corresponding HBD descriptors, α^H_2. Using data in cyclohexane, it was shown [46] that equation 12 also applies to K_c values in this solvent. This is not unexpected in view of the fact that equation 10 and its extension, equation 11, were deduced from data in this solvent.

3.3. HB interactions in the gas phase

3.3.1. Neutral species. We present in Table 2 a series of log K_c values for reaction 1a involving a variety of systems for which α^H_2 and β^H_2 values are available. Treatment of representative subsets of data leads to the following results:

a. Figure 6 is a plot of log K_c vs. $\alpha^H_2 \cdot \beta^H_2$ for associations between CF_3CH_2OH and all the bases of the set (only the somewhat abnormal [23] datum for CF_3CH_2OH dimerization is excluded). The relationship, equivalent to a plot of log K_c vs. β^H_2 (on account of the constant value of α^H_2) is linear and of excellent statistical quality ($r^2=0.999$; sd=0.04 log units).

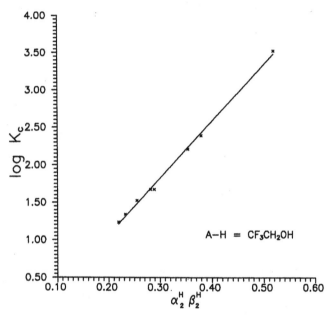

Figure 6. log K_{ij} vs. $\alpha^H_2 \cdot \beta^H_2$ for A-H = CF_3CH_2OH

b. Of the same high quality is the linear relationship between log K_c and α^H_2 for associations between $(CH_3)_3N$ and the set of HBDs, Figure 7.

c. Finally, we present in Figure 8 a plot of log K_c vs. $\alpha^H_2 \cdot \beta^H_2$ for the full set of data.

146

Table 2
HB interactions between neutral species A-H and B in the gas phase (reaction 1a)

A-H	B	$K_p^{(a)}$	$K_c^{(b)}$	Ref.	$\alpha_2^H(A\text{-}H)^{(c)}$	$\beta_2^H(B)^{(d)}$	$\alpha_2^H \cdot \beta_2^H$
$t\text{-}C_4H_9OH$	$t\text{-}C_4H_9OH$	0.322 0.178	6.06	1 2	0.320	0.49	0.16
$i\text{-}C_3H_7OH$	$i\text{-}C_3H_7OH$	0.314 0.120 0.106 0.117	3.88	1 3 4 5	0.325	0.47	0.15
C_2H_5OH	C_2H_5OH	0.0771 0.0974 0.135	2.42	6 3 1	0.333	0.44	0.15
CH_3OH	CH_3OH	0.060 0.062 0.073 0.061 0.077	1.62	7 8 3,9 1 6	0.367	0.41	0.15
	$(CH_3)_3N$	0.78	18.9	10		-	-
H_2O	H_2O	0.038 0.073 0.040	1.21	11 8,12 13	0.353	0.38	0.13
C_4H_5N (pyrrole)	$(CH_3)_3N$	1.2	29.1	10	0.408	-	-

147

CF₃CH₂OH	CF₃CH₂OH	0.263	6.30	14	0.567	0.18	0.10
	C₂H₅CHO	0.87	21.1	10		0.39	0.22
	CH₃OH	1.0	24.2	15		0.41	0.23
	(C₂H₅)₂O	1.7	41.2	10		0.45	0.26
	(CH₃)₂CO	2.2	53.3	10		0.50	0.28
	C₄H₈O (THF)	3.4	82.3	10		0.51	0.29
	NH₃	2.3	55.7	10		-	-
	C₅H₅N	7.8	189	10		0.62	0.35
	(CH₃)₃N	16.6	402	10		-	-
	(C₂H₅)₃N	11.8	286	10		0.67	0.38
	(CH₃)₂NC=NH	158	3.83×10³	10		0.91	0.52
(CF₃)₂CHOH	CH₃OH	3.5	84.8	16	0.771	0.41	0.32
	(C₂H₅)₂O	7.7	186	10		0.45	0.35
	(CH₃)₂CO	10.4	252	10		0.50	0.39
	C₄H₈O (THF)	23	557	10		0.51	0.39
	NH₃	24	581	10		-	-
	(CH₃)₃N	333	8.07×10³	10		-	-
	(C₂H₅)₃N	156	3.78×10³	10		0.67	0.52
(CF₃)₃COH	(C₂H₅)₂O	28	678	10	0.862	0.45	0.39
	(CH₃)₂CO	45	1.09×10³	10		0.50	0.43
	C₄H₈O (THF)	156	3.78×10³	10		0.51	0.44
	NH₃	210	5.09×10³	10		-	-
	(CH₃)₃N	6.67×10³	1.62×10⁵	10		-	-
	(C₂H₅)₃N	1.42×10³	3.44×10⁴	10		0.67	0.58
HF	CH₃CN	0.8	19.4	16	-	0.44	-
	H₂O	96	2.33×10³	16		0.38	-

148

(a) in atm^{-1}, at 295 K
(b) in l mol^{-1}, at 295 K
(c) values taken from M.H. Abraham, P.L. Grellier, D.V. Prior, P.P. Duce, J.J. Morris and P.J. Taylor, J. Chem. Soc., Perkin Trans. 2, (1989) 699
(d) values taken from M.H. Abraham, P.L. Grellier, D.V. Prior, J.J. Morris and P.J. Taylor, J. Chem. Soc., Perkin Trans. 2, (1990) 521

REFERENCES

1. D.J. Frurip, L.A. Curtiss and M. Blander, Int. J. Thermophys., 2 (1981) 115
2. E.T. Beynon and J.J. McKetta, J. Phys. Chem., 67 (1963) 2761
3. C.B. Kretschmer and R.Wiebe, J. Am. Chem. Soc., 76 (1954) 2579
4. N.S. Berman, C.W. Larkman and J.J. McKetta, J. Chem. Eng. Data, 9 (1964) 218
5. J.L. Hales, J.D. Cox and E.B. Lees, Trans. Faraday Soc., 59 (1963) 1544
6. G.M. Barrow, J. Chem. Phys., 20 (1952) 1739. See L.A. Curtiss and M. Blander, Chem. Rev., 88 (1988) 827
7. W. Weltner and K.S. Pitzer, J. Am. Chem. Soc., 73 (1951) 2606
8. G.S. Kell and G.E. McLaurin, J. Chem. Phys., 51 (1969) 4345
9. See L.A. Curtiss and M. Blander, Chem. Rev., 88 (1988) 827
10. J. Marco, J.M. Orza and J.-L.M. Abboud, Vibrational Spectroscopy, in press
11. J.S. Rowlinson, Trans. Faraday Soc., 45 (1949) 974
12. G.S. Kell, G.E. McLaurin and E. Whalley, J. Chem. Phys., 48 (1968) 3805
13. L.A. Curtiss, D.J. Frurip and M. Blander, J. Chem. Phys., 71 (1979) 2703
14. L.A. Curtiss, D.J. Frurip and M. Blander, J. Am. Chem. Soc., 100 (1978) 79
15. S.B. Farnhan, Ph. D. Dissertation, University of Oklahoma, 1970, quoted in ref.9
16. S.L.A. Adebayo, A.C. Legon and D.J. Millen, J. Chem. Soc., Faraday Trans., 87 (1991) 443

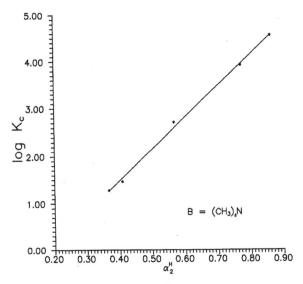

Figure 7. log K_c *vs.* α^H_2 for B = $(CH_3)_3N$

The regression equation we obtain is:

$$\log K_c = 8.85 \ (\pm\ 0.42)\ \alpha^H_2 \cdot \beta^H_2 - 0.86 \ (\pm\ 0.27) \tag{15}$$

n=21; r^2=0.994 and sd = 0.18 log units.

Equation 15 is important in that it **shows that the bilinear formalism that succeeds in CCl_4 or c-C_6H_{12} solutions also applies to the gas phase.** Moreover, if we back-calculate β^H_2 values for NH_3 and $(CH_3)_3N$ from equation 15, we obtain respectively 0.54 and 0.71, in nearly perfect agreement with recent, unpublished results from Profs. Berthelot and Laurence (for CCl_4 solutions). The intercept of equation 15 is very close to that in equation 12. **This strongly suggests that the envelope point** (also called "magic point" by Abraham [44 - 45]) **is endowed with a deep physical meaning,** likely defining the onset of "true" HB interactions. The slope is some 20% larger than in CCl_4 solution, indicating a modest attenuation of the HB interactions in solution.

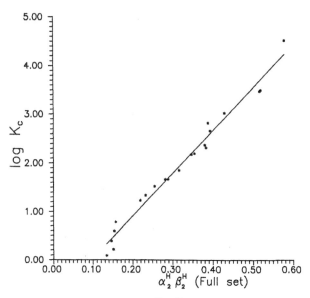

Figure 8. log K_c vs. $\alpha^H_2 \cdot \beta^H_2$ for the full set of data

3.3.2. HB complexes involving ions in the gas phase (the last challenge).

HB interactions, even in the absence of solvent span a wide range of structures and energies. Thus, their onset merges with "pure van der Waals" interactions, as those involved in complexes between rare gases and hydrogen halides [47]. On the other hand, an appreciable degree of proton-transfer has been recently found to exist in the gas phase in the case of the very strong complex $(CH_3)_3N\cdots HCl$ [48].

This suggests an increase in the weight of mesomeric structures II with the thermodynamic stability of the complex:

$$A\text{-}H\cdots B \leftrightarrow A^-\cdots H^+\text{-}B$$
$$\text{I} \qquad\qquad \text{II}$$

From the preceeding discussion, it seems that equations 12 and 15 hold irrespective of the extent of proton transfer. It is then tempting to extend this study to HB complexes between neutral HBD molecules and HBA anions (III) as well as between neutral HBAs and cationic HBDs (IV):

$$A-H\cdots B^- \longleftrightarrow A^-\cdots H-B$$

IIIa   IIIb

$$A-H^+\cdots B \longleftrightarrow A\cdots H^+-B$$

IVa   IVb

In the case of complexes III, involving a constant anion, B^-, we consider $\Delta G^\circ_{298}(HB)$, the standard Gibbs energy change for reaction 1c:

$$A-H(g) + B^-(g) \rightleftharpoons (A-H\cdots B)^-(g) \tag{1c}$$

It seems reasonable to analyze $\Delta G^\circ_{298}(HB)$ as a linear combination of the following contributions:
(a) $\Delta G^\circ_{\alpha\beta}$, the change in Gibbs energy associated to the "general" HB contribution defined by equation 15.
(b) ΔG°_{Ac}, the change in Gibbs energy for the ionization of A-H in the gas phase, reaction 16:

$$A-H(g) \rightleftharpoons A^-(g) + H^+(g) \tag{16}$$

(c) ΔG°_{elec} , the change in Gibbs energy originating in the electrostatic interaction between A-H(g) and B^-(g).
   In short:

$$\Delta G^\circ_{298}(HB) = \text{constant} + \rho_{\alpha\beta} \Delta G^\circ_{\alpha\beta} + \rho_{Ac} \Delta G^\circ_{Ac} + \rho_{elec} \Delta G^\circ_{elec} \tag{17}$$

Analogously, $\Delta G^\circ_{298}(HB)$ for reaction 1b involving a constant $A-H^+$ can be decomposed as in equation 18:

$$\Delta G^\circ_{298}(HB) = \text{constant} + \rho_{\alpha\beta} \Delta G^\circ_{\alpha\beta} + \rho_B \Delta G^\circ_B + \rho_{elec} \Delta G^\circ_{elec} \tag{18}$$

ΔG°_B is the standard Gibbs energy change for the deprotonation of BH^+ in the gas phase, equation 19:

$$BH^+(g) \rightleftharpoons B(g) + H^+(g) \tag{19}$$

As regards the $\Delta G^\circ_{\alpha\beta}$ the term in equation 17, one expects it to be proportional to α^H_2, the descriptor of the HB "acidity" of A-H and, in the case of equation 18, to be proportional to β^H_2, the descriptor of the HB "basicity" of B.
   The main drawback in the analysis of experimental data by means of equations 17 and 18 lies in the fact that the three sets of explanatory variables must be linearly independent. We present in Tables 3 and 4 representative sets of $\Delta G^\circ_{298}(HB)$ values

Table 3
HB interactions between neutral, A-H, and anionic, B⁻, species in the gas phase (reaction 1c)

A-H	B⁻	$-\Delta G^\circ(HB)$[a]	$-\Delta H^{\circ}$[b]	$-\Delta S^{\circ}$[c]	Ref.	K_c[d]	$\Delta G^\circ_{acid}(AH)$[e]	$\alpha_2^H(AH)$[f]
CH HB donors								
HCN	F	32.9	39.5	22.2	1	3.19×10^{25}	343.8	-
	Cl⁻	(16.2)	21.8	18.9	2	2.86×10^{12}		
		13.9	21.0	23.7	3,4			
	CN⁻	13.7	21.7	26.9	5	6.27×10^{11}		
		(14.6)	20.7	20.6	2,6			
	CH₃CO₂⁻	(16.9)	23.5	22.2	6	5.97×10^{13}		
	SH⁻	(15.0)	21	(20)	2	2.42×10^{12}		
	Br⁻	(13.5)	19.4	19.8	6	1.92×10^{11}		
	I⁻	(10.5)	16.8	21.2	6	1.21×10^{9}		
CH₃CN	F⁻	12.0	16.0	13.4	10	1.53×10^{10}	365.2	0.09
	Cl⁻	9.2	13.4	14.3	10			
		(8.9)	13.6	15.7	11			
		9.4	15.8	21.4	3	1.14×10^{8}		
		(9.0)	13.5	15.0	2			
		8.4 (300K)	-	-	9			
	CN⁻	9.1	16.4	24.3	5	1.90×10^{8}		
		(9.6)	13.8	14.2	2			
	CH₃CO₂⁻	(9.4)	14.8	18.0	2	1.90×10^{8}		
	Br⁻	8.0	12.9	16.5	10	1.79×10^{7}		
	I⁻	6.4	11.9	18.2	10	1.20×10^{6}		
C₆H₆	F⁻	(9.5)	15.3	19.5	13	2.25×10^{8}	390.9	≤0.05
	Cl⁻	(4.1)	9.4	17.9	14			

Table 3 (continued)
HB interactions between neutral, A-H, and anionic, B⁻, species in the gas phase (reaction 1c)

A-H	B⁻	$-\Delta G°(HB)$[a]	$-\Delta H°$[b]	$-\Delta S°$[c]	Ref.	K_c[d]	$\Delta G°_{acid}(AH)$[e]	$\alpha_2^H(AH)$[f]
C_6H_6	Cl⁻	(3.8)	10.4	22	15	2.48×10^4		
		4.8	9.9	17.1	3,4			
		3.6	9.5	19.9	22			
	Br⁻	(3.9)	9.0	17.0	14	1.77×10^4		
		0.0 (423K)	-	-	12			
$C_6H_5CH_3$	I⁻	(1.9)	6.1	14.2	14	6.04×10^2	373.7	≤0.05
	Cl⁻	4.0 (300K)	-	-	9	2.09×10^4		
	Br⁻	0.1 (423K)	-	-	12			
CHF_3	F⁻	19.6	27.1	25.1	1	5.69×10^{15}	369.2	-
	Cl⁻	9.8	16.7	22.9	3,4	3.73×10^8		
	CN⁻	9.6	17.0	24.4	5	2.66×10^8		
$CHCl_3$	Cl⁻	11.2	18.1	23.2	3	3.35×10^9	349.3	0.197
		10.3	-	-	17			
		11.8	19.1	24.5	7			
	CN⁻	10.8	18.2	24.8	5	2.02×10^9		
CH_2Cl_2	Cl⁻	8.9	15.5	22.0	7	1.14×10^8	366.8	0.129
		9.2	15.8	22.1	3			
	CN⁻	9.0	16.3	24.2	5	9.67×10^7		
CH_3F	Cl⁻	5.4	11.5	20.3	3	2.22×10^5	379.0	-
CH_3Cl	Cl⁻	4.1	8.6	15.3	7		389.1	-
		6.1	12.2	20.5	3,4	3.31×10^6		
		10.8	15.2	14.8	19			

154

Compound	Ion							
CH₃Br	Cl⁻	7.1	10.9	12.8	25	3.92×10^{6}	385.8	-
	Br⁻	5.0	9.2	14.0	25	1.13×10^{5}		
CH₃I	Cl⁻	7.6	9.8	7.3	25	9.10×10^{6}	379.4	-
	I⁻	4.1	9.0	16.4	25	2.48×10^{4}		
CF₃CF₂H	F⁻	22.5	30.4	26.6	1	7.60×10^{17}	366.8	-
	Cl⁻	11.8	18.8	23.3	3,4	1.09×10^{10}		
1,4-pentadiene	Cl⁻	3.7 (300K)	-	-	9	1.26×10^{4}	-	0
cyclopentadiene	Cl⁻	<2.5 (300K)	-	-	9	$<1.66\times10^{3}$	371.8	0
CH₂(C₆H₅)₂	Cl⁻	7.4 (300K)	-	-	9	6.50×10^{6}	358.2	≤0.05
CH(C₆H₅)₃	Cl⁻	4.1 (300K)	-	-	9	2.48×10^{4}	350.7	≤0.05
(CH₃)₂CO	Cl⁻	(7.9)	13.7	19.6	9	2.12×10^{7}	361.2	0.04
		8.2	14.1	19.6	3,4			
	CN⁻	8.0	14.7	22.5	5	1.79×10^{7}		
	CH₃CO₂⁻	(9.2)	15.7	21.9	2	1.36×10^{8}		
C₂H₅COCH₃	Cl⁻	8.5	14.8	21.0	3	4.16×10^{7}	363.4	≤0.04
(C₂H₅)₂CO	Cl⁻	8.2	14.1	19.6	3	2.51×10^{7}	360.5	~0
CH₃CONHCH₃	CH₃CO₂⁻	17.2	25.4	27.4	2	9.91×10^{13}	-	0.383
(CH₃)₂SO	Cl⁻	(12.5)	18.6	20.4	26	3.56×10^{10}	366.4	-
	Br⁻	(10.9)	17.3	21.4	26	2.39×10^{9}		
	I⁻	(9.2)	15.7	21.7	26	1.36×10^{8}		

Table 3 (continued)
HB interactions between neutral, A-H, and anionic, B⁻, species in the gas phase (reaction 1c)

A-H	B⁻	$-\Delta G^\circ(HB)$[a]	$-\Delta H^\circ$[b]	$-\Delta S^\circ$[c]	Ref.	K_c[d]	$\Delta G^\circ_{acid}(AH)$[e]	$\alpha_2^H(AH)$[f]
NH HB donors								
NH₃	F⁻	-	<23	-	21		396.1	-
	Cl⁻	3.6	8.2	15.4	22	2.48×10^4		
		4.5	10.5	19.9	3			
	Br⁻	2.0	7.7	19.1	22	7.15×10^2		
	I⁻	1.2	7.4	20.9	22	1.85×10^2		
CF₃CH₂NH₂	F⁻	20.3	28.1	26.0	1	1.85×10^{16}	-	-
	Cl⁻	10.8	18.0	24.0	3	2.02×10^9		
C₆H₅NH₂	F⁻	23.4	31.2	26.2	1	3.47×10^{18}	359.1	0.26
	Br⁻	6.1 (423K)	-	-	12			
C₄H₅N (pyrrole)	F⁻	26.6	34.2	25.5	1	7.69×10^{20}	350.9	0.408
	Cl⁻	~14.0(300K)	-	-	9	6.98×10^{10}		
	CN⁻	11.8	18.8	23.4	3			
		12.3	19.5	23.8	5	3.19×10^{11}		
	CH₃CO₂⁻	(15.3)	23.4	27.1	2			
		(16.5)	24.0	25.1	2	3.04×10^{13}		
	SH⁻	(15.7)	23.0	24.4	2	7.88×10^{12}		
OH HB donors								
H₂O	F⁻	18.1	23.3	17.4	17	4.52×10^{14}	384.1	0.353
	Cl⁻	9.0	14.9	19.7	23			

Base						Ref	k
Br⁻	8.2	13.1	16.5			17	4.16×10^{7}
	8.4	14.4	20.1			3,4	
	8.2 (296K)	-	-			27	
I⁻	8.8	14.8	20.1			28	
	7.0	12.6	18.4			17	1.79×10^{7}
	8.9	14.8	19.8			28	
	5.4	10.2	16.3			17	
	5.3	11.1	19.3			23	2.22×10^{5}
	5.6	-	-			29	
CN⁻	(5.4)	10.1	15.8			30	
	7.9	13.8	19.8			24,5	3.51×10^{7}
	(8.8)	14.6	19.6			2	
CH₃CO₂⁻	(9.3)	16.0	22.5			31	2.66×10^{8}
	(9.8)	15.8	20.3			2	
SH⁻	(8.6)	14.2	18.7			2,32	4.92×10^{7}
C₆H₅S⁻	(5.6)	11.4	19.6			32	3.11×10^{5}
CH₃O⁻	13.3	19.9	(22)			34	3.39×10^{12}
t-C₄H₉O⁻	(17.1)	23.9	22.9			31	2.86×10^{12}
C₆H₅O⁻	(15.1)	23.4	(28)			31	2.97×10^{7}
HCO₂⁻	(8.3)	15.4	23.9			31	1.14×10^{8}
	(9.1)	16.0	23.0			31	
HCO₂H							
F⁻	38.1	43.3	24.2	338.3	-	1	2.07×10^{29}
Cl⁻	18.4	25.6	24.1			3,4	1.00×10^{17}
I⁻	25.4	37.2	39.6			19	4.98×10^{10}
	(20.1)	27.4	24.5			9	
	(12.7)	18.9	20.7			30	
C₆H₅S⁻	9.6 (488K)	-	-			32	6.12×10^{19}
HCO₂⁻	(25.1)	36.8	39.1			31	
CH₃CO₂H							
F⁻	36.5	44.1	25.6	341.5	0.55	1	1.39×10^{28}
Cl⁻	15.8	21.6	19.3			19	2.17×10^{13}

Table 3
HB interactions between neutral, A-H, and anionic, B^-, species in the gas phase (reaction 1c)

A-H	B^-	$-\Delta G^\circ(HB)$[a]	$-\Delta H^\circ$[b]	$-\Delta S^\circ$[c]	Ref.	K_c[d]	$\Delta G^\circ_{acid}(AH)$[e]	$\alpha_2^H(AH)$[f]
CH_3CO_2H	Cl^-	16.7	23.9	24.0	3,4			
	I^-	(10.6)	16.9	21.3	30	1.44×10^9		
	$CH_3CO_2^-$	(20.5)	29.3	29.6	31	2.60×10^{16}		
	$C_6H_5S^-$	(12.5)	20.3	26.2	32	3.56×10^{10}		
	$C_6H_5O^-$	(20.2)	(27.4)	(24.0)	31	1.57×10^{16}		
$C_2H_5CO_2H$	I^-	(10.5)	16.6	20.4	30	1.22×10^9	340.4	-
	$C_6H_5S^-$	(12.4)	20.0	25.6	32	3.00×10^{10}		
CH_3OH	F^-	22.8	29.6	22.6	1	1.26×10^{18}	374.0	0.367
	Cl^-	(10.2)	17.4	24.1	11			
		9.8	14.2	14.8	8,9	4.42×10^8		
		9.9	16.8	22.9	3,4			
		9.7	14.1	14.8	19			
	I^-	(6.0)	11.3	17.8	30	6.12×10^5		
	CN^-	9.2	16.5	24.3	5	3.73×10^8		
		(10.4)	16.6	20.8	2			
	$CH_3CO_2^-$	(10.6)	17.6	23.6	2	1.44×10^9		
	SH^-	(11.0)	17	(20)	2	2.83×10^9		
	$C_6H_5S^-$	(6.5)	13.4	23.0	32	1.42×10^6		
	CH_3O^-	15.3	21.8	(22)	34	4.52×10^{14}		
		(20.8)	28.8	26.7	31			
	$C_2H_5O^-$	13.6	20.2	(22)	34	2.28×10^{11}		
	$n\text{-}C_3H_7O^-$	13.2	19.8	(22)	34	1.16×10^{11}		
	$t\text{-}C_4H_9O^-$	12.3	18.9	(22)	34	1.72×10^{12}		
		(17.2)	25.5	27.9	31			
	$C_6H_5C\equiv C^-$	7.7	13.3	(22)	34	1.08×10^7		

Solvent	Anion						
C_2H_5OH	$t\text{-}C_5H_{11}O^-$	12.0	18.6	(22)	34	1.53×10^{10}	370.8	0.33
	HCO_2^-	(10.6)	17.6	23.6	31	1.44×10^{9}		
	F	24.1	31.5	24.9	1	1.13×10^{19}		
	Cl^-	10.4	17.3	23.1	3	1.03×10^{9}		
	I^-	(6.5)	12.1	18.9	30	1.42×10^{6}		
	CN^-	10.0	17.4	24.5	5	1.03×10^{9}		
		(10.7)	17.4	22.5	2			
	$CH_3CO_2^-$	(12.0)	20.7	29.2	31	1.53×10^{10}		
	SH^-	(10.3)	16.2	19.8	2	1.22×10^{9}		
		(10.6)	16.3	19.0	32			
	$C_2H_5O^-$	14.0	20.6	(22)	34	5.04×10^{13}		
		(19.6)	27.6	26.8	31			
	$n\text{-}C_3H_7O^-$	13.7	20.3	(22)	34	2.69×10^{11}		
	$t\text{-}C_4H_9O^-$	12.9	19.5	(22)	34	6.98×10^{10}		
	$t\text{-}C_5H_{11}O^-$	12.6	19.2	(22)	34	4.21×10^{10}		
	$C_6H_5O^-$	(11.3)	19.3	27.0	31	4.69×10^{9}		
$n\text{-}C_3H_7OH$	F	24.7	32.3	25.4	1	3.11×10^{19}	369.4	0.33
	Cl^-	10.8	17.7	23.2	3	2.02×10^{9}		
	$n\text{-}C_3H_7O^-$	14.4	21.0	(22)	34	8.78×10^{11}		
$n\text{-}C_3H_7OH$	$t\text{-}C_4H_9O^-$	13.6	20.2	(22)	34	2.28×10^{11}		
	$t\text{-}C_5H_{11}O^-$	13.2	19.8	(22)	34	1.16×10^{11}		
	$C_6H_5C\equiv C^-$	8.8	15.4	(22)	34	6.90×10^{7}		
$i\text{-}C_3H_7OH$	F	24.7	32.3	25.6	1	3.11×10^{19}	368.8	0.32
	Cl^-	10.7	17.6	23.2	3	1.70×10^{9}		
	I^-	(6.5)	12.2	19.1	30	1.42×10^{6}		
	CN^-	10.7	18.1	24.8	5	1.70×10^{9}		
	$C_6H_5S^-$	(7.3)	15.0	26.0	32	5.49×10^{6}		
$n\text{-}C_4H_9OH$	F	24.5	32.2	25.9	1	2.22×10^{19}	368.8	0.33

Table 3 (continued)
HB interactions between neutral, A-H, and anionic, B⁻, species in the gas phase (reaction 1c)

A-H	B⁻	$-\Delta G°(HB)$ [a]	$-\Delta H°$ [b]	$-\Delta S°$ [c]	Ref.	K_c [d]	$\Delta G°_{acid}(AH)$ [e]	$\alpha_2^H(AH)$ [f]
n-C₄H₉OH	Cl⁻	10.7	17.6	23.2	3	1.70×10^9		
t-C₄H₉OH	F⁻	25.5	33.3	26.1	1	1.20×10^{20}	368.0	0.319
	Cl⁻	11.1	19.2	27.0	9			
		11.1	14.2	10.3	19	3.35×10^9		
		11.1	18.1	23.4	3,4			
	I⁻	(6.5)	12.1	18.7	30	1.42×10^6		
	CN⁻	10.7	18.1	24.8	5	1.70×10^9		
	HS⁻	(10.9)	16.8	19.9	32	2.39×10^9		
	C₆H₅S⁻	(7.2)	14.6	25.0	32	4.63×10^6		
	t-C₄H₉O⁻	13.8	20.4	(22)	34	3.19×10^{11}		
	t-C₅H₁₁O⁻	13.7	20.3	(22)	34	2.69×10^{11}		
t-C₅H₁₁OH	C₆H₅C≡C⁻	10.5	17.1	(22)	34	1.22×10^9	366.0	-
	t-C₅H₁₁O⁻	14.9	21.5	(22)	34	2.04×10^{12}		
C₆H₅CH₂OH	C₆H₅C≡C⁻	12.9	19.5	(22)	34	6.98×10^{10}	363.4	-
CF₃CH₂OH	F⁻	31.1	39.1	26.8	1	1.53×10^{24}	354.1	0.567
	Cl⁻	16.5	~24.0	~25.0	3	3.04×10^{13}		
	CN⁻	16.4	24.6	26.1	5	2.57×10^{13}		
	HS⁻	(20.1)	26.8	22.6	32	1.32×10^{16}		
	C₆H₅S⁻	(13.5)	21.0	25.1	32	1.92×10^{11}		
(CF₃)₂CHOH	Cl⁻	>19.0	>26.5	~25.0	3	$>2.07 \times 10^{15}$	338.3	0.771
	CN⁻	16.8	25	26	5	5.04×10^{13}		

Compound	Anion				Ref.			
CH_2FCH_2OH	F^-	8.9	11.5	8.8	1	8.17×10^7	363.9	0.396
	Cl^-	13.0	~20.5	~25.0	3,4	8.27×10^{10}		
	CN^-	12.9	20.4	25.3	5	6.98×10^{10}		
$(CH_2F)_2CHOH$	F^-	11.8	14.5	9.1	1	1.09×10^{10}	356.7	-
	Cl^-	16.1	~23.6	~25.0	3	1.55×10^{13}		
	CN^-	15.8	23.5	26.0	5	9.33×10^{12}		
$(CF_3)_2(CH_3)COH$	Cl^-	>19.0	>26.5	>25.0	3	$>2.07 \times 10^{15}$	340.6	-
	CN^-	17.8	25.7	26.4	5	2.73×10^{14}		
$ClCH_2CH_2OH$	Cl^-	14.0	~21.5	~25.0	3	4.47×10^{11}	-	0.346
	CN^-	13.5	21.1	25.5	5	1.92×10^{11}		
C_6H_5OH	F^-	33.5	41.4	26.3	1	8.78×10^{25}	342.3	0.596
	Cl^-	14.8	19.4	15.5	19	1.39×10^{14}		
		(20.0)	27.4	25	20			
		(17.3)	25.0	26.0	9			
	$CH_3CO_2^-$	15.9 (423K)	-	-	35			
	Br^-	(18.9)	26.1	24.0	31	1.75×10^{15}		
		11.1 (423K)	-	-	35			
	I^-	8.4 (423K)	-	-	35			
$4\text{-}FC_6H_4OH$	Cl^-	18.0 (423K)	-	-	35		339.9	0.629
	Br^-	19.3 (300K)	-	-	9,36	3.43×10^{15}		
		13.6 (423K)	-	-	35			
	I^-	10.5 (423K)	-	-	35			
$4\text{-}ClC_6H_4OH$	Cl^-	19.5 (423K)	-	-	35		336.2	0.67
	Br^-	20.8 (300K)	-	-	9,36	4.31×10^{16}		
		14.7 (423K)	-	-	35			

Table 3 (continued)
HB interactions between neutral, A-H, and anionic, B⁻, species in the gas phase (reaction 1c)

A-H	B⁻	-ΔG°(HB)[a]	-ΔH°[b]	-ΔS°[c]	Ref.	K_c[d]	ΔG°acid(AH)[e]	α_2^H(AH)[f]
4-ClC₆H₄OH	I⁻	11.6 (423K)	-	-	35			
4-CH₃C₆H₄OH	Cl⁻	15.3 (423K)	-	-	35		343.4	0.569
		16.6 (300K)	-	-	9,36	3.60×10^{13}		
	Br⁻	10.6 (423K)	-	-	35			
	I⁻	7.5 (423K)	-	-	35			
4-CNC₆H₄OH	Cl⁻	24.7 (423K)	-	-	35		325.3	0.787
		26.0 (300K)	-	-	9,36	2.79×10^{20}		
	Br⁻	18.1 (423K)	-	-	35			
	I⁻	14.9 (423K)	-	-	35			
4-NH₂C₆H₄OH	Br⁻	9.8 (423K)	-	-	35		345.6	-
4-NO₂C₆H₄OH	Br⁻	18.7 (423K)	-	-	35		320.9	0.824
4-OCH₃C₆H₄OH	Br⁻	11.3 (423K)	-	-	35		343.5	0.573
4-CH₃COC₆H₄OH	Br⁻	15.4 (423K)	-	-	35		328.6	-
4-CHOC₆H₄OH	Br⁻	16.1 (423K)	-	-	35		326.1	-
4-CF₃C₆H₄OH	Br⁻	16.9 (423K)	-	-	35		336.1	-
3-FC₆H₄OH	Br⁻	13.7 (423K)	-	-	35		336.8	0.676
	I⁻	10.7 (423K)	-	-	35			

3-ClC$_6$H$_4$OH	Br⁻	14.4 (423K)	-	-	35	335.0	0.693
	I⁻	12.2 (423K)	-	-	35		
3-CH$_3$C$_6$H$_4$OH	Br⁻	10.9 (423K)	-	-	35	343.4	-
3-CNC$_6$H$_4$OH	Br⁻	17.7 (423K)	-	-	35	328.9	0.772
	I⁻	14.1 (423K)	-	-	35		
3-NH$_2$C$_6$H$_4$OH	Br⁻	10.6 (423K)	-	-	35	343.7	-
3-NO$_2$C$_6$H$_4$OH	Br⁻	17.7 (423K)	-	-	35	327.5	0.785
	I⁻	14.3 (423K)	-	-	35		
3-OCH$_3$C$_6$H$_4$OH	Br⁻	12.4 (423K)	-	-	35	341.1	-
3-CHOC$_6$H$_4$OH	Br⁻	15.5 (423K)	-	-	35	333.6	-
3-CF$_3$C$_6$H$_4$OH	Br⁻	16.0 (423K)	-	-	35	332.4	-
2-FC$_6$H$_4$OH	Br⁻	11.2 (423K)	-	-	35	339.0	-
2-ClC$_6$H$_4$OH	Br⁻	11.5 (423K)	-	-	35	337.1	0.650
2-CH$_3$C$_6$H$_4$OH	Br⁻	10.9 (423K)	-	-	35	-	0.519
2-CNC$_6$H$_4$OH	Br⁻	17.3 (423K)	-	-	35	-	0.738
2-CF$_3$C$_6$H$_4$OH	Br⁻	14.4 (423K)	-	-	35	-	-
C$_6$F$_5$OH	Br⁻	15.9 (423K)	-	-	35	-	0.764

Table 3 (continued)
HB interactions between neutral, A-H, and anionic, B⁻, species in the gas phase (reaction 1c)

A-H	B⁻	$-\Delta G°(HB)$[a]	$-\Delta H°$[b]	$-\Delta S°$[c]	Ref.	K_c[d]	$\Delta G°_{acid}(AH)$[e]	$\alpha_2^H(AH)$[f]
SH HB donors								
H₂S	F⁻	29.0	34.7	18.7	1	4.42×10^{22}	344.8	≤0.1
	Cl⁻	-	21	-	2			
	CN⁻	12.4	19.8	23.8	5	4.98×10^{10}		
		(12.9)	18.9	20.1	2			
	SH⁻	(7.3)	13.2	19.7	2	5.49×10^{6}		
CH₃SH	F⁻	27.3	34.2	23.3	1	2.51×10^{21}	350.6	≤0.1
	CH₃CO₂⁻	(8.1)	14.9	22.8	2	2.12×10^{7}		
Others HB donors								
HF	F⁻	32.0	38.5	21.9	1	3.00×10^{24}	365.5	-
		(30.9)	38.6	25.7	16			
	Cl⁻	15.1	21.8	22.5	3	2.86×10^{12}		
HCl	Cl⁻	16.7	23.7	23.5	33		328.0	-
		13.6	20.4	22.8	23	1.10×10^{13}		
		16.0	23.1	23.5	3,4			
		(17.2)	23.8	22.3	16			
	Br⁻	(13.0)	19.6	22.0	16	8.27×10^{10}		
	I⁻	(8.8)	14.8	20.0	16			
		(7.4)	14.2	22.7	18	2.12×10^{7}		

HBr	Br⁻	(14.0)	20.6	22.3	16	4.47×10^{11}	317.0	-
	I⁻	(10.3)	16.1	19.6	16	8.68×10^{8}		
HI	I⁻	(9.7)	17.0	24.4	16	3.15×10^{8}	309.3	-

(a) in kcal mol⁻¹, at 298 K. Values given in parentheses are calculated from the given ΔH^0 and ΔS^0 values.
(b) in kcal mol⁻¹. Values in parentheses are estimated.
(c) in cal mol⁻¹ K⁻¹. Values in parentheses are estimated.
(d) in l mol⁻¹.
(e) in kcal mol⁻¹. Values taken from S.G. Lias, J.E. Bartmess, J.F. Liebman, J.L. Holmes, R.D. Levin and W.G. Mallard, NIST Standard Reference Database, Computerized Version 1.1. NIST, Gaitherburg, MD 20899, 1989.
(f) Values taken from M.H. Abraham, P.L. Grellier, D.V. Prior, P.P. Duce, J.J. Morris and P.J. Taylor, J. Chem. Soc., Perkin Trans. 2, (1989) 699.

REFERENCES

1. J.W. Larson and T.B. McMahon, J. Am. Chem. Soc., 105 (1983) 2944
2. M. Meot-Ner (Mautner), J. Am. Chem. Soc., 110 (1988) 3854
3. J.W. Larson and T.B. McMahon, J. Am. Chem. Soc., 106 (1984) 517
4. J.W. Larson and T.B. McMahon, Can. J. Chem., 62 (1984) 675
5. J.W. Larson and T.B. McMahon, J. Am. Chem. Soc., 109 (1987) 6230
6. M. Meot-Ner (Mautner), S.M. Cybulski, S. Scheiner and J.F. Liebman, J. Phys. Chem., 92 (1988) 2738
7. R.C. Dougherty, J. Dalton and J.D. Roberts, Org. Mass. Spectrom., 8 (1974) 77
8. R. Yamdagni, J.D. Payzant and P. Kebarle, Can. J. Chem., 51 (1973) 2507
9. M.A. French, S. Ikuta and P. Kebarle, Can. J. Chem., 60 (1982) 1907
10. R. Yamdagni and P. Kebarle, J. Am. Chem. Soc., 94 (1972) 2940
11. S. Yamabe, Y. Furumiya, K. Hiraoka and K. Morise, Chem. Phys. Lett., 131 (1986) 261
12. G.J.C. Paul and P. Kebarle, J. Am. Chem. Soc., 113 (1991) 1148
13. K. Hiraoka, S. Mizuse and S. Yamabe, J. Chem. Phys., 86 (1987) 4102
14. K. Hiraoka, S. Mizuse and S. Yamabe, Chem. Phys. Lett., 147 (1988) 174
15. J. Sunner, K. Nishizawa and P. Kebarle, J. Phys. Chem., 85 (1981) 1814
16. G. Caldwell and P. Kebarle, Can. J. Chem., 63 (1985) 1399

17. M. Arshadi, R. Yamdagni and P. Kebarle, J. Phys. Chem., 74 (1970) 1475
18. R. G. Keesee and A. W. Castleman, Jr.,J. Am. Chem. Soc., 102 (1980) 1446
19. R. Yamdagni and P. Kebarle, J. Am. Chem. Soc., 93 (1971) 7139
20. P. Kebarle, Ann. Rev. Phys. Chem., 28 (1977) 445
21. K. G. Spears and E. E. Ferguson, J. Chem. Phys., 59 (1973) 4174
22. R. G. Keesee and A. W. Castleman, Jr.,J. Phys. Chem. Ref. Data, 15 (1986) 1011
23. R. G. Keesee and A. W. Castleman, Jr.,Chem. Phys. Lett., 74 (1980) 139
24. J.D. Payzant, R. Yamdagni and P. Kebarle, Can. J. Chem., 49 (1971) 3308
25. R.C. Dougherty and J.D. Roberts, Org. Mass. Spectrom., 8 (1974) 81
26. T. F. Magnera,G. Caldwell, J. Sunner, S. Ikuta and P. Kebarle, J. Am. Chem. Soc., 106 (1984) 6140
27. F.C. Fehsenfeld and E. E. Ferguson, J. Chem. Phys., 61 (1974) 3181
28. N.A. Burdett and A.N. Hayhurst, J. Chem. Soc. Faraday Trans. 1, 78 (1982) 2997
29. P. Kebarle, M. Arshadi and J. Scarborough, J. Chem. Phys., 49 (1968) 817
30. G. Caldwell and P. Kebarle, J. Am. Chem. Soc., 106 (1984) 967
31. M. Meot-Ner (Mautner) and L.W. Sieck,J. Am. Chem. Soc., 108 (1986) 7525
32. L.W. Sieck and M. Meot-Ner (Mautner), J. Phys. Chem., 93 (1989) 1586
33. R. Yamdagni and P. Kebarle, Can. J. Chem., 52 (1974) 2449
34. G. Caldwell, M.D. Rozeboom, J.P. Kiplinger and J.E. Bartmess, J. Am. Chem. Soc., 106 (1984) 4660
35. G.J.C. Paul and P. Kebarle, Can. J. Chem., 68 (1990) 2070
36. J.B. Cumming, M.A. French and P. Kebarle, J. Am. Chem. Soc., 99 (1977) 6999

Table 4
HB interactions between neutral, B, and cationic, (A-H)$^+$, species in the gas phase (reaction 1b)

B	(A-H)$^+$	$-\Delta G°(HB)$ [a]	$-\Delta H°$ [b]	$-\Delta S°$ [c]	Ref.	K_c [d]	$\Delta G°_{base}(B)$ [e]	$\beta_2^H(B)$ [f]
C HB acceptors								
CH_4	H_3O^+	(1.9)	8.0	20.4	1	6.04×10^2	124.0	0
	NH_4^+	(-1.0)	3.59	15.5	2	1.85×10^{-1}		
C_2H_4	NH_4^+	(4.0)	10	(20)	3	2.09×10^4	155.2	0.07
$c\text{-}C_6H_{12}$	NH_4^+	(<3.0)	(<9)	(20)	3	$<3.87 \times 10^3$	161.4	0
C_6H_6	NH_4^+	(12.4)	19.3	23.3	3	3.00×10^{10}	175.3	0.14
	$CH_3NH_3^+$	(11.3)	18.8	25.1	3	4.69×10^9		
	$(CH_3)_3NH^+$	(7.6)	15.9	27.7	3	9.10×10^6		
	$C_2H_5OH_2^+$	(13.6)	21	(25)	3	2.28×10^{11}		
cyclohexene	$CH_3NH_3^+$	(6.6)	11.6	16.9	3	1.68×10^6	181.4	0.07
N HB acceptors								
HCN	H_3O^+	25.3	32.5	24.3	4,5	7.24×10^{19}	164.2	0.44
		25.0	32.3	24.9	8			
	NH_4^+	(14.5)	20.5	20.2	15	1.04×10^{12}		
	$CH_3NH_3^+$	(14.0)	20.8	22.9	15	4.47×10^{11}		
	$(CH_3)_3NH^+$	(9.9)	16.8	23.0	15	4.42×10^8		
	$(CH_3)_2COH^+$	(13.4)	20.4	23.6	15	1.62×10^{11}		

167

Table 4 (continued)
HB interactions between neutral, B, and cationic, (A-H)$^+$, species in the gas phase (reaction 1b)

B	(A-H)$^+$	-ΔG$^{o(a)}$	-ΔH$^{o(b)}$	-ΔS$^{o(c)}$	Ref.	K$_c^{(d)}$	ΔG$^o_{base}$(B)$^{(e)}$	β$_2^H$(B)$^{(f)}$
CH$_3$CN	H$_3$O$^+$	(38.0)	46.7	29.3	8	1.75×10^{29}	180.1	0.44
	NH$_4^+$	(20.4)	27.6	24.2	15	2.20×10^{16}		
	CH$_3$NH$_3^+$	(16.8)	24.5	25.8	9	2.30×10^{14}		
		(18.5)	26.2	25.7	15			
	(CH$_3$)$_2$OH$^+$	23.0	(29.9)	(23)	6,15	1.77×10^{18}		
	(CH$_3$)$_2$COH$^+$	20.8	(27.7)	(23)	6,15	4.31×10^{16}		
n-C$_3$H$_7$CN	NH$_4^+$	(22.0)	28.4	21.6	15	3.27×10^{17}	186.1	0.45
	CH$_3$NH$_3^+$	(19.8)	28.1	29	9	7.97×10^{15}		
C$_6$H$_5$CN	CH$_3$NH$_3^+$	(20.1)	29.4	31.2	9	1.32×10^{16}	188.0	0.42
NH$_3$	H$_3$O$^+$	(48.0)	54.8	22.7	11	3.73×10^{36}	195.6	-
	NH$_4^+$	(17.2)	24.8	25.8	7,9			
		17.5	27	32	12	8.37×10^{13}		
		18.1	25.4	24.3	13			
		15.5	21.5	20	14			
	CH$_3$NH$_3^+$	(13.7)	21.4	(26.0)	7	2.69×10^{11}		
	(CH$_3$)$_3$NH$^+$	(10.2)	17.3	23.9	9	7.33×10^8		
	C$_5$H$_5$NH$^+$	(10.6)	17.3	22.5	9	1.44×10^9		
CH$_3$NH$_2$	NH$_4^+$	(24.3)	(-32)	(26)	7	1.59×10^{19}	205.4	0.70
	CH$_3$NH$_3^+$	(14.7)	21.7	23.6	7	1.31×10^{13}		
		(17.3)	25.4	(27.3)	26			
(CH$_3$)$_2$NH	NH$_4^+$	(30.5)	(38.9)	(28.2)	7	5.55×10^{23}	212.7	0.70
	CH$_3$NH$_3^+$	(20.1)	27.5	24.9	7	1.32×10^{16}		

Base	Species				Ref			
(CH₃)₃N	(CH₃)₃NH⁺	(12.0)	20.5	(28.5)	7	1.53×10^{10}		
	NH₄⁺	(34.6)	(43.3)	(29.1)	7	5.62×10^{26}	217.6	–
	CH₃NH₃⁺	(24.1)	(32.5)	(28.2)	7	1.13×10^{19}		
	(CH₃)₃NH⁺	(13.0)	22.5	32.0	7	1.92×10^{11}		
		(13.9)	22.0	(27.2)	26			
C₄H₅N (pyrrole)	CH₃NH₃⁺	(12.3)	18.6	21.0	3	2.54×10^{10}	200.0	–
C₅H₅N	C₅H₅NH⁺	16.7	26.3	32.1	16		213.2	0.62
		(15.4)	23.7	(28)	18	1.83×10^{13}		
		(16.2)	24.6	28.2	17			
		(16.4)	25.2	29.6	26			

O HB acceptors

Base	Species				Ref			
H₂O	H₃O⁺	(24.3)	31.6	24.3	5		159.5	0.38
		(24.2)	31.5	24.4	19			
		(26.0)	35.0	30.2	28,29	3.11×10^{19}		
		(26.2)	36	33.3	30			
		(23.0)	33	33.6	31			
	NH₄⁺	11.4	17.3	19.7	35	2.14×10^{10}		
		(13.0)	19.9	23.1	19			
	CH₃NH₃⁺	(11.0)	18.8	26.3	20	1.70×10^{9}		
		(10.3)	16.8	21.8	19			
	(CH₃)₃NH⁺	(7.3)	14.5	24.1	17,19	5.49×10^{6}		
	(CH₃)₂OH⁺	(14.7)	22.6	26.5	21	2.86×10^{11}		
		(15.4)	24.0	29.0	19			
	(CH₃)₂COH⁺	(12.8)	20.5	26.0	19	5.90×10^{10}		
	C₂H₅OH₂⁺	(17.4)	26.5	30.5	29	1.39×10^{14}		
	C₅H₅NH⁺	(7.4)	15.0	25.5	22	1.28×10^{26}		
		(8.1)	16.1	27.0	17			

Table 4 (continued)
HB interactions between neutral, B, and cationic, $(A-H)^+$, species in the gas phase (reaction 1b)

B	$(A-H)^+$	$-\Delta G^\circ(HB)^{[a]}$	$-\Delta H^{o[b]}$	$-\Delta S^{o[c]}$	Ref.	$K_c^{[d]}$	$\Delta G^\circ_{base}(B)^{[e]}$	$\beta_2^H(B)^{[f]}$
CH_3OH	H_3O^+	(33.6)	40.8	24.0	8	1.04×10^{26}	174.9	0.41
	$CH_3NH_3^+$	(11.8)	19.0	24.2	9	1.09×10^{10}		
	$(CH_3)_2OH^+$	(18.2)	26.3	27.1	21	5.36×10^{14}		
C_2H_5OH	$CH_3NH_3^+$	(13.9)	21.3	25	9	3.78×10^{11}	179.8	0.44
	$C_2H_5OH_2^+$	23.5	32.0	28.5	8	4.86×10^{18}		
		23.7	32.2	28.5	8			
$n-C_3H_7OH$	$CH_3NH_3^+$	(14.4)	22.0	25.6	9	8.78×10^{11}	182.0	0.45
	$(CH_3)_2OH^+$	21.8	30.3	28.4	8	2.33×10^{17}		
	$C_2H_5OH_2^+$	24.6	32.8	27.4	8	2.63×10^{19}		
$i-C_3H_7OH$	$C_2H_5OH_2^+$	25.5	33.6	27.2	8	1.20×10^{20}	183.6	0.47
	$(CH_3)_2OH^+$	22.6	31.0	28.2	8	9.00×10^{17}		
$n-C_4H_9OH$	$CH_3NH_3^+$	(15.8)	23.5	(26)	9	9.33×10^{12}	182.6	0.46
	$C_2H_5OH_2^+$	24.8	33.1	27.7	8	3.69×10^{19}		
$t-C_4H_9OH$	$CH_3NH_3^+$	(15.2)	22.9	(26)	9	3.39×10^{12}	186.1	0.49
CF_3CH_2OH	H_3O^+	25.7	33.0	24.6	8	1.68×10^{20}	161.7	0.18
	$CH_3NH_3^+$	(10.6)	19.1	28.5	9	1.44×10^9		
HCO_2H	$CH_3NH_3^+$	(11.8)	19.0	24.2	9	1.09×10^{10}	169.3	0.38
CH_3CO_2H	$CH_3NH_3^+$	(14.8)	22.0	24.3	9	1.72×10^{12}	180.8	-

Compound	Ion							
CF₃CO₂H	(CH₃)₂OH⁺	20.8	29.3	28.4	8	4.31×10^{16}	163.1	-
	C₂H₅OH₂⁺	23.2	31.4	27.6	8	2.48×10^{18}		
	H₃O⁺	23.4	30.8	24.7	8	3.47×10^{18}		
HCO₂CH₃	CH₃NH₃⁺	(14.2)	21.4	(24)	9	6.27×10^{11}	180.9	0.38
HCO₂C₂H₅	(CH₃)₂OH⁺	22.6	31.2	28.8	8	9.00×10^{17}	184.6	0.38
HCO₂n-C₄H₉	CH₃NH₃⁺	(16.8)	24.5	(26)	9	5.04×10^{13}	186.0	0.38
CH₃CO₂CH₃	CH₃NH₃⁺	(16.1)	23.5	24.8	9	1.55×10^{13}	189.1	0.40
	(CH₃)₂COH⁺	22.5	31.4	29.9	8	7.60×10^{17}		
CH₃CO₂i-C₃H₇	CH₃NH₃⁺	(19.5)	30.0	35.2	9	4.81×10^{15}	-	0.45
n-C₄H₉CO₂n-C₃H₇	CH₃NH₃⁺	(19.6)	30.0	34.8	9	5.69×10^{15}	-	0.45
CF₃CO₂C₂H₅	CH₃NH₃⁺	(13.7)	21.4	(26)	9	2.69×10^{11}	175.8	-
H₂CO	H₃O⁺	26.4	32.9	21.6	8		164.8	0.40
		25.5	32.0	21.7	8			
		25.2	-	-	8			
		25.3 (299K)	-	-	8	1.20×10^{20}		
		25.1	32.2	23.8	8			
CH₃CHO	C₂H₅OH₂⁺	23.2	31.2	26.9	8	2.48×10^{18}	177.6	0.40
(CH₃)₂CO	CH₃NH₃⁺	(17.1)	24.0	23.2	9	8.37×10^{13}	187.9	0.50
	(CH₃)₂COH⁺	22.3	31.5	30.9	8	1.66×10^{17}		
		(22.3)	30.7	28.2	27			
		(21.0)	30.1	30.4	23			

Table 4 (continued)
HB interactions between neutral, B, and cationic, $(A\text{-}H)^+$, species in the gas phase (reaction 1b)

B	$(A\text{-}H)^+$	$-\Delta G°(HB)$ [a]	$-\Delta H°$ [b]	$-\Delta S°$ [c]	Ref.	K_c [d]	$\Delta G°_{base}(B)$ [e]	$\beta_2^H(B)$ [f]
$(CH_3)_2CO$	$(CH_3)_2COH^+$	(20.9)	29.6	29.3	28			
$C_2H_5COCH_3$	$CH_3NH_3^+$	(17.8)	25.2	(25)	9	2.73×10^{14}	190.8	0.48
	$(CH_3)_2COH^+$	23.8	32.7	29.8	8	6.82×10^{18}		
$(C_2H_5)_2CO$	$CH_3NH_3^+$	(18.2)	25.9	(26)	9	5.36×10^{14}	193.0	0.48
	$(CH_3)_2COH^+$	24.4	33.4	30.1	8	1.88×10^{19}		
$C_2H_5COn\text{-}C_3H_7$	$CH_3NH_3^+$	(19.0)	27.0	27.0	9	2.07×10^{15}	-	0.48
$(CH_3)_2O$	H_3O^+	(39.9)	48.2	27.7	8	9.44×10^{29}	183.9	0.43
		(38.1)	45.4	24.5	8			
	$CH_3NH_3^+$	(12.8)	21.5	29.3	9	5.90×10^{10}		
	$(CH_3)_2OH^+$	(21.5)	29.5	27.0	9			
		(21.9)	30.7	29.6	25	3.27×10^{17}		
		(22.5)	32.0	31.9	27			
$(C_2H_5)_2O$	$CH_3NH_3^+$	(14.6)	22.0	25.0	10	1.23×10^{12}	191.3	0.45
	$(CH_3)_3NH^+$	(10.7)	19.5	29.4	10	1.70×10^{9}		
	$(CH_3)_2COH^+$	24.0	32.9	29.7	8	9.56×10^{18}		
	$C_5H_5NH^+$	(12.7)	22.5	32.9	10	4.98×10^{10}		
$(n\text{-}C_3H_7)_2O$	$CH_3NH_3^+$	(16.0)	24.0	26.7	9	1.31×10^{13}	193.9	0.44
	$C_5H_5NH^+$	(14.3)	23.5	(31)	9	7.42×10^{11}		
$(n\text{-}C_4H_9)_2O$	$CH_3NH_3^+$	(16.6)	25.0	28.0	10	3.60×10^{13}	196.1	0.42

172

(n-C₆H₁₃)₂O	$CH_3NH_3^+$	(17.8)	27.2	31.4	9	2.73×10^{14}	-	0.45
C₄H₈O (THF)	$(CH_3)_2COH^+$	24.5	33.4	29.9	8	2.22×10^{19}	190.4	0.51
1,4-dioxane	$(CH_3)_2OH^+$	23.1	31.7	28.9	8	2.09×10^{18}	185.1	0.41
	$(CH_3)_2COH^+$	21.0	30.0	30.2	8	6.04×10^{16}		
HCONH₂	$CH_3NH_3^+$	(21.1)	30.0	30.0	9	3.08×10^{16}	192.0	0.66
		(20.0)	30.0	33.5	24			
CH₃CON(CH₃)₂	$(CH_3)_3NH^+$	(20.0)	27.2	24.1	24	1.12×10^{16}	208.6	0.73
CH₃NO₂	$CH_3NH_3^+$	(13.6)	20.5	23.0	9	2.28×10^{11}	173.6	0.25
(CH₃)₂SO	$(CH_3)_2COH^+$	(32.3)	41.1	29.6	23	1.16×10^{25}	203.7	0.78

S HB acceptors

H₂S	H_3O^+	(17.3)	24.9	25.5	32,5	2.57×10^{13}	163.3	-
		(15.5)	20.7	17.3	31,8			
	NH_4^+	6.5	12.0	18.5	8	1.42×10^6		
		(6.4)	11.4	16.7	34			
	$CH_3NH_3^+$	(4.7)	11.3	22	8	8.07×10^4		
		(4.8)	10.8	(20)	34			
CH₃SH	H_3O^+	-	34.5	-	8		178.7	0.16
	$CH_3NH_3^+$	7.1	14.5	24.7	8	3.31×10^6		
		(6.8)	13.4	22.1	34			
C₂H₅SH	$CH_3NH_3^+$	9.0	15.5	21.8	8	8.17×10^7	183.2	0.16

Table 4 (continued)
HB interactions between neutral, B, and cationic, (A-H)$^+$, species in the gas phase (reaction 1b)

B	(A-H)$^+$	$-\Delta G°(HB)$[a]	$-\Delta H°$[b]	$-\Delta S°$[c]	Ref.	K_c[d]	$\Delta G°_{base}(B)$[e]	$\beta_2^H(B)$[f]
C_2H_5SH	$CH_3NH_3^+$	(8.8)	14.6	19.6	34		184.0	0.16
$n\text{-}C_3H_7SH$	$CH_3NH_3^+$	10.1	17.5	24.9	8	6.19×10^8	195.9	0.29
$C_2H_5SCH_3$	$CH_3NH_3^+$	(12.4)	19.8	25	33	3.00×10^{10}		
Other HB acceptors								
PH_3	H_3O^+	-	34.5	-	8		179.6	-
CH_3F	$CH_3NH_3^+$	(4.9)	11.8	23.3	9	9.55×10^4	137.4	-
C_6H_5F	NH_4^+	(9.0)	14.4	18.0	3	9.67×10^7	175.0	0.10
CH_3Cl	$CH_3NH_3^+$	(4.6)	10.7	20.6	9	5.76×10^4	155.4	0.15
CH_3Br	$CH_3NH_3^+$	(4.9)	11.2	21.0	9	9.55×10^4	158.1	0.17

(a) in kcal mol^{-1}, at 298 K. Values given in parentheses are calculated from the given $\Delta H°$ and $\Delta S°$ values.
(b) in kcal mol^{-1}. Values in parentheses are estimated.
(c) in cal mol^{-1} K^{-1}. Values in parentheses are estimated.
(d) in l mol^{-1}.
(e) in kcal mol^{-1}. Values taken from S.G. Lias, J.E. Bartmess, J.F. Liebman, J.L. Holmes, R.D. Levin and W.G. Mallard, NIST Standard Reference Database, Computerized Version 1.1. NIST, Gaitherburg, MD 20899, 1989.
(f) Values taken from M.H. Abraham, P.L. Grellier, D.V. Prior, J.J. Morris and P.J. Taylor, J. Chem. Soc., Perkin Trans. 2, (1990) 521.

174

REFERENCES

1. S.L. Bennet and F.H. Field, J. Am. Chem. Soc., 94 (1972) 5188
2. S.L. Bennet and F.H. Field, J. Am. Chem. Soc., 94 (1972) 6305
3. C.A. Deakyne and M. Meot-Ner (Mautner), J. Am. Chem. Soc., 107 (1985) 474
4. D.W. Berman and J.L. Beauchamp, J. Phys. Chem., 84 (1980) 2233
5. A.J. Cunningham and J.D. Payzant, P. Kebarle, J. Am. Chem. Soc., 94 (1972) 7627
6. J. Bromilow, J.-L.M. Abboud, C.B. Lebrilla, R.W. Taft, G. Scorrano and V. Lucchini, J. Am. Chem. Soc., 103 (1981) 5448
7. R. Yamdagni and P. Kebarle, J. Am. Chem. Soc., 95 (1973) 3504
8. R.G. Keesee and A.W. Castleman, Jr., J. Phys. Chem. Ref. Data, 15 (1986) 1011
9. M. Meot-Ner (Mautner), J. Am. Chem. Soc., 106 (1984) 1257
10. M. Meot-Ner (Mautner), J. Am. Chem. Soc., 105 (1983) 4912
11. S.K. Searles and P. Kebarle, Can. J. Chem., 47 (1969) 2619
12. S.K. Searles and P. Kebarle, J. Phys. Chem., 72 (1968) 742
13. I.N. Tang and A.W. Castleman, Jr., J. Chem. Phys., 62 (1975) 4576
14. M.R. Arshadi and J.H. Futrell, J. Phys. Chem., 78 (1974) 1482
15. C.V. Speller and M. Meot-Ner (Mautner), J. Phys. Chem., 89 (1985) 5217
16. P.M. Holland and A.W. Castleman, Jr., J. Chem. Phys., 76 (1982) 4195
17. M. Meot-Ner (Mautner) and L.W. Sieck, J. Am. Chem. Soc., 105 (1983) 2956
18. M. Meot-Ner (Mautner), J. Am. Chem. Soc., 101 (1979) 2396
19. M. Meot-Ner (Mautner), J. Am. Chem. Soc., 106 (1984) 1265
20. Y.K. Lau and P. Kebarle, Can. J. Chem., 59 (1981) 151
21. K. Hiraoka, E.P. Grimsrud and P. Kebarle, J. Am. Chem. Soc., 96 (1972) 3359
22. W.R. Davidson, J. Sunner and P. Kebarle, J. Am. Chem. Soc., 101 (1979) 1675
23. Y.K. Lau, P.P.S. Saluja and P. Kebarle, J. Am. Chem. Soc., 102 (1980) 7429
24. M. Meot-Ner (Mautner), J. Am. Chem. Soc., 106 (1984) 278
25. E.P. Grimsrud and P. Kebarle, J. Am. Chem. Soc., 95 (1973) 7939
26. M. Meot-Ner (Mautner), J. Am. Chem. Soc., 114 (1992) 3312
27. M. Meot-Ner (Mautner) and L.W. Sieck, J. Am. Chem. Soc., 113 (1991) 4448
28. K. Hiraoka, H. Takimoto and S. Yamabe, J. Phys. Chem., 90 (1986) 5910
29. K. Hiraoka, H. Takimoto and K. Morise, J. Am. Chem. Soc., 108 (1986) 5683

175

30. P. Kebarle, S.K. Searles, A. Zolla, J. Scarborough and M. Arshadi, J. Am. Chem. Soc., 89 (1967) 6393
31. M. Meot-Ner (Mautner) and F.H. Field, J. Am. Chem. Soc., 99 (1977) 998
32. K. Hiraoka and P. Kebarle, Can. J. Chem., 55 (1977) 24
33. M. Meot-Ner (Mautner) and C.A. Deakyne, J. Am. Chem. Soc., 107 (1985) 469
34. M. Meot-Ner (Mautner) and L.W. Sieck, J. Phys. Chem., 89 (1985) 5222
35. J.D. Payzant, A.J. Cunningham and P. Kebarle, Can. J. Chem., 51 (1973) 3242

for 1:1 HB associations between neutral molecules and ionic species. We have carefully examined these data and, found that the requirement of perfect orthogonality is never satisfied. This originates in the fact that, *while in general β^H_2 and $\Delta G°_B$ as well as α^H_2 and $\Delta G°_{Ac}$ are not correlated, substantial correlations exist within families of compounds (AH or B).* This fact is shown in figure 9 (kindly provided by Prof. R.W. Taft), a plot of protonic acidities in DMSO (where the phenomenon also appears and more data are available) *vs.* α^H_2.

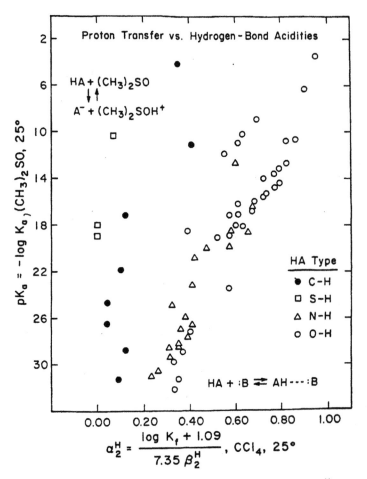

Figure 9. Protonic acidity in DMSO *vs.* α^H_2

Under these circumstances, only truncated forms of equations 17 and 18 were used. Some relevant results obtained using the best truncated forms are summarized as follows:

1. *Associations of $CH_3CO_2^-$.* A set of seven HBDs: MeCN, H_2O, CH_3OH, C_2H_5OH, CH_3CO_2H, CH_3SH and pyrrole was found for which α^H_2 and ΔG°_{Ac} values are available. The condition of orthogonality of both sets is well satisfied ($r^2 = 0.022$).

For the sake of consistency with previous sections, log K_c is used instead of ΔG°_{298}(HB). The best correlation equation found is 20:

$$\log K_c = 52 \,(\pm 19) + 13.2 \,(\pm 4.2)\; \alpha^H_2 - 0.125 \,(\pm 0.052)\; \Delta G^\circ_{Ac} \qquad (20)$$

with $r^2=0.985$ and sd=0.7 log units. Uncertainties at the 95% level.

Notice that equation 20 applies to a set of K_c values spanning a range of over nine powers of ten. The sd corresponds to 0.96 kcal mol^{-1}, a satisfactory value on account of the large effects found in ion-molecule reactions in the gas phase.

For $\alpha^H_2 = 0$, log K_c=0 for ΔG°_{Ac} = 416 kcal mol^{-1}. This is the order of magnitude found in saturated hydrocarbons. Aromatic compounds with α^H_2 values as low as 0.1 should lead to detectable associations (say, log $K_c \geq 4.3$) with acetate ion in the gas phase.

Let us take $\rho_{\alpha\beta} = 1$ in equation 17. That is, let us assume that the "general" HB interaction embodied in equation 15 is also fully operative in the case of equilibria 1b and 1c. Equating $\partial(\log K_c) / \partial\alpha^H_2$ in equations 15 and 20 we obtain $8.85\beta_{CH_3CO_2^-}$ = 13.2, from which we derive $\beta^H_2(CH_3CO_2^-) = 1.5$, in excellent agreement with the value recently found in solution [50]. Last, consideration of the values of α^H_2 and ΔG°_{Ac} shows that in the case of these associations, the contributions from the "general" HB effect and charge-transfer to ΔG°_{298}(HB) (or log K_c) are quite comparable.

2. *Associations of Cl^- and other anions.* A "minimal" set of five equilibrium constants pertaining to associations of CH_3CN, H_2O, $CHCl_3$, pyrrole, CH_3COOH and 4-cyanophenol has been selected on the basis of the low correlation between their α^H_2 and ΔG°_{Ac} values. They lead to the following regression equation:

$$\log K_c = 46.8 - 0.111\; \Delta G^\circ_{Ac} + 9.86\; \alpha^H_2 \qquad (21)$$

This equation applies to the entire set of associations for which K_c, ΔG°_{Ac} and α^H_2 are available. From the data presented in Table 3 we have selected 19 K_c values ranging from 10^4 to 10^{18}. They are described by equation 21 with sd=0.66 log units and r^2=0.985. Only four phenols (phenol, 4-chlorophenol, 4-cresol and 4-cyanophenol) were included in the data set in order not to artificially increase the correlation coefficient. Equation 21 nicely accounts for the significant interaction between aromatic hydrocarbons and Cl^-. The case of cyclopentadiene is important: taking ΔG°_{Ac} for this compound (350.9 kcal mol^{-1}) at face value leads to log K_c= 11.9, way larger than the experimental value. Since the relative low value of ΔG°_{Ac} originates in the stability of the conjugate anion, one is lead to suspect that the HB interaction between Cl^- and the hydrocarbon involves the ethylenic hydrogens. Indeed, the experimental log K_c can be obtained using $\alpha^H_2 \approx 0$ and $\Delta G°_{Ac} \approx 390$ kcal mol^{-1} (essentially the value for benzene). As in the case of acetate ion, associations with saturated hydrocarbons are predicted to be extremely weak and possibly undetectable. A value of α^H_2 between 0.05 and 0.1 can be estimated for benzene (0.05 was used in the correlations). A treatment similar to that for the associations of acetate ion leads to $\beta^H_2(Cl^-) = 0.9$, in good agreement with solution data.

The associations of CN$^-$ can be treated similarly, with a similar outcome.

The case of associations involving F$^-$ is noteworthy: First, K_c values are possibly the largest known, 10^{32} for the formation of the paradigmatic FHF$^-$ complex and 10^8 for the "modest" interaction with benzene. Second, the quantitative treatment is less successful, even if $\Delta G°_{\alpha\beta}$ and $\Delta G°_{Ac}$ are able to account for some 90% of the covariance. There is no possibility of introducing $\Delta G°_{elec}$ because of orthogonality problems. Also, the enormous electric field created by "bare" F$^-$ is also likely to induce significant electronic perturbations in A-H. More experimental data are thus necessary. Correlation between $\Delta G°_{\alpha\beta}$ and $\Delta G°_{Ac}$ also precludes the analysis of the database for RO$^-$, RS$^-$ and Br$^-$. In the case of I$^-$, there are just enough data to obtain equation 22:

$$\log K_c = 61.8 - 0.154\,\Delta G°_{Ac} + 6.09\,\alpha^H_2 \tag{22}$$

n=9; r^2=0.987; sd=0.57 log units.

To avoid bias, equation 22 was obtained using data for only two phenols (phenol and 4-cyanophenol). It leads to $\beta^H_2(I^-) = 0.7$.

3. *Associations of $CH_3NH_3^+$ and other cations.* The treatment of the database for reaction 1b, Table 4, meets the same difficulties in terms of correlations between the explanatory variables. This notwithstanding, several significant conclusions can be reached:

In the case of $CH_3NH_3^+$, the statistically preponderant terms were found to be $\Delta G°_{\alpha\beta}$ and $\Delta G°_{elec}$. The latter was estimated by using the dipolarity - polarizability parameter π^* [49].

Equation 23 obtains for the set of nine associations for which all the necessary data were available or could be safely estimated, namely: C_6H_6, CH_3CN, NH_3, HCO_2CH_3, $(CH_3)_2CO$, $(C_2H_5)_2O$, $CH_3SC_2H_5$, DMF and CH_3NO_2.

$$\log K_c = 6.9\,(\pm 1.6)\,\pi^* + 16.3\,(\pm 4.8)\,\beta^H_2 + 4.1\,(\pm 1.1) \tag{23}$$

n=9; r^2=0.975; sd=1.0 log units. Uncertainties at the 95% level.

While the uncertainties on the various coefficients are large, the correlation is satisfactory considering that *it spans a range of 17 powers of 10 in K_c.* The most remarkable feature is the fact that $\Delta G°_{\alpha\beta}$ and $\Delta G°_{elec}$ account for most of the covariance. This possibly reflects the relatively high proton affinity of CH_3NH_2. NH_4^+ and $(CH_3)_3NH^+$ display a similar behavior.

HB interactions of $(CH_3)_2COH^+$, the conjugated acid of a weaker base seem to be better described in terms of π^*, β^H_2 and $\Delta G°_B$. Unfortunately, the database is too small to warrant a more detailed discussion.

We are aware of the fact that, at least in some of these correlations, contributions from $\Delta G^{\circ}{}_{elec}$ might well be hidden behind the $\Delta G^{\circ}{}_{Ac}$ or $\Delta G^{\circ}{}_{B}$ terms, but as indicated above, a more complete partitioning can not be carried out at this time. Also, some systems may have complex (say bidentate) structures. However, the excellent agreement between the $\beta^{H}{}_{2}$ values for anions obtained in the gas phase (with $\rho_{\alpha\beta}=1$) and in solution, very strongly supports the treatment. This and the good correlations involving $\alpha^{H}{}_{2}$ and $\beta^{H}{}_{2}$ for the neutral species show a consistent adherence to a pattern of "general HB reactivity". This pattern appears in solution, where thousands of examples have been found [35, 43 - 46], and in associations between neutral species in gas phase [21 - 22, 51].

4. CONCLUSION

This work presents what we believe to be the first unified treatment of Gibbs energy changes for 1:1 HB associations in solution and in the gas phase for both neutral and charged species. *Its most important conceptual consequence is the unveiling of a "general pattern" of HB reactivity that seems quite independent of the medium (in the absence of specific interactions) and of the neutral or charged character of the reagents.*

These results also show the need for more work:
1. Regarding interactions in solution of neutral species, it is most desirable to go beyond bilinear forms (equations 11 and 12). The implementation of the full "vectorial" model requires more experimental data.
2. Little is known about 1:1 HB associations involving ionic species in solution, particularly cations. A database is needed.
3. Data for associations between neutral species in the gas phase are of great importance and, as seen above, they are scarce.
4. The available information on 1:1 complexes between neutral molecules and ions in the gas phase is substantial but its full exploitation is hampered by the lack of a relatively small number of experimental values.
5. These treatments are phenomenological and "a posteriori". As usual, the burden of rationalization, prediction and critical evaluation rests on theoreticians.

5. ACKNOWLEDGMENTS

We thank Profs. M.H. Abraham (University College, London), M. Berthelot and C. Laurence (University of Nantes), J.M. Orza (C.S.I.C., Madrid), J. Murray and P. Politzer (University of New Orleans) and R.W. Taft (University of California, Irvine) for their interest in this study. Valuable discussions with Drs. T. del Río and G. Chirico are most appreciated.

REFERENCES

1. W.M. Latimer and W.H. Rodebush, J. Am. Chem. Soc., 42 (1920) 1419
2. Fascinating examples are given in the the chapters by M.F. Perutz (p.17), A. Rich (p.31) and F. Crick (p.87) in " The Chemical Bond. Structure and Dynamics", A. Zewail, Ed., Academic Press, San Diego, 1992.
3. A.C. Legon, D.J. Millen and S.C. Rogers, Proc. Roy. Soc. London, A370 (1980) 213
4. (a) T.R. Dyke, Top. Curr. Chem., 120 (1984) 85 (b) Z. Kisiel, A.C. Legon and D.J. Millen, J. Chem. Phys., 78 (1983) 2910
5. T.R. Dyke and J.S. Muenter, in International Reviews of Science, Physical Chemistry Series Two, Vol.2, A.D. Buckingham, Ed., Butterworths, London, 1975, Ch.2.
6. (a) T.J. Balle, E.J. Campbell, M.R. Keenan and W.H. Flygare, J. Chem. Phys., 71 (1979) 2723. (b) Ibid., J. Chem. Phys., 72 (1980) 922 (c) A.C. Legon, Ann. Rev. Phys. Chem., 34 (1983) 275
7. A.C. Legon and D.J. Millen, Chem. Rev., 86 (1986) 635
8. A.C. Legon, Chem. Soc. Rev., 19 (1990) 197
9. K.B. Borisenko and I. Hargittai, J. Phys. Chem., 97 (1993) 4080
10. J.W. Bevan, A.C. Legon, D.J. Millen and S.C. Rogers, Proc. Roy. Soc. London, A370 (1980) 239
11. D.J. Millen, J. Mol. Struct., 237 (1990) 1
12. (a) A.J.B. Cruickshank, Faraday Discuss. Chem. Soc., 73 (1982) 127 (b) A.C. Legon, D.J. Millen and H.M. North, Chem. Phys. Lett., 135 (1987) 303
13. S.L.A. Adebayo, A.C. Legon and D.J. Millen, J. Chem. Soc. Faraday Trans., 87 (1991) 443
14. See, e.g., Z. Slanina, Z. Phys. Chem. Leipzig, 271 (1990) 109, and references therein
15. (a) G.C. Pimentel and A.L. McClellan, "The Hydrogen Bond", Freeman, New York, 1960. (b) "Spectroscopy and Structure of Molecular Complexes", J. Yarwood, Ed., Plenum Press, London, 1973. (c) "The Hydrogen Bond", Vols. I to III, P. Schuster, G. Zundel and C. Sandorfy, Eds., North Holland, Amsterdam, 1976. (d) "Molecular Interactions", Vols. I to III. M. Ratajczak and W.J. Orville-Thomas, Eds. Wiley, New York, 1980. (e) "Hydrogen Bonds", Topics in Current Chemistry, Vol. 120, P. Schuster, Ed., Springer, Berlin, 1984.
16. The vibrational spectroscopy of HB complexes in the gas-phase is dealt with more specifically in the following: (a) D.J. Millen, J. Mol. Struct., 100 (1983) 351 (b) J.C. Lassègues and J. Lascombe, in "Vibrational Spectra and Structure", J.R. Durig, Ed., Elsevier, Amsterdam, 1982. Vol.11. (c) Y. Maréchal in "Vibrational Spectra and Structure", J.R. Durig, Ed., Elsevier, Amsterdam, 1987. Vol.16.
17. R.D. Hunt and L. Andrews, J. Phys. Chem., 96 (1992) 6945
18. S.R. Davis and L. Andrews, J. Am. Chem. Soc., 109 (1987) 4768
19. M.D. Joesten and L.J. Schaad, "Hydrogen Bonding", Marcel Dekker, New York, 1974.

20. P.R. Griffiths and J.A. de Haseth, " Fourier Transform Infrared Spectrometry" (Chemical Analysis, Vol. 83). Wiley, New York, 1986.
21. J. Marco, Ph.D. Thesis, Universidad Complutense de Madrid, 1994.
22. J. Marco, J.M. Orza and J.-L.M. Abboud, Vibrational Spectroscopy, in press.
23. L.A. Curtiss and M. Blander, Chem. Rev., 88 (1988) 827.
24. (a) R.G. Keesee and A.W. Castleman, Jr., J. Phys. Chem. Ref. Data, 15 (1986) 1011. (b) "Gas Phase Ion Chemistry", M.T. Bowers, Ed. Academic Press, New York, Vols. 1 and 2, 1979 and Vol. 3, 1984. (c) "Fundamentals of Gas Phase Ion Chemistry", K.R. Jennings, Ed. NATO ASI Series C, Vol. 347, Kluwer, Dordrecht, 1991.
25. (a) R.T. McIver, Jr., in "Kinetics of Ion-Molecule Reactions". P. Ausloos, Ed., Plenum, New York , 1978. (b) R.T. McIver, Jr., Sci. Am., 243 (1980) 186
26. (a) M.B. Comisarow, in "Ion Cyclotron Resonance Spectrometry", H. Hartmann and K.P. Wanczek, Eds., Lect. Notes Chem., Springer, Berlin and New York, 1978. (b) R.T. McIver, Int. Lab., 17 (1981) 10 (c) M.B. Comisarow in "Fourier, Hadamard, Hilbert Transforms in Chemistry". A.G. Marshall, Ed., Plenum, New York, 1982. (d) J.-F. Gal, Actual. Chim., (1985) 15
27. (a) P. Kebarle and A.M. Hogg, J. Chem. Phys., 42 (1965) 798 (b) P. Kebarle, Ann. Rev. Phys. Chem., 28 (1977) 445 (c) Y.K. Dan, P.P.S. Saluja, P. Kebarle and R.W. Alder, J. Am. Chem. Soc., 100 (1978) 7328. (d) M. Meot-Ner (Mautner), J. Am. Chem. Soc., 101 (1979) 2396
28. See, e.g., (a) M. Meot-Ner (Mautner), J. Am. Chem. Soc., 114 (1992) 3312 and references therein and (b) Ibid.,J. Am. Chem. Soc., 108 (1986) 6189
29. (a) E.E. Ferguson, F.C. Fehsenfeld and A.L. Schmeltekopf, Adv. At. Mol. Phys., 5 (1969) 1 (b) K. Tanaka, C.I. Mackay and D.K. Bohme, Can. J. Chem., 57 (1976) 193
30. M.T. Molina and A. Dafali, manuscript in preparation
31. J. Hine, "Structural Effects on Equilibria in Organic Chemistry", Wiley, New York, 1975
32. (a) R.W. Taft, Jr., J. Phys. Chem., 64 (1960) 1805 (b) R.W. Taft, Jr., and I.C. Lewis, J. Am. Chem. Soc., 81 (1959) 5343 (c) R.W. Taft, Jr., S. Ehrenson, I.C. Lewis and R.E. Glick, J. Am. Chem. Soc., 82 (1960) 4877. See also, H. van Bekkum, P.E. Verkade and B.M. Wepster, Rec. Trav. Chim., 78 (1952) 815
33. R.W. Taft and R.D. Topsom, Prog. Phys. Org. Chem., 16 (1987) 1
34. R.W. Taft, D. Gurka, L. Joris, P. von R. Schleyer and J.W. Rakshys, J. Am. Chem. Soc., 91 (1969) 480
35. J.-L.M. Abboud and L. Bellon, Ann. Chim. (Paris), 5 (1970) 63
36. R.W. Taft, Jr., in "Steric Effects in Organic Chemistry", E. Newman, Ed. Wiley, New York, 1956, Ch.13.
37. J.E. Loffler and E. Grunwald, "Rates and Equilibria of Organic Reactions". Wiley, New York, 1965.
38. E.M. Arnett, Prog. Phys. Org. Chem., 1 (1963) 289
39. See, e.g., N. Abbagnano, "Historia de la Filosofía", Montaner y Simón, Barcelona, 1978. Vol. 3.

40. Ref. 35, page 72.
41. P.-C. Maria, J.F. Gal, J. de Franceschi and E. Fargin, J. Am. Chem. Soc., 109 (1987) 483
42. M.H. Abraham, P.L. Grellier, D.V. Prior, J.J. Morris, P.J. Taylor, P.-C. Maria and J.-F. Gal, J. Phys. Org. Chem., 2 (1989) 243
43. M.H. Abraham, P.L. Grellier, D.V. Prior, R.W. Taft, J.J. Morris, P.J. Taylor, C. Laurence, M. Berthelot, R.M. Doherty, M.J. Kamlet, J.-L.M. Abboud, K. Sraïdi and G. Guihéneuf, J. Am. Chem. Soc., 110 (1988) 8534
44. M.H. Abraham, P.L. Grellier, D.V. Prior, J.J. Morris and P.J. Taylor, J. Chem. Soc. Perkin Trans. 2, (1990) 521
45. M.H. Abraham, P.L. Grellier, D.V. Prior, P.P. Duce, J.J. Morris and P.J. Taylor, J. Chem. Soc. Perkin Trans. 2, (1989) 699
46. J.-L.M. Abboud, K. Sraïdi, M. Abraham and R.W. Taft, J. Org. Chem., 55 (1990) 2230
47. D.J. Nesbitt, Chem. Rev., 88 (1988) 843
48. A.C. Legon and C.A. Rego, J. Chem. Phys., 90 (1989) 6867
49. (a) M.J. Kamlet, J.-L.M. Abboud and R.W. Taft, J. Am. Chem. Soc., 99 (1977) 6027. (b) M.J. Kamlet, J.-L.M. Abboud and R.W. Taft, Prog. Phys. Org. Chem., 13 (1981) 485. (c) M.J. Kamlet, J.-L.M. Abboud, M.H. Abraham and R.W. Taft, J. Org. Chem., 48 (1983) 2877. (d) C. Laurence, P. Nicolet, M. Tawfik Dalati, J.-L.M. Abboud and R. Notario, J. Phys. Chem., in press.
50. Private communication from Profs. M. Berthelot and C. Laurence
51. This work

P. Politzer and J.S. Murray
Quantitative Treatments of Solute/Solvent Interactions
Theoretical and Computational Chemistry, Vol. 1
© 1994 Elsevier Science B.V. All rights reserved.

Solvent Acidity and Basicity in Polar Media and Their Role in Solvation

W. Ronald Fawcett

Department of Chemistry, University of California, Davis, CA 95616, USA

1. INTRODUCTION

The solvation of ions and molecules in polar solvents has been the subject of considerable interest during the past 70 years [1-4]. The early work of Born [5] on ion solvation relied upon continuum concepts in which the solvent was represented as a dielectric with a uniform permittivity. However, it rapidly became apparent that this model over estimates the thermodynamic solvation properties of simple ions because it ignores the local chemical interactions between the ion and individual solvent molecules. Various other models for ionic solvation which are based on continuum concepts have been presented [6], but they are not popular because they neglect the molecular nature of the solvent and the specific way it interacts with cations and anions.

As far as neutral solutes are concerned, the solvation of the noble gases in polar solvents has been studied in some detail [7]. In this case, the solvation parameters are estimated on the basis of the molecular properties of individual solvent molecules including molecular polarizability, magnetic susceptibility, and diameter. In the case of a polyatomic solute with a dipole moment, the description of solvation is more complex, and includes consideration of dipole-dipole interactions induced dipole-dipole interactions and interactions involving multipoles. Thus, the description of solute-solvent interactions on the basis of a hard sphere model becomes increasingly more difficult as the electrostatic description of the components becomes more complex. Moreover, these models neglect chemical interactions such as hydrogen bonding, which are often important in determining the properties of the system. In order to include these, one must consider the quantum mechanical description of the system.

A quite different approach to describing the solvation of polar molecules and ions is based upon the Lewis acidic and basic properties of the solvent [3,4]. Accordingly, the ability of a solvent to solvate a cation depends on its donicity, that is, its ability to donate a pair of electrons or to act as a Lewis base. On the other hand, the solvation of anions depends upon the solvent's ability to act as a Lewis acid, that is, to accept a pair of electrons. Empirical scales measuring solvent acidity and basicity have been developed [3,4] and can be used to assess the change in experimental quantities related to solvation with the nature of the solvent. When solvent acidity and basicity are the dominant features leading to the change in an experimental parameter Q with solvent nature, these changes can be expressed using the simple linear relationship [8]

$$Q = Q_0 + \alpha A + \beta B \tag{1}$$

where A is the parameter measuring solvent acidity, B, that measuring basicity, α and β, the corresponding response factors, and Q_0, the value of Q when A and B are both zero. By analyzing experimental data on the basis of this relationship one may assess the relative roles of solvent acidity and basicity in solvation, and rationalize the variation in solubility, ion pairing and other phenomena with the nature of the solvent.

Quite recently [9], a connection between the fundamental theoretical approach to ion solvation and that based on linear solvation energy relationships (LSER) was demonstrated on the basis of a non-primitive statistical mechanical model of the solution. This model which uses the mean spherical approximation (MSA) is derived from the integral equation approach to estimate the thermodynamic properties of electrolyte solutions. Accordingly, the system is described as a collection of hard spheres with point dipoles corresponding to the solvent molecules, and hard spheres with points charges for the ions. The resulting expression for the Gibbs solvation energy is similar to that derived by Born [5] on the basis of the primitive model, but with the important difference that the ionic radius is corrected by a quantity which depends on the nature of the solvent and whether the ion is a cation or an anion. This feature of the MSA result is the same as that found much earlier on an empirical basis by Latimer, Pitzer, and Slansky [10]. Moreover, the reciprocal of the correction term, which is called the polarization parameter, measured in various solvents is linearly related to empirical measures of solvent acidity in the case of anions, and to solvent basicity in the case of cations [9]. These relationships provide the connection between fundamental theory and the empirical description of two important chemical properties relevant to solvation.

In the present review, the appropriate parameters for solvent acidity and basicity are considered for electrolyte solutions. Since the discussion is limited to polar solvents, it is appropriate to note that a polar solvent is considered to be one with a relative dielectric permittivity greater than 15. This criterion was chosen on the basis of Bjerrum's model [11] for ion pairing applied to a 0.1M solution of a 1-1 electrolyte. In such a solution the ions are 2 nm apart from one another on the average. The Bjerrum cut off distance for ion pairing in this solution is 2 nm if its relative permittivity is 15. This simple calculation provides a convenient, although somewhat arbitrary, means of defining what dielectric properties are required to consider a solvent as polar.

This review is organized in the following way. First, the appropriate parameters for solvent acidity and basicity are considered. Other solvent parameters needed to assess solvation are also tabulated. Then, data relevant to ionic solvation are examined using linear solvation energy relationships (LSER). These data include Gibbs transfer energies for 1-1 electrolytes and single ions, and formal potentials for simple redox couples measured as a function of solvent. Finally, spectroscopic data relevant to the solvation of polar molecules are discussed and analyzed with respect to the roles of solvent acidity and basicity.

2.   ACIDITY AND BASICITY SCALES FOR POLAR SOLVENTS

Solvent acidity may be viewed as the ability of the solvent to accept a pair of electrons or as its ability to donate a hydrogen bond. In the case of protic solvents, the latter function is generally considered to be more important, protic solvents being strong Lewis acids. Several empirical parameters for measuring solvent acidity have been developed and they are summarized in Table 1.

Mayer et al. [12] formulated the acceptor number AN on the basis of the relative ^{31}P-NMR chemical shifts produced by a given solvent with a strong Lewis base, triethylphosphine oxide. The data were normalized so that the acceptor number of hexane is zero, and that for the 1:1 adduct with the strong Lewis acid $SbCl_5$, 100 when dissolved in 1,2-dichloroethane. The attractive feature of this scale is that it varies over a wide range for the polar solvents normally considered, for example, from 10.6 for hexamethylphosphoramide (HMPA) to 54.8 for water (W).

Another acidity scale is based on the Dimroth-Reichardt E_T parameter [3]. This is obtained by measuring the wavelength of the longest wavelength band in the spectrum of a dilute solution of a betaine dye in the given solvent. This dye, namely, 4-(2,4,6-triphenylpyridinium)-2,6-diphenylphenoxide undergoes a $\pi-\pi^*$ transition in the visible region which is accompanied by a large decrease in molecular dipole moment. E_T was developed as a measure of solvent polarity [3] but clearly is also related to solvent acidity. Its variation for the solvents considered is about half of that of AN so that it is statistically less useful in LSERs (see Table 1). A related acidity scale is the Kosower Z which is based on the absorption bond associated with an intermolecular electron transfer in an ion-pair complex [13]. However, this parameter is not available for a significant number of polar solvents, and is not considered further.

Recently, an acidity scale for polar solvents was introduced on the basis of the MSA expression for the Gibbs solvation energy of a monoatomic monovalent ion [14]. Considering ion-dipole interactions only, the MSA expression for the Gibbs solvation energy of such an ion is

Table 1

Acidity Scales for Polar Solvents

Solvent	AN[a]	E_T[b]	A_p[c]
Protic			
1. water (W)	54.8	63.1	48.0
2. methanol (MeOH)	41.5	55.4	41.0
3. ethanol (EtOH)	37.9	51.9	37.2
4. n-propanol (PrOH)	37.3	50.7	34.7
5. n-butanol (BuOH)	36.8	50.2	34.5
6. formamide (F)	39.8	56.6	34.1
7. N-methylformamide (NMF)	32.1	54.1	31.5
Aprotic			
8. acetone (AC)	12.5	42.2	22.0
9. acetonitrile (AN)	18.9	45.6	24.4
10. benzonitrile (BzN)	15.5	41.5	23.3
11. butyronitrile (BuN)	-	43.1	-
12. dimethylacetamide (DMA)	13.6	43.7	20.1
13. dimethylformamide (DMF)	16.0	43.8	22.2
14. dimethylsulfoxide (DMSO)	19.3	45.1	25.6
15. hexamethylphosphoramide (HMPA)	10.6	40.9	19.2
16. N-methylpyrrolidinone (NMP)	13.3	42.2	22.1
17. nitrobenzene (NB)	14.8	41.2	26.4
18. nitromethane (NM)	20.5	46.3	25.5
19. propylene carbonate (PC)	18.3	46.6	23.9
20. tetramethylene sulphone (TMS)	19.2	44.0	22.0
21. tetramethylurea (TMU)	-	41.0	-

[a]Gutmann acceptor number [12]
[b]Dimroth-Reichardt polarity parameter [3]
[c]Fawcett polar acidity [14]

$$\Delta G_s = - \frac{N_0 e_0^2}{8\pi\varepsilon_0} \left(1 - \frac{1}{\varepsilon_s}\right) \left(\frac{1}{r_i + \delta_s}\right) \tag{2}$$

where N_0 is the Avogadro constant, e_0, the fundamental electronic charge, ε_0, the permittivity of free space, ε_s, the relative permittivity of the pure solvent, r_i, the ionic radius, and δ_s, the MSA distance parameter [9, 14]. The latter quantity depends on the nature of the solvent and also on whether the solvated ion is an anion or cation, reflecting the fact that the mechanism of ion solvation is very different for the two types of ions. It was also found that the reciprocal of δ_s determined in a variety of solvents on the basis of solvation data for the halide ions was linearly related to parameters measuring solvent acidity such as AN and E_T. This suggests that one may define a new acidity scale A_p for polar solvents based on the value of $1/\delta_s$ appropriate for the halide ions in a given solvent. The parameter A_p is defined by the equation

$$\Delta G_s \, (X^-) = - \frac{N_0 e_0^2}{8\pi\varepsilon_0} \left(1 - \frac{1}{\varepsilon_s}\right) \left(\frac{A_p}{1 + r_i A_p}\right) \tag{3}$$

where $\Delta G_s(X^-)$ is the Gibbs solvation energy of the halide ion with radius r_i in a given solvent. This relationship demonstrates that the Gibbs solvation energy of a monovalent anion is linear in the solvent acidity A_p provided that variation in $(1 - 1/\varepsilon_s)$ with solvent is not large, and that the ratio $(1 + r_i A_p)^{-1}$ is also linear with respect to A_p. A plot of the Gibbs solvation energy of the Cl^- anion in 19 polar solvents against the acidity parameter A_p is shown in Figure 1. A good linear correlation is obtained with a correlation coefficient r of 0.956. The scatter is mainly due to variation in $(1 - 1/\varepsilon_s)$ which changes from 0.943 to 0.995 for the solvents considered.

A quite different approach to the assessment of solvent acidity was taken by Taft and Kamlet [15,16]. They defined a hydrogen bond donating ability α on the basis of the solvatochromic comparison method. On this scale, α is only significant for protic solvents and is zero or close to zero for aprotic solvents [16]. In order to account for solvation effects related to solvent polarity, Taft and Kamlet use the solvatochromic parameter π^*. Thus, in order to describe solvent acidity one must use two parameters, namely, α and π^*. Although this separation may be appealing from a fundamental point of view, it is not from a practical one. When one is examining the dependence of a physico-chemical parameter on solvent acidity and basicity on the basis of a LSER, the number of solvents for which data are available is usually limited so that expansion of the description from one involving two independent variables (equation (1)) to one with more independent variables is often unjustified.

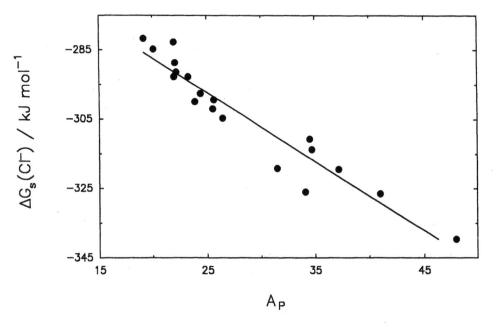

Figure 1. The Gibbs solvation energy of the Cl⁻ anion as a function of the acidity parameter A_p.

Marcus [17] has analyzed the relationship between the parameters AN and E_T and the Taft-Kamlet parameters α and π^*, and showed that the AN parameter is more sensitive to hydrogen bond donating ability α than the E_T parameter. As pointed out earlier [14] correlations considering a wide range of solvents are greatly influenced by the protic solvents. When only the aprotic solvents were considered it was concluded on the basis of correlations with A_p that both AN and E_T provide acceptable descriptions of solvent acidity in the absence of hydrogen bonding [14]. The correlations between these quantities considering the 19 solvents for which A_p is available are

$$A_p = 13.3 + 0.604 \text{ AN} \qquad (r = 0.975) \qquad (4)$$

$$\text{and} \quad E_T = 36.0 + 0.467 \text{ AN} \qquad (r = 0.964) \qquad (5)$$

The fact that A_p is derived from the Gibbs solvation energies of the halide ions, which are very strong Lewis bases, suggests it should be the preferred parameter for estimating solvent acidity for polar solvents. However, the acceptor number AN also performs this function very well, and has the added advantage that it varies over a wide range for the solvents considered here.

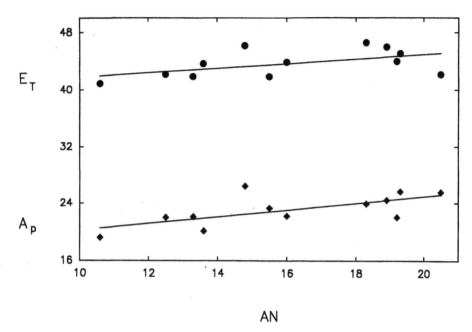

Figure 2. Plot of the acidity parameters $E_T(\bullet)$ and $A_p(\blacklozenge)$ against the acceptor number AN for the aprotic solvents.

It is important to recognize that the quality of the correlations found between the acceptor number AN and the parameters A_p and E_T is determined to a great extent by the protic solvents which are strongly acidic. However, when the data for the protic solvents are removed quite good correlations are also found between AN and these parameters for the aprotic solvents alone (see Figure 2). These scales distinguish between the acidities of common aprotic solvents such as DMF, DMSO, and HMPA, ordering them HMPA<DMF<DMSO with respect to increasing acidity. On the other hand, the Kamlet-Taft parameter α does not distinguish between the acidities of these solvents. However, their polarity parameter π^* does show that DMSO is a more effective Lewis acid than HMPA or DMF, the relative ordering being HMPA ~ DMF < DMSO. All of the above considerations lead to the conclusion that AN is the best parameter for assessing solvent acidity with A_p and E_T being acceptable alternatives.

A number of parameters have been introduced to estimate solvent basicity, the important ones being given in Table 2. These have been compared and analyzed in detail by Persson et al. [18, 19]. The most popular basicity scale is based on the donor number (DN) introduced by Gutmann [20, 21]. This quantity is obtained by measuring the heat of reaction of the solvent with the strong Lewis acid $SbCl_5$ when these reactants are dilute solutes in 1,2-dichloromethane. An important defect of this parameter is that the DN for protic solvents cannot be measured

Table 2

Basicity Scales for Polar Solvents

	Solvent	DN^a	DNs^b	B_{sc}^c	B_p^d
Protic					
1.	W	18.0	17	591	12.1
2.	MeOH	19.0	18	589	12.1
3.	EtOH	19.2[e]	19	589	11.9
4.	PrOH	19.8[e]	18	-	12.05
5.	BuOH	19.5[e]	18	589	11.3
6.	F	24	21	598	12.5
7.	NMF	~27	22	604	12.6
Aprotic					
8.	AC	17.0	15	569	12.55
9.	AN	14.1	12	573	11.47
10.	BzN	11.9	12	572	11.2
11.	BuN	16.6	13	-	11.7
12.	DMA	27.8	24	608	13.5
13.	DMF	26.6	24	602	13.2
14.	DMSO	29.8	27.5	613	13.3
15.	HMPA	38.8	34	633	13.8
16.	NMP	27.3	27	-	14.0
17.	NB	4.4	9	522	10.6
18.	NM	2.7	9	530	10.7
19.	PC	15.1	12	554	11.4
20.	TMS	14.8	15	562	12.5
21.	TMU	31	24	596	14.2

[a]Gutmann acceptor number [20,21]
[b]Persson soft donicity [18,19]
[c]Persson solvatochromic basicity [18,19]
[d]Fawcett polar basicity [14]
[e]Estimates by Kanevsky and Zarubin [22]

directly because of the instability of SbCl$_5$ in these systems. However, values of DN for protic solvents have been estimated by a variety of other techniques [14,22,23], and values are available for all solvents considered here. It is also interesting that values of a bulk DN have been estimated for many solvents [23]. Thus, it was considered that the Lewis basicity of a solvent can be considerably different when the solvent molecules act in concert, rather than when a single molecule is involved as is the case in the Gutmann definition of DN. Significantly higher estimates of the bulk donicity are obtained for protic solvents which are highly structured because of hydrogen bonding. However, on the basis of the polar basicity scale estimated from the Gibbs solvation energies of alkali metal cations, the higher "bulk values" of DN are not relevant for the quantities considered in this review. This may be a reflection of the fact that the structure of protic solvents is very much disrupted near strong Lewis acids such as cations, so that local solvent properties rather than bulk solvent properties are important.

Persson et al. [18] have recently introduced a donor scale for soft acceptors, designated here as DN$_S$. This scale is defined as the shift in the symmetric stretching frequency in HgBr$_2$ when it is in the gas phase compared to when it is a solute in a given solvent. These parameters correlate in an approximate way with the Gutmann DN, but deviations from the best linear fit are seen for solvents with soft donating atoms. Persson et al. [18] also devised a scale for hard acceptors based on the Gibbs energy of transfer of the Na$^+$ ion from water to another solvent [18]. Keeping in mind the fact that the DN is defined with respect to SbCl$_5$, an acceptor with properties on the borderline between hard and soft, DN is preferable to the Persson DN$_S$ except in cases where only soft acceptors are considered.

Persson et al. [18,19] also reported values of the maximum wavelength for the absorption band of a solvatochromic Cu^{2+} complex, namely Cu(II) N,N,N',N'–tetramethylethylenediamine acetoacetonate. This parameter, which is designated B$_{sc}$, is especially convenient because it may be measured directly for all the solvents considered here, both protic and aprotic. The relationship between B$_{sc}$ and DN is

$$B_{sc} = 525.2 + 2.89 \, DN \qquad (r = 0.942) \qquad (6)$$

This relationship may be used to check values of the DN for protic solvents which were obtained by indirect methods. For instance, the DN for N-methylformamide, which is given as ~27 [24], is estimated to be 27.3 on the basis of the corresponding value of B$_{sc}$.

The last parameter listed in Table 2 is the polar basicity B$_p$ determined from the Gibbs solvation energies of the alkali metal cations on the basis of the MSA. From the previous discussion and equation (2), one may write

$$\Delta G_s\ (C^+) = -\frac{N_0 e_0^2}{8\pi\varepsilon_0}\left(1 - \frac{1}{\varepsilon_s}\right)\left(\frac{B_p}{1 + r_i B_p}\right) \tag{7}$$

Values of B_p were extracted from the available data using the procedure described previously [9]. As argued above for the parameter A_p, the Gibbs solvation energy for a given cation should be linear with respect to B_p provided that variation in $(1 - 1/\varepsilon_s)$ is not large, and the ratio $(1 + r_i B_p)^{-1}$ is also linear with respect to B_p. The relationship between ΔG_s and B_p is illustrated in Figure 3. A good linear correlation is found ($r = 0.941$), most of the observed scatter being due to variation in $(1 - 1/\varepsilon_s)$. B_p is also linear with respect to the DN, the relationship being

$$B_p = 10.14 + 0.108\ DN \qquad\qquad (r = 0.896) \tag{8}$$

The range of variation of B_p for the solvents considered here is much smaller than that of DN so the latter parameter is preferred in LSERs.

Several other basicity scales should be mentioned here. Maria and Gal [25] proposed a donicity scale based on the heat of reaction of BF_3 with a given solvent when these are dilute solutes in dichloromethane. Thus, its basis is very similar to that of DN. The Lewis acid BF_3 and solvent dichloromethane were chosen because there are fewer side reactions with this system than with $SbCl_5$ and 1,2-dichloroethane [25].

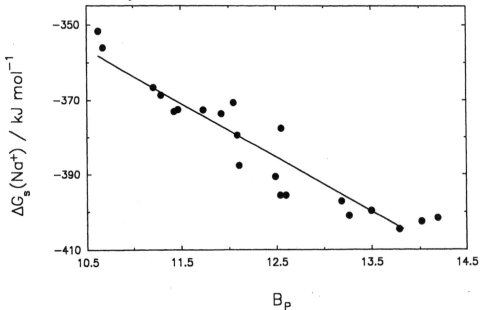

Figure 3. A plot of the Gibbs solvation energy of the Na^+ ion against the basicity parameter B_p.

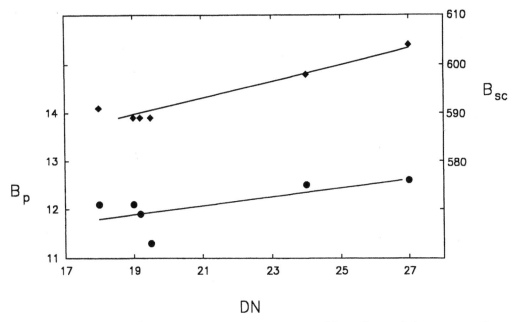

Figure 4. Plot of the basicity parameters B_p(•) and B_{sc} (♦) against the donor number DN for the protic solvents only. The left hand ordinate scale is for B_p and the right hand scale for B_{sc}.

However, the Maria-Gal parameter is not available for the protic solvents considered here, and therefore is not considered further. Another parameter B was introduced by Koppel and Palm [26,27] on the basis of the red shift of the O-D stretching vibration for CH_3OD dissolved in a given solvent. The value of B for water is anomalously low on this scale when one compares it with other basicity scales such as DN and B_p [14]. This may be related to problems in resolving overlapping bands in strongly associated media. Finally, Taft and Kamlet [15,16] introduced the basicity parameter β using the solvatochromic comparison method. This parameter was described as less certain for the alcohols and water in [16] but improved values were given later by Marcus et al. [28]. Because the Taft-Kamlet parameters require three independent variables to describe local solvation effects, they are considered to be less practical than other scales as discussed above.

In order to illustrate the question of the donicity of protic solvents, the values of B_p and B_{sc} for six solvents are plotted against estimates of their donor number in Figure 4. First of all, it is clear that the donor number of water is close to 18 on the basis of both the polar basicity B_p and the solvatochromic parameter B_{sc}. The donicity of the alcohols is not much different from that of water on all three scales. The estimates of DN given for formamide and N-methyl formamide [24] appear to be quite reasonable on the basis of Persson's B_{sc} [18, 19]. The excellent correlation between B_{sc} and DN clearly provides the best method of estimating the DN in protic media.

In conclusion, the best solvent parameters describing acidity and basicity for polar solvents are AN and DN, respectively. This choice is based on the fact that they are available for all solvents considered and vary most widely within this group. Furthermore, values of DN which are unavailable directly for protic solvents, are supported by the values of B_{sc} and B_p obtained by quite different methods. Since the range of variation in the variables used is important in determining the quality of the statistical analysis required to obtain LSERs, it was the overriding consideration in making the present choice. Clearly, scales such as B_{sc} and B_p should be expanded so that this question can be reconsidered.

3. METHOD OF DATA ANALYSIS

In previous analyses of solvent effects [8, 14, 29] attention was focussed on ion-solvent interactions. Under these circumstances local or specific solvation effects dominate and the two parameter LSER involving an acidity and basicity parameter often suffices. However, in general, one should also consider non-specific effects which depend on the bulk dielectric properties of the solvents. Such an approach is originally due to Koppel and Palm [3,27,30]. These authors examined many sets of experimental data [27] using four independent variables with an acidity and basicity parameter for specific effects, and with solvent polarity and polarizability for non-specific effects. Solvent polarity is defined as

$$Y = (\varepsilon_s - 1) / (\varepsilon_s + 2) \tag{9}$$

and polarizability as

$$P = (\varepsilon_{op} - 1) / (\varepsilon_{op} + 2) \tag{10}$$

where ε_{op} is the solvent's relative permittivity at optical frequencies. The equation describing the solvent effect on quantity Q is then

$$Q = Q_0 + \alpha AN + \beta DN + \gamma Y + \delta P \tag{11}$$

where the coefficients α, β, γ, and δ describe the response of Q to the respective solvent parameter, and Q_0 is the value of Q when AN, DN, Y and P are all zero. Application of this equation to experimental data requires that Q be measured in at least six solvents and that proper statistical analysis be made to test the significance of each of these coefficients.

In order to discuss the application of equation (11) to experimental data it is written in a more general form:

$$Q = Q_0 + \sum_{i=1}^{n} k_i X_i \qquad (12)$$

where X_i is the ith solvent parameter, k_i, the corresponding coefficient and n, the number of independent variables. The coefficients k_i are calculated from the normal equations for linear regression using well known matrix diagonalization techniques [31]. Assessment of the quality of the fit is based on the scatter of the observed values of Q from those calculated on the basis of the fit, and the correlation coefficient R. However, these quantities provide no indication of the importance of an individual solvent parameter X_i or the validity of its inclusion in the analysis of a specific set of data. In order to carry out a complete analysis, one must calculate quantities related to the variance of each independent variable, and the dependent variable, namely

$$v_i = m \sum_{k=1}^{m} X_i^2 - \left(\sum_{k=1}^{m} X_i \right)^2 \qquad (13)$$

where m is the number of values of Q_i. In addition, the quantities related to the covariance between any two independent variables, or between an independent variable and the observed quantity Q_i are needed:

$$v_{ij} = m \sum_{k=1}^{m} X_i X_j - \left(\sum_{k=1}^{m} X_i \right) \left(\sum_{k=1}^{m} X_j \right) \qquad (14)$$

On the basis of the variances one may define partial regression coefficients k_i' which are normalized to remove dependence of k_i on the range of variation of X_i. Thus, the partial regression coefficients are [8,14].

$$k_i' = k_i \left(v_i / v_Q \right)^{1/2} \qquad (15)$$

An even better way of presenting these coefficients is in terms of relative partial regression coefficients defined as

$$\bar{k}_i = k_i' / \sum_{i=1}^{n} k_i' \qquad (16)$$

A given $\bar{k}_i$ represents the fraction of the explained variation in Q due to the independent variable X_i. The relative partial regression coefficients are much more informative than the normal regression coefficients usually reported, and are given in the tables of results below.

The covariances are also important in assessing the quality of the fit. From these one may calculate the correlation coefficient between any two variables. Thus, the correlation coefficient for variables X_i and X_j is

$$r_{ij} = \frac{v_{ij}}{v_i^{1/2} \, v_j^{1/2}} \qquad (17)$$

When these are independent variables, the value of r_{ij} should be zero or close to zero. This is certainly the case for the variables used in the Koppel-Palm equation, namely, AN, DN, Y and P as applied here. The value r_{ij} depends on the specific solvents used in the experimental study so that the values of r_{ij} should be calculated in each analysis to check for fortuitous correlations. Values of r_{ij} are also calculated for the correlation between the dependent variable and each of the independent variables. In this way, one immediately finds which of the independent variables is most important, and can then perform the analysis adding one independent variable at a time to the LSER in order of increasing importance.

Finally, one needs to calculate the overall correlation coefficient at each stage in the analysis. This is given by [31]

$$R^2 = \sum_{i=1}^{n} k_i' \, r_{iQ} \qquad (18)$$

By comparing the value of R using n independent variables with that for n-1 independent variables one may assess the importance of adding the nth variable. Using this criterion and the other usual statistical criteria, one may avoid descriptions of the solvent effect which are unnecessarily detailed.

In order to carry out the analyses presented in the following section, one also needs values of the relative permittivities ε_s and ε_{op}. These are listed in Table 3 for the 21 solvents considered. Also listed in this table are values of the molecular dipole moment and polarizability. It was shown recently [32] on the basis of the MSA, that the bulk dielectric properties of polar solvents may be derived from their molecular properties provided one includes in the electrostatic description of the system a stickiness parameter which accounts for interactions other than dipole-dipole interactions. An understanding of interactions between solvent molecules at a molecular level is essential to the development of a picture of the mechanism of solvation.

Before presenting the results of analyzing some experimental data, the approach taken here to the LSER is illustrated with the example of infrared data for the solvent induced frequency shift for the C≡N stretching vibration in acetonitrile [33]. Studies of this system in both

Table 3

Bulk and Molecular Dielectric Properties for Polar Solvents

Solvent	Relative Permittivity		Dipole Moment	Polarizability[c]
	Static ε_s	Optical[a] ε_{op}	p/Debye[b]	$10^3\alpha/nm^3$
Protic				
1. W	78.3	1.7756	1.83	1.48
2. MeOH	32.7	1.7596	1.66	3.29
3. EtOH	24.6	1.8480	1.66	5.21
4. PrOH	20.3	1.9146	1.66	7.10
5. BuOH	17.5	1.9525	1.66	8.79
6. F	111.0	2.0932	3.82	4.26
7. NMF	182.4	2.0449	3.82	6.20
Aprotic				
8. AC	20.7	1.8387	2.87	6.51
9. AN	37.5	1.7999	3.47	4.48
10. BzN	25.2	2.3284	4.54	13.05
11. BuN	22.7	1.9099	3.50	8.29
12. DMA	37.8	2.0609	3.80	9.91
13. DMF	36.7	2.0398	3.80	8.12
14. DMSO	46.7	2.1824	3.96	8.24
15. HMPA	30.0	2.1228	4.47	19.6
16. NMP	32.2	2.1550	4.09	11.0
17. NB	34.8	2.4025	4.28	13.5
18. NM	35.8	1.9033	3.46	5.04
19. PC	66.1	2.0190	4.98	8.80
20. TMS	43.3	2.1963	4.81	11.2
21. TMU	23.1	2.1005	3.40	13.1

[a]The square of the refractive index measured at the sodium D line.
[b]1 Debye is equal to 3.335×10^{-30} Cm.
[c]Calculated on the basis of the MSA as described in [32]

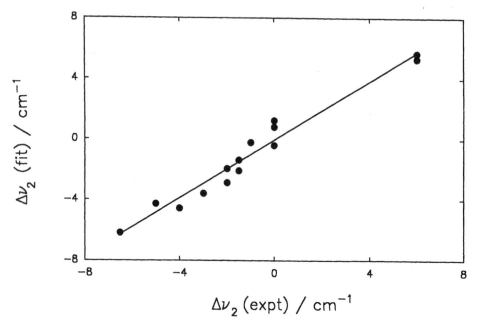

Figure 5. The frequency shift for the C≡N stretching mode in d-acetonitrile calculated by the Koppel-Palm equation (see text) plotted against the experimentally observed value.

polar and non-polar solvents have shown that the v_2 band is blue shifted when the solvent is a stronger Lewis acid than acetonitrile, and red shifted when it is a stronger Lewis base. When the LSER analysis is applied to data obtained in 14 polar solvents from those listed in Tables 1-3, the strongest correlation is with the acceptor number (R = 0.746), correlations with each of the other three parameters being much weaker. The standard deviation for this simple linear fit is 2.5 cm^{-1}. Considering the fact that the correlation with AN is not very good, addition of a second parameter is expected to improve the description of the solvent effect.

Each of the remaining three parameters is now added as a second parameter in a least squares fit, the best results being obtained with solvent polarity Y. The correlation coefficient increases significantly to 0.905 and the standard deviation drops to 1.6 cm^{-1}. The next parameter in order of importance is the donor number DN. When this is added in a least squares fit involving three independent variables the correlation coefficient increases to 0.968 and the standard deviation decreases to 1.0 cm^{-1}. Finally, addition of solvent polarizability P in a least squares fit with four independent variables results in a correlation coefficient of 0.981 with a standard deviation of 0.8 cm^{-1}. It is clear that all four independent parameters are significant in the description of the solvent effect. The quality of the fit obtained is illustrated in Figure 5 where the value of the C≡N stretching frequency shift calculated by the four parameter least squares analysis is plotted against the experimentally observed value. The level of error estimated by the least squares fit corresponds rather well with the precision with which the frequency shift can be determined (0.5 cm^{-1}).

If during the course of the analysis the correlation coefficient does not increase significantly or the standard deviation remains constant or increases, then one has a clear indication that addition of the next parameter is not justified. As will be seen below, this is often the case. Thus, the statistical analysis should be performed in a sequential fashion in order to avoid addition of meaningless correlations in the multiparameter fit.

4.   ANALYSIS OF EXPERIMENTAL DATA

4.1   The Gibbs Energy of Transfer of 1-1 Electrolytes
The standard Gibbs energy of transfer between water and a non-aqueous solvent has been measured for a number of 1-1 electrolytes [34-41]. These data have then been used to extract values for single ion Gibbs energies of transfer on the basis of the equality of solvation of two very large ions, namely, the tetraphenylarsonium cation and the tetraphenylborate anion (TATB assumption) [4,42]. Analysis of the data for transfer of the whole electrolyte on the basis of equation (11) provides an opportunity to assess the relative importance of cationic and anionic solvation, and specific and non-specific solvation for several simple electrolytes.

Examination of the data in Table 4 reveals that very good fits to the Koppel-Palm LSER are obtained for all electrolytes considered. In three cases, the contribution from solvent polarizability is negligible, and in one case, that from solvent basicity is unimportant. Under these circumstances, the parameters quoted are those from a fit with three independent variables.

In the case of the alkali metal halides, solvent acidity is the most important parameter when ionic size is small. As ion size increases, solvent acidity becomes less important, whereas solvent polarity becomes more important. For the perchlorate salts, solvent polarity is the dominant factor. In the case of the tetraalkylammonium and tetraphenylarsonium salts, the relative importance of specific and non-specific effects varies from one system to another, no particular trend being apparent.

It was pointed out previously [14] that the analysis of the data for tetraphenylarsonium tetraphenylborate is especially significant with respect to the TATB assumption used to separate the cationic and anionic contributions for thermodynamic transfer properties [42]. If this

200

Table 4

LSER Analysis of Data for Gibbs Energy of Transfer of 1-1 Electrolytes

Relative Partial Regression Coefficients

Electrolyte	Number of Solvents	$\bar{\alpha}$	$\bar{\beta}$	$\bar{\gamma}$	$\bar{\delta}$	stand. Dev. kJmol^{-1}	R
LiCl	9	0.53	0.44	0.03	-	2.5	0.997
NaCl	10	0.58	0.32	0.04	0.06	1.3	0.999
KCl	8	0.55	0.30	0.12	0.03	1.7	0.999
KBr	10	0.38	0.43	0.19	-	4.4	0.984
CsBr	10	0.20	0.32	0.32	0.16	6.6	0.967
KClO$_4$	10	0.19	0.25	0.40	0.16	3.9	0.977
RbClO$_4$	10	0.22	0.22	0.40	0.16	4.4	0.967
(CH$_3$)$_4$NClO$_4$	8	0.33	0.04	0.43	0.20	4.2	0.950
(C$_2$H$_5$)$_4$NI	10	0.04	-	0.70	0.26	2.3	0.968
(C$_6$H$_5$)$_4$As(C$_6$H$_5$)$_4$B	12	0.87	0.12	0.01	-	7.3	0.959

assumption is correct, the values of $\bar{\alpha}$ and $\bar{\beta}$ for this salt should be equal. Since they are clearly not equal, the TATB assumption can be questioned. On the other hand, if one removes the data for the protic solvents namely, water, methanol and formamide, and performs the analysis using the data for aprotic solvents alone, the values of $\bar{\alpha}$ and $\bar{\beta}$ are equal [14]. This suggests that the TATB assumption is valid only in the absence of hydrogen bonding. It is clear that the TATB assumption needs to be reexamined and a better reference solvent other than water be used to tabulate thermodynamic transfer quantities.

A large body of data is also available for the enthalpy of transfer of simple 1-1 electrolytes from water to non-aqueous solvents [34-36, 38, 43-47]. When these are combined with results for the Gibbs energy of transfer for the same system, one may calculate the entropy of transfer. The present analysis was applied to the enthalpic data for 12 electrolytes including seven alkali metal halides, three alkali metal perchlorates, and two tetraalkylammonium bromides. In the case of the alkali metal salts, successful fits of the Koppel-Palm equation were obtained in all cases with correlation coefficients greater than 0.9 for data sets containing not less than 9 points. The standard deviation of the fits was approximately 6 kJmol^{-1}, that is, somewhat poorer than the corresponding fits for the Gibbs energy of transfer (Table 4). However, in the case of the tetraalkylammonium salts, the data could not be fitted to the LSER used here.

According to the Born [5, 6] model or MSA model [9, 48], the enthalpy of transfer depends on the temperature coefficients of the solvent permittivity and in the latter case on the temperature coefficients of the MSA distance parameter δ. This follows from the fact that ΔH_{tr} is calculated by adding $T\Delta S_{tr}$ to ΔG_{tr}. The parameters used in the present analysis are all temperature dependent and there is no reason to expect the LSER to follow the solvent dependence of a temperature coefficient such as ΔS_{tr}. Thus, if the entropic contribution to ΔH_{tr} is small, one can still obtain an acceptable fit to the solvent dependence of ΔH_{tr} because one is really following the corresponding solvent dependence of ΔG_{tr}. On the other hand, when the entropic contribution is large as in the case for tetraalkylammonium salts [38], the present LSER fails.

The above analysis suggests on the basis of theory and available experimental data that one should not apply the Koppel-Palm equation to analyzing thermodynamic data for enthalpy changes. A proper analysis of such data should consider both the specific and non-specific parameters discussed here, as well as their temperature coefficients.

4.2   Standard Potentials of Simple Electrode Reactions

The simplest systems which can be examined in terms of the acid/base properties of the solvent are redox reactions involving the one-electron reduction or oxidation of an organic molecule to form the corresponding anion or cation radical. Data from the literature for eight systems [49-53] are summarized in Table 5. Very good to excellent fits to the Koppel-Palm equation are obtained in all cases, some of the relative partial regression coefficients being negligibly small. When cation radicals are formed in an oxidation reaction (1,4-diaminobenzene and phenothiazene), solvent basicity is the most important parameter. On the other hand, when the molecule is reduced to form an anion radical as in the case for the other six systems, solvent acidity plays the most important role. The importance of the remaining parameters varies from system to system. In two cases, namely, 1,4-benzoquinone and phenazine, only solvent acidity is important in determining the variation in standard potential with solvent.

Another group of electrode reactions for which solvent effects have been studied involve the reduction of various cations at a dropping mercury electrode. Gritzner [54,55] has reported half-wave potentials for a number of metal cations in a wide variety of solvents. The results of applying the Koppel-Palm analysis to data for the monovalent cations in the solvents considered in this review are summarized in Table 6. As one would expect, solvent basicity plays the predominant role. As the size of the cation increases in the alkali metal cation series, the relative importance of this factor decreases, and that of solvent acidity increases.

Table 5

LSER Analysis of Data for the Standard Potential for Simple Redox Reactions involving Organic Molecules

Reactant	Number of Solvents	$\bar{\alpha}$	$\bar{\beta}$	$\bar{\gamma}$	$\bar{\delta}$	stand. dev. mV	R
		Relative Partial Regression Coefficients					
9,10-anthraquinone (0/-)	7	.50	.17	.09	.24	8	.998
benzophenone (0/-)	7	.47	.26	.08	.19	14	.988
1,4-benzoquinone (0/-)	6	1.00	-	-	-	9	.998
1,4-diaminobenzene (0/+)	9	.08	.63	.13	.16	46	.972
1,4-naphthoquinone (0/-)	6	.89	-	.11	-	4	.999
9,10-phenanthene-quinone (0/-)	7	.49	.24	-	.27	5	.999
phenazine (0/-)	8	1.00	-	-	-	10	.993
phenothiazene (0/+)	11	.04	.77	.14	.05	15	.982

Table 6

LSER Analysis of Data for the Polarographic Half-Wave Potential for Reduction of Monovalent Metal Ions at Mercury

Reactant	Number of Solvents	$\bar{\alpha}$	$\bar{\beta}$	$\bar{\gamma}$	$\bar{\delta}$	stand. dev. mV	R
		Relative Partial Regression Coefficients					
Li^+	12	.08	.84	.08	-	43	.988
Na^+	14	-	.96	.04	-	48	.955
K^+	13	.21	.72	.08	-	44	.920
Rb^+	13	.26	.60	.14	-	33	.926
Cs^+	13	.31	.58	.12	-	29	.936
Tl^+	17	.13	.83	.04	-	41	.962

Solvent polarity plays a minor role, but in no case is there a significant contribution from solvent polarizability. The role of solvent acidity in the solvent effect is difficult to rationalize but it may be connected with ion pairing which depends on both solvent acidity and basicity. It should also be noted that the level of error in fitting these data is significantly higher than that for the redox reactions of organic molecules. This is undoubtedly because the half-wave potential is often not simply related to the standard potential for the corresponding electrode reaction. In some cases, the polarographic current-potential dependence does not correspond to a reversible (fast) electrode reaction so that the half-wave potential reflects not only the thermodynamics of the process but also its kinetics. Nevertheless, the correlation between the half-wave potential and solvent basicity is very strong.

Other systems which have been considered involve transition metal complexes which are often highly charged ions [14,56]. These systems are usually cationic so that solvent basicity plays a predominant role. When anions are involved, solvent acidity predominates. Transition metal ion redox potentials are undoubtedly complicated by ion pairing especially when the ions are highly charged. In the analyses reported previously, at least three parameters are involved in the LSER, namely, solvent acidity and basicity, and either solvent polarity or polarizability [14]. The quality of the fit to the experimental data using the Koppel-Palm equation was better than that obtained by Lay et al. [56] on the basis of Kamlet-Taft parameters. This not only reflects the fact that the Kamlet-Taft parameters are often poorly defined for some of the polar solvents used in electrochemical studies but also that the hydrogen bond donating ability of the solvent was not considered in their analysis.

All of the electrochemical data analyzed here involve an extrathermodynamic assumption. More specifically, the electrode potentials are reported with respect to either the ferrocenium$^{+/0}$ or bis(biphenyl) chromium$^{+1/0}$ redox couples [57] in the same solvent. In this way, problems with changing liquid junction potentials were avoided. The assumption made is that the standard potential of the reference couple is independent of solvent. If the assumption is not valid, there should be some contribution to the correlation with solvent basicity. On the basis of the data presented in Table 5 this cannot be very important since in three cases involving reduction reactions, no contribution from solvent basicity was found.

In conclusion, the electrochemical data for standard potentials are especially interesting for assessing the role of solvent acidity and basicity in solvation. However, ideally these data should be obtained at zero ionic strength in order to avoid problems associated with ion pairing. Unfortunately, all available data were obtained in the presence of an inert electrolyte so that ion pairing effects are undoubtedly present.

4.3 Solvent Induced Frequency Shifts for Polar Solutes

A particularly interesting subject is the role of solvent acidity and basicity in the solvation of polar solutes. Since these molecules are dipolar one might expect the solvent to be involved both as an acid and a base at the negative and positive ends of the molecular dipole, respectively. Usually polar molecules have a characteristic vibrational frequency associated with the electronegative end of the molecular dipole. Interaction of a Lewis acid with this part of the molecule results in a shift

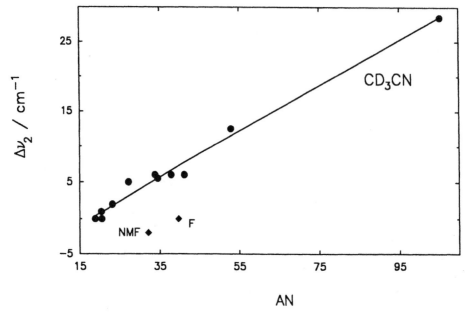

Figure 6. The shift in the C≡N stretching frequency for d-acetonitrile, $\Delta\nu_2$ in solvents which are more acidic plotted against the solvent's acceptor number AN. The data for formamide and N-methyl formamide (♦) were not included in the linear correlation (see text).

in the frequency of the associated infrared band which can be seen using IR spectroscopy. Thus, in ketones such as acetone one can follow the change in C=O stretching frequency with nature of the Lewis acid, and with acetonitrile, the change in the C≡N stretching frequency. These frequency shifts have been used effectively to study cation solvation in a number of aprotic solvents [58-61].

In the experiments discussed here the polar molecule, which is often used as a solvent, is a dilute solute in other solvents both polar and non-polar. The system which has been studied most extensively is acetonitrile [33, 62]. By using deuterated acetonitrile, the effect of the medium on the CD_3 symmetrical and asymmetrical stretching modes could be examined as well as the frequency shift for the C≡N stretching mode [33]. In solvents which are more acidic than acetonitrile itself, the C≡N stretching frequency is shifted in the blue direction by an amount which depends on the acceptor number of the solvent (Figure 5). This clearly shows that solvent acidity plays a major role in determining solvation of the electronegative end of the molecular dipole. The correlation shown in Figure 6 does not include formamide and N-methylformamide. These solvents are moderately strong as both a Lewis acid and base. As a result, the solvent induced frequency shift does not

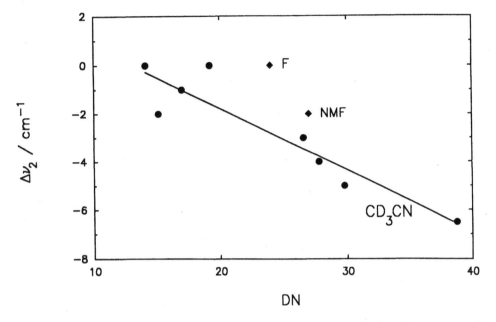

Figure 7. The shift in the C≡N stretching frequency for d-acetonitrile, $\Delta\nu_2$, in solvents which are more basic plotted against the solvent's donor number DN. The data for formamide and N-methylformamide (♦) were not included in the linear correlation (see text).

correlate with those of the other solvents which are stronger acids than acetonitrile but whose basicity is approximately the same or less.

In the case of solvents which are more basic than acetonitrile, the C≡N stretching frequency is shifted in the red direction to an extent which depends on solvent donicity (Figure 7). These observations were rationalized as involving interaction of the electronegative end of the solvent molecule with the CD_3 group in d-acetonitrile, which in turn affects the C≡N stretching frequency by weakening this bond. This explanation is supported by the observation that the symmetrical and asymmetrical stretching frequencies for the CD_3 group are also shifted in the same direction [33].

The LSER analysis based on the Koppel-Palm equation was applied to the acetonitrile data and data for six other polar solutes [62-68]. The solvents included both polar and non-polar liquids with the polar media being limited to those considered above. In the case of acetonitrile, the dominant parameter is solvent acidity which accounts for 42 percent of the explained frequency shift for the C≡N stretching frequency using data for 22 different solvents (Table 7). The second parameter in order of importance is solvent polarity, a non-specific parameter. The quality of the fit is quite good with an overall correlation coefficient of 0.950. If the correlation is limited to the polar solvents considered in this review

(Tables 1-3), the fit improves considerably with a correlation coefficient of 0.987 and standard deviation of 0.8 cm^{-1} for 14 solvents. However, the relative importance of the four solvent parameters remains approximately the same.

A related solvent for which shifts in the C≡N stretching frequency have been studied is benzonitrile [62]. For this system, the dominant parameter in explaining the observed frequency shifts is solvent polarizability, which accounts for 32 percent of the explained variation. The remaining parameters are of approximately equal importance. The changes observed from acetonitrile to benzonitrile undoubtedly reflect the presence of the phenyl ring with its very polarizable electron density.

Another group of molecules including acetone [64] cyclopentanone [65] and tetramethylurea [63] contain the C=O group whose stretching frequency depends on the nature of the solvent. These systems are moderately strong to strong bases. As a result, solvent acidity plays a dominant role in accounting for the observed frequency shifts. For each of these systems, one of the remaining three parameters is of negligible importance. Obviously, the details of molecular structure are important in determining the relative roles of the other parameters.

The remaining molecules considered here are the nitro compounds, nitromethane and nitrobenzene. Both are very weak bases and not very strong acids. The medium effect may be examined by studying the vibrational modes of the NO_2 group [67]. On the basis of the asymmetric stretching frequency for this group in nitromethane, the predominant factor in solvation of this molecule is the solvent's basicity. In fact, most solvents considered in the study carried out by Nyquist [67] are stronger Lewis bases than nitromethane itself. Thus, it is not at all surprising that no significant correlation with the acceptor number alone was found in his study [67]. The parameter of second importance is solvent polarity, indicating that non-specific effects are important for this system. When nitrobenzene is considered, the relative importance of the solvent parameters changes considerably. Because of the polarizable electron density associated with the phenyl ring, solvent polarizability predominates. Solvent polarity is also important, and solvent acidity plays a negligible role.

The data summarized in Table 7 confirm that one needs to consider both specific and non-specific effects in order to understand the solvation of polar molecules. It is particularly striking that solvent acidity or basicity often plays the major role. The vibrational spectroscopic data allow one to examine the electronegative end of the molecular dipole for most polar molecules, and in some cases the electropositive end. Since many polar organic solvents contain methyl or phenyl groups, it is helpful to use the deuterated solute so that the solute's vibrational modes associated with these groups may be distinguished from those of the solvent.

Table 7

LSER Analysis of Data for the Solvent Induced Frequency Shift of the Band Associated with the Electronegative Group in Polar Molecules

Solute	Vibrational Band	Number of Solvents	Relative Partial Regression Coefficients				stand. Dev./in V	R
			$\bar{\alpha}$	β	$\bar{\gamma}$	δ		
AC	C=O stretch	10	0.73	0.10	-	0.17	1.6	0.980
AN	C≡N stretch	22	0.42	0.17	0.24	0.17	1.4	0.950
BzN	C≡N stretch	11	0.22	0.24	0.22	0.32	0.9	0.914
CPN[a]	C=O stretch	13	0.52	-	0.25	0.23	2.5	0.924
NB	asym. NO_2 stretch	10	0.11	-	0.39	0.50	1.4	0.932
NM	asym. NO_2 stretch	10	0.16	0.47	-	0.37	2.2	0.891
TMU	C=O stretch	14	0.72	0.12	0.17	-	2.4	0.981

[a]CPN = cyclopentanone.

Most of the previous studies [62-68] focused on the role of solvent acidity in determining the frequency shift. For this reason, the solvents chosen in this work covered a wide range of acidities. However, there are a significant number of other polar and non-polar solvents which could be included. Extension of the number of solvents for systems such as benzonitrile and the nitrocompounds would undoubtedly improve the correlations observed for these systems and clarify the role of the less important parameters in the LSER.

5. CONCLUSION

The important conclusion of the studies reviewed here is that one must consider both specific and non-specific solvation effects in order to understand the related physical characteristics. Attempts to rationalize solvation on the basis of the continuum dielectric properties of the solvent can only be partially successful because they do not recognize the chemical interactions due to the specific composition of the solute and solvent. Solvent acidity and basicity estimated through empirical parameters provide a very effective way of evaluating specific effects.

The analysis presented in this review emphasizes the relative importance of four solvent parameters namely solvent acidity, basicity, polarity and polarizability. Rather than reporting regression coefficients, only relative partial regression coefficients are given in the tables. This approach is quite different from that usually taken in reporting linear free energy relationships and is meant to emphasize that the main purpose of the analysis is to assess the relative importance of the independent variables rather than to predict a value for the quantity under investigation for an unstudied solvent. When a sufficient number of solvents are included in the analysis, the relative partial regression coefficients do not change significantly. In the absence of this information, one has no reliable way of assessing the relative importance of each parameter on the basis of the overall fit. This follows from the fact that the range of variation of each parameter is not the same.

As far as solvent acidity and basicity are concerned, the parameters developed by Gutmann [24] appear to be the best. This is largely due to the fact that they change over a wide range and are available for most polar solvents. Strong support for these parameters is obtained from analyzing thermodynamic data for the Gibbs solvation energy of simple monoatomic monovalent cations and anions [9, 14]. The fact that the donor number cannot be measured for protic solvents is then circumvented by estimating it using the value of B_p calculated from the Gibbs solvation energy of the alkali metal cations, or B_{sc} estimated from Persson's solvatochromic parameter.

In conclusion, much can be learnt about the mechanism of solvation of both ionic and polar solutes by applying the above analysis provided data are available in a sufficiently large number of solvents. Since application of the Koppel-Palm requires estimation of five parameters, the experimental study must involve at least six solvents, and preferably, at least ten. This limitation often prevents analysis of existing data in the literature for which only a few solvents have been involved in the study. Considering the importance of solvation to reactions in solution, further experimental work to study solvation as a function of solvent nature seems well worthwhile.

ACKNOWLEDGMENT

Recent work from this laboratory discussed in this review was supported by the Office of Naval Research, Washington and by the National Science Foundation, Washington (Grant No. CHE 9008171).

REFERENCES

1. R.R. Dogonadze, E. Kalman, A.A. Kornyshev and J. Ulstrup, The Chemical Physics of Solvation, Part A, Theory of Solvation, Elsevier, Amsterdam (1985).

2. A. Ben-Naim, Solvation Thermodynamics, Plenum Press, New York (1987).

3. C. Reichardt, Solvents and Solvent Effects in Organic Chemistry, 2nd edition, VCH Publishers, New York (1988).

4. Y. Marcus, Ion Solvation, Wiley-Interscience, New York (1985).

5. M. Born, Z. Physik, 1 (1920) 45.

6. For a review see J.E., Desnoyers and C. Jolicoeur, Comprehensive Treatise of Electrochemistry, B.E. Conway, J.O.'M. Bockris and E. Yeager, editors, Plenum Press, New York (1983), Vol. 5, Chap. 1.

7. R.A. Pierotti, Chem. Rev., 76 (1976) 717.

8. T.M. Krygowski and W.R. Fawcett, J. Am. Chem. Soc., 97 (1975) 2143.

9. L. Blum and W.R. Fawcett, J. Phys. Chem., 96 (1992) 408.

10. W.M. Latimer, K.S. Pitzer and C. Slansky, J. Chem. Phys., 7 (1939) 108.

11. N. Bjerrum, Koninklinge Dans. Vidensk. Selsk., 7 (1926).

12. U. Mayer, V. Gutmann, and W. Gerger, Monatsh. Chem., 106 (1975) 1235.

13. E.M. Kosower, J. Am. Chem. Soc., 80 (1958) 5253.

14. W.R. Fawcett, J. Phys. Chem., 97 (1993) 9540.

15. R.W. Taft and M.J. Kamlet, J. Am. Chem. Soc., 98 (1976) 2886.

16. M.J. Kamlet, J.-L.M. Abboud, M.H. Abraham and R.W. Taft, J. Org. Chem., 48 (1983) 2877.

17. Y. Marcus, J. Solution Chem., 20 (1991) 929.

18. I. Persson, M. Sandström, and P.L. Goggin, Inorg. Chim. Acta, 129 (1987) 183.

19. M. Sandström, I. Persson, and P. Persson, Acta Chem. Scand., 44 (1990) 653.

20. V. Gutmann and E. Wychera, Inorg. Nucl. Chem. Lett., 2 (1966) 257.

21. V. Gutmann, Coord. Chem. Rev., 19 (1976) 225.

22. E.A. Kanevsky and A.I. Zarubin, Zh. Org. Khim., 45 (1975) 130.

23. Y. Marcus, J. Solution Chem., 13 (1984) 599.

24. V. Gutmann, G. Resch and W. Linert, Coord. Chem. Rev., 43 (1982) 133.

25. P.-C. Maria and J-F Gal, J. Phys. Chem., 89 (1985) 1296.

26. I.A. Koppel, and V.A. Palm, Reakts. Sposobn. Org. Soedin., 11 (1974) 121.

27. I.A. Koppel and V.A. Palm, in Advances in Linear Free Energy Relationships, N.B. Chapman and J. Shorter, editors, Plenum, London (1972) Chap. 5.

28. Y. Marcus, M.J. Kamlet and R.W. Taft, J. Phys. Chem., 92 (1988) 3613.

29. W.R. Fawcett and T.M. Krygowski, Can. J. Chem., 54 (1976) 3283.

30. I.A. Koppel and V.A. Palm, Reakts. Sposobn. Org. Soedin., 8 (1971) 291.

31. P.R. Bevington, Data Reduction and Error Analysis for the Physical Sciences, McGraw-Hill, New York (1969).

32. L. Blum and W.R. Fawcett, J. Phys. Chem., 97 (1993) 7185.

33. W.R. Fawcett, G. Liu and T.E. Kessler, J. Phys. Chem., 97 (1993) 9293.

34. B.G. Cox, G.R. Hedwig, A.J. Parker, and D.W. Watts, Aust. J. Chem., 27 (1974) 477.

35. B.G. Cox, Annu. Rep. Chem. Soc. A (1973) 249.

36. B.G. Cox and W.E. Waghorne, Chem. Soc. Rev., 9 (1980) 381.

37. C.L. deLigny, H.J.M. Denessen, and M. Alfenaar, Recl. Trav. Chim. 90 (1971) 1265.

38. M.H. Abraham, J. Chem. Soc., Faraday Trans., 71 (1975) 1375.

39. A.F. Danil de Namor and T.J. Hill, J. Chem. Soc., Faraday Trans. 1, 79 (1983) 2713.

40. A.F. Danil de Namor, and L. Ghousseini, J. Chem. Soc., Faraday Trans. 1, 80 (1984) 2843.

41. A.F. Danil de Namor, and H. Berroa de Ponce, J. Chem. Soc., Faraday Trans 1, 83 (1987) 1577.

42. B.G. Cox and A.J. Parker, J. Am. Chem. Soc., 95 (1973) 408.

43. G. Somsen and L. Weeda, J. Electroanal. Chem., 29 (1971) 375.

44. M.H. Abraham, A.F. Danil de Namor, and R.A. Schultz, J. Solution Chem., 6 (1977) 491.

45. M.H. Abraham, E. Ah-Sing, A.F. Daniel de Namor, T. Hill, A. Nasehzadeh and R.A. Schultz, J. Chem. Soc. Faraday Trans. 1, 74 (1978) 359.

46. M. Castagnolo, G. Petrella, A. Inglese, A Sacco, and M. della Monica, J. Chem. Soc. Faraday Trans. 1, 79 (1983) 2211.

47. A.F. Danil de Namor and L. Ghousseini, J. Chem. Soc. Faraday Trans. 1, 82 (1986) 3275.

48. W.R. Fawcett and L. Blum, J. Chem. Soc. Faraday Trans., 88 (1992) 3339.

49. J.S. Jaworski, E. Lesniewska, and M.K. Kalinowski, J. Electroanal. Chem., 105 (1979) 329.

50. W.R. Fawcett and M. Fedurco, J. Phys. Chem., 97 (1993) 7075.

51. M. Opallo, J. Chem. Soc., Faraday Trans. 1, 82 (1986) 339.

52. B. Paduszek and M.K. Kalinowski, Electrochim. Acta, 28 (1983) 639.

53. M. Opallo and A. Kapturkiewicz, Electrochim. Acta, 30 (1985) 1301.

54. G. Gritzner, J. Phys. Chem., 90 (1986) 5478.

55. G. Gritzner, Pure Appl. Chem., 62 (1990) 1839.

56. P.A. Lay, N.S. McAlpine, J.T. Hupp, M.J. Weaver, and A.M. Sargeson, Inorg. Chem., 29 (1990) 4322.

57. G. Gritzner, V. Gutmann, and R. Schmid, Electrochim. Acta, 13 (1968) 919.

58. D.E. Irish and M.H. Booker, Adv. Infrared Raman Spectros., 2 (1976) 212.

59. I.S. Perelygin in Ionic Solvation, G.A. Krestov, editor, Nauka, Moscow (1987) Chap. 3.

60. W.R. Fawcett and G. Liu, J. Phys. Chem., 96 (1992) 4231.

61. W.R. Fawcett, G. Liu, P.W. Faguy, C.A. Foss, Jr. and A.J. Motheo, J. Chem. Soc. Faraday Trans., 89 (1993) 811.

62. R.A. Nyquist, Appl. Spectrosc., 44 (1990) 1405.

63. M.M. Wohar, J.K. Seehra and P.W. Jagodzinski, Spectrochim. Acta, 44A (1988) 999.

64. R.A. Nyquist, T.M. Kirchner and H.A. Fouchea, Appl. Spectrosc., 43 (1989) 1053.

65. R.A. Nyquist, Appl. Spectrosc., 43 (1989) 1208.

66. R.A. Nyquist, Appl. Spectrosc., 44 (1990) 426.

67. R.A. Nyquist, Appl. Spectrosc., 44 (1990) 433.

68. R.A. Nyquist, Appl. Spectrosc., 44 (1990) 594.

P. Politzer and J.S. Murray
Quantitative Treatments of Solute/Solvent Interactions
Theoretical and Computational Chemistry, Vol. 1
© 1994 Elsevier Science B.V. All rights reserved.

213

Using Theoretical Descriptors in Linear Solvation Energy Relationships

George R. Famini[a], and Leland Y. Wilson[b]

[a]U.S. Army Edgewood Research, Development and Engineering Center, Aberdeen Proving Ground, MD 21010

[b]Department of Chemistry, La Sierra University, Riverside, CA 92515

1. INTRODUCTION

1.1. QSAR, LFER

Inherent to chemistry is the concept that there is a relationship between bulk properties of compounds and the structure of the molecules of those compounds. This provides a connection between the macroscopic and the microscopic properties of matter. For example, compounds with carboxyl groups are known to be acids; they have the characteristic sour taste, form red litmus solutions and neutralize bases. Quantitative structure-activity (property) relationships (QSAR, QSPR) go a step further by assuming that there is a quantitative relation between microscopic (molecular structural) features and a macroscopic (empirical) property of a compound. QSAR have been used to correlate molecular structural features of compounds with their known biological, chemical and physical properties for many systems of compounds and properties; many bulk properties do involve solvent-solute interactions.

Two main approaches can be used to find quantitative relationships between molecular structure and bulk properties. A direct attack would be to use quantum mechanics and statistical mechanics to calculate the property. This is extremely complex since it requires models for the molecular interactions involved (solute-solute, solvent-solvent and solvent-solute) and requires great amounts of computer time. However, the increase in computer speed and data storage capacity makes this direct approach possible in principle. The more conventional QSAR approach is to perform a (statistical) correlation analysis. The bulk property of interest is measured for each member of a representative set ("training set") of compounds; then the set of molecular parameters for those compounds is obtained through experiment or calculation. Some type of correlation analysis is then used to find a relation between these two sets of values; often this analysis involves a "curve fit" using multilinear regression.

This technique provides a theoretical probe. Generally the molecular parameters have particular physical meaning; consequently, when that parameter is statistically significant one may infer something about the process at the molecular level. For example, if a molecular parameter for solvent acidity is significant then this implies that there is interaction with a solute basicity function.

Originally, activity and QSAR primarily referred to a biological property; however, the QSAR concepts apply equally well to physicochemical properties. Once a relation is found for a particular property, it can be used to predict that property for any compound from its molecular structure [1]. One such equation is based on the *linear free energy relationship* (LFER). In 1935 Burkhardt [2] and Hammett [3] reviewed the existence of LFER's; in 1937 Hammett [4] proposed the equation that bears his name. Exner [5] provided a more recent (1988) survey of LFER and a clear discussion of the background for its use.

1.2. LSER

An enormous number of descriptors has been used by researchers to increase the ability to correlate biological, chemical and physical properties. One of the most successful sets has been used in the correlations of Hansch [6] and Kamlet, Taft and coworkers [7] who extended the LFER of earlier workers [8] to involve solute/solvent interactions [9]. This *linear solvation energy relationship* (LSER) model has the general form shown in equation (1).

property = bulk/cavity term(s) + dipolarity/polarizability term(s) +
hydrogen bonding term(s) + constant   (1)

The property is often the logarithm of a measured quantity (involving solute-solvent interactions) related to some rate or equilibrium constant which, in turn, can be related to a free energy consistent with the LFER concept. Specifically, the relations, $\Delta G° = -RT \ln K = \Delta H° - T \Delta S°$ and/or the analogous activation energy relations from kinetics, are appealed to. Many properties can be modelled by a process which in turn involves a rate constant, equilibrium constant or an energy transition.

On a microscopic (molecular, bonding) scale, terms in equation (1) can be interpreted in the following manner. The bulk and cavity terms model the energy (endoergic) to overcome the solvent cohesive forces to make a solute molecule sized hole. Furthermore, size effects of the solvent, such as those influencing solvent separation of reactants may occur; the solvent separated ion pair (SSIP) concept would be an example here. Dipole-polarizability terms model van der Waals interactions, essentially electrostatic, which include combinations involving dipole and induced dipole interactions and contribute to the formation of the solute-solvent complex. Dispersive interactions make up a subset of these interactions that involve induced dipole-induced dipole effects. Other than ion-ion effects the strongest electrostatic interactions involve hydrogens and are specifically referred to as the hydrogen bonding terms; these include both acceptor hydrogen-bond basicity (HBB) and donor hydrogen-bond acidity (HBA) within the Bronsted-Lowery context. These terms model the contribution to the formation of the solvent-solute complex through interactions (exoergic) of solute HBA-solvent HBB and solute HBB-solvent HBA.

Kamlet, Taft, Abraham and coworkers have employed the empirically based solvatochromic (LSER) descriptor set for the terms in equation (1). For a given property of solutes in a given solvent, the bulk term uses the solute intrinsic volume, V_I, the dipolarity terms use the solute dipolarity parameter, π^*, and the solute polarizability correction, δ, while the hydrogen bonding terms employ a solute acidity descriptor, α, and a solute basicity descriptor, β. Early work used the molar volume, V_m, while V_I is computed. For a property of solvents with a given solute, the Hildebrand solubility parameter, δ_H^2, replaces the volume in the bulk term while the other parameters, now pertaining to the solvent, are retained. Subscripts 1 and 2 refer to solvent and solute respectively. Recently the characteristic volume, V_x, has replaced V_I and the excess molar refraction has replaced, δ, the polarizability correction [10]. When the parameters are applied to the LSER model equation (2) results. LogSSP represents the logarithm of some property effected by solvent-solute interactions.

$$\log SSP = a\, V_{I2}\, \delta_{H1}^2 + b\, \pi_2^*\pi_1^* + c\, \beta_2\, \alpha_1 + d\, \alpha_2\, \beta_1 + \log SSP_0 \qquad (2)$$

Usually not all the terms in equation (2) are statistically significant for a given property and set of compounds.

Thermodynamic interpretation (a macroscopic scale viewpoint) can be inferred from work by Abraham and coworkers [11] who correlated thermodynamic quantities for the process, X(water) -> X(hexadecane), with the LSER descriptors. The volume term (bulk) is related to the difference in energy needed to create a solute molecule sized cavity in the two solvents; it is endoergic in each solvent. If the energy is greater in the water it will make the standard enthalpy change for the process more exothermic. The cavity formation involves the general dipole and induced dipole interactions also; these will be more exoergic and exothermic in a nonpolar solvent than in water again contributing to a more exothermic overall change. The dipolarity and polarizability terms seem not to be as easily interpreted thermodynamically; however, greater dipolarity implies a greater tendency to form solute-water dipole-dipole interactions which are expected to be exoergic. The hydrogen bonding terms involve the difference between the exothermic solute-water interaction and the much less exothermic solute-hexadecane interaction resulting in an overall endothermic enthalpy of transfer. However, the entropy change from the formation of solute-water bonds will be less than that for solute-hexadecane bonds resulting in an overall positive entropy change.

A strong point of these solvatochromic (LSER) descriptors is their very successful correlation of more than 250 biological, chemical and physical properties involving solute-solvent interactions for a large number of compounds [12]. The coefficients of the descriptors in the correlation equation also can provide insight into the nature of the solute/solvent interactions as typified by the discussion in the previous paragraphs. However, the LSER descriptors are somewhat limited in their ability to make a *priori* predictions because they are empirical. Although there are tables of LSER parameters and predictive relations to help in their estimation LSER values for complex molecules are not as easily found. Hickey and Passino-Reader have provided "rules of thumb" for LSER parameter estimation [13].

Closely related to the LSER approach is a four parameter correlation recently reported by Gajewski [14] who indicates good success for solvent rate effects using the KOPMH (Kirkwood, Onsager, Parker, Marcus, Hildebrand) equation. The four solvent parameters are the Kirkwood-Onsager function, $(\epsilon-1)/(2\epsilon+1)$ [15], the Hildebrand solvent cohesive energy density, $(\Delta H_{vap}-RT)/V_m$ [16], (this is the δ^2 mentioned earlier) and anion and cation relative stabilization parameters, α' and β', respectively. For chloride and potassium ions, values for these last two parameters are derived from the Parker and Marcus [17] compilations of relative free energy of solvation in several solvents. The molar refraction can be related to ϵ, the electrical permittivity in this case, which is related to the polarizability of the molecule.

1.3. Theoretical Descriptors in QSAR

In the past, theoretical chemistry has been used to provide descriptors for QSAR. Ford and Livingstone [18] point out advantages of computationally derived descriptors over extra-thermodynamically derived descriptors such as pi and sigma. They are not restricted to closely related compounds as is often the case with group theoretical, topological and other variables. They describe clearly defined molecular properties making the interpretation of QSAR equations more straightforward. Furthermore, their values are easily obtained; no laboratory measurements are needed thus saving time, space, materials, equipment and alleviating safety (toxicity) and disposal concerns.

There have been attempts to relate the experimental solvatochromic (LSER) parameters to theoretical, computationally derived structural and electronic molecular parameters. Lewis [19] did so with moderate success. More recently Politzer and coworkers have found good correlations for parameters based on the molecular electrostatic potential (MEP) to the LSER dipolarity/polarizability index [20] and solute hydrogen bonding descriptors [21].

However, there has been considerable effort to find theoretical, molecular descriptors to be used in and of themselves regardless of their correlation with the LSER parameters. This section presents some of the work regarding theoretical descriptors. Most often these descriptors are used in correlation equations; but, some computational work takes on a more direct approach to evaluating properties perturbed by solvent-solute interactions. Loew and coworkers [22] have used Mulliken net charges on atoms, orbital energies (including E(LUMO)) and enthalpies. These parameters were calculated with semi-empirical quantum mechanical MO methods. Pedersen [23] provides an overview of the use of computational methods, with emphasis on *ab initio* methods, to examine conformational properties of molecules; however, little mention is made of the complications from including solvent molecules. Chastrette and colleagues [24] combined the empirical and theoretical parameter approach. They used five macroscopic descriptors and three molecular descriptors to classify 83 solvents. The macroscopic properties were the Kirkwood function, molecular refraction, Hildebrand parameter, index of refraction, and boiling point; some of these have been mentioned earlier. The molecular parameters were the E(HOMO), E(LUMO) and dipole moment.

Lewis [25] gives a more recent, extensive summary of molecular orbital calculations applied to QSAR (MO-QSAR) for a variety of activities. Molecular parameters used in these structure-activity relationships include interatomic distances, surface areas, molecular volume, electronic charges, molecular orbital coefficients, resonance integrals, an orbital weighted electron density parameter, polarizability, as well as orbital coefficients and energies for the E(HOMO) and E(LUMO). Kier and Hall [26] have used molecular connectivity parameters, topological graph theoretical descriptors, to correlate a number of activities. Somewhat related to this last parameter is the molecular transform applied by King and coworkers [27]. This parameter is the square root of the integrated area under a squared curve resulting from a Fourier Transform operation on a topological representation of a molecule. The latter representation can be in the form of a neighboring atom bond distance matrix for the molecule as well as other bond distance or bond count matrices. These distances can be found through a molecular orbital optimization of the geometry. In addition to the comments made earlier, Politzer and coworkers have correlated many properties with the MEP and its associated parameters. Their work is discussed in Chapter 8 of this book.

A more direct approach to solvent effects is that of Wiberg and coworkers [28] who use the Onsager reaction field model and *ab initio* algorithms. They have applied this model to the study of conformational equilibria, charge distributions, dipole moments, spectral frequencies and intensities of representative molecules. There is good qualitative agreement with experiment. Peradejordi and colleagues also used a more direct approach for rate constants based on partition functions and quantum mechanical quantities [29]. Their relations involved three type of descriptors: net charges, electrophilic and nucleophilic delocalizabilities. These parameters were combinations of net charges on atoms, orbital coefficients and energies, and coulomb and resonance integrals.

It is important to note that while good correlation or other quantitative and qualitative agreement has been possible with these theoretical efforts, very few sets of parameters have been applied across a wide range of properties and compounds.

1.4. TLSER

Based on the LSER philosophy and general structure a new, theoretical set of parameters for correlating a wide variety of properties has been developed [30]. These theoretical linear solvation energy relationship (TLSER) descriptors have shown good correlations and physical interpretations for a wide range of properties. These include the following: five nonspecific toxicities [31]; activities of some local anesthetics and the molecular transform [32]; opiate receptor activity of some fentanyl-like compounds [33]; and six physicochemical

Table 1
Some systems studied with TLSER

Property	number of compounds	Ref.[a]
Microtox Test Toxicity	41	31
Golden Orfe Fish Toxicity	32	"
Tadpole Narcosis	41	"
Könemann's Industrial Pollutants Test	28	"
Frog Muscle Activity Inhibition	21	"
Minimum Blocking Concentration (Anesthetics)	36	"
Molecular Transform	36	32
Opiate Receptor Activity of Fentanyls	56	33
Charcoal Absorption	33	34
HPLC Retention Index	21	"
Octanol-water Partition Coefficient	67	"
Rate Constant for Phosphonthiolate Hydrolysis	35	"
pKa	42	"
Electronic Absorption Spectra of Pyridinium Ylides	22	"
Gas Phase Acidity[b]	110	35
Solubility in Supercritical CO_2	22	36
Cytochrome P450 Mediated Acute Nitrile Toxicity	26	37

[a]See reference section. [b]Does not involve solvent-solute interactions.

properties: charcoal absorption, HPLC retention index, octanol-water partition coefficient, phosphonothiolate hydrolysis rate constant, aqueous acid equilibrium constant, and electronic absorption of some ylides [34]. Table 1 lists some properties that have been studied and results published using these theoretical descriptors. Also included are TLSER studies of gas phase acidity, which does not involve solvent-solute interactions [35]; solubility in supercritical CO_2 [36]; and cytochrome P450 mediated acute nitrile toxicity, a specific toxicity in contrast with the nonspecific toxicities mentioned earlier, has been examined [37].

These TLSER parameters are determined solely from computational methods thus permitting nearly *a priori* prediction of properties. Modelled after the LSER parameters TLSER descriptors were developed to correlate closely with those descriptors; to give equations with correlation coefficients, R, and standard deviations, SD, close to those for LSER; and to be as widely applicable to solute-solvent interactions as the LSER set. Table 2 gives a summary of these TLSER descriptors as used in this paper and described in the next paragraphs.

The TLSER bulk/steric term is described by the molecular van der Waals volume, V_{mc}, in units of 100 cubic angstroms. The dipolarity/polarizability term uses the polarizability index, π_I, obtained by dividing the polarizability volume by the molecular volume to produce a unitless, size independent quantity which indicates the ease with which the electron cloud may be moved

Table 2
TLSER descriptors

Symbol	Name	Definition	Units	Meaning
V_{mc}	molecular volume	molecular volume	$100A^3$	cavity/steric
π_I	polarizability index	polarizability/V_{mc}	none	polarizability
ε_B	covalent basicity	$0.30- |\Delta E(h,lw)|/100$	hev	acceptor HBB
q_-	electrostatic basicity	maximum $|(-)$ charge$|$	acu	acceptor HBB
ε_A	covalent acidity	$0.30-|\Delta(E(l,hw)|/100$	hev	donor HBA
q_+	electrostatic acidity	maximum $H(+)$ charge	acu	donor HBA

A = Angstrom; hev = hecto-electronvolt; acu = atomic charge unit ; HBB or A = hydrogen bond basicity or acidity; $\Delta E(h,lw) = E(h)-E(lw)$; $E(h)$ = HOMO energy; $E(l)$ = LUMO energy; $E(lw)$ and $E(hw)$ refer to the $E(LUMO)$ and $E(HOMO)$ for water, respectively; $|\ |$ indicate absolute magnitudes.

or polarized. For example, aromatics and chlorine rank high while alkanes and fluorine rank low on the scale.

The acceptor hydrogen bond basicity (HBB) is composed of covalent, ε_b , and electrostatic, q_-, basicity terms. Analogously, the donor hydrogen bond acidity (HBA) is made up of covalent, ε_a, and electrostatic, q_+, acidity terms. The covalent HBB parameter, ε_b, is the magnitude of the difference between the energy of the highest occupied molecular orbital (HOMO) of the solute and the lowest unoccupied molecular orbital (LUMO) of water. The result is divided by 100 for convenience in presentation and comparison of coefficients; the units are in hectoelectronvolts (heV). Analogously, the covalent HBA parameter, ε_a, is the magnitude of the difference between the energies of the LUMO of the solute and the HOMO of water, again scaled like the covalent HBB with the same units. The water energies are included for aesthetic reasons; the smaller these differences the greater is the ability to form a hydrogen bond with water. According to molecular orbital theory when the solute HOMO energy is close to the water LUMO energy the solute can more readily share its electrons with the water; thus, the solute serves as a Lewis base. Analogously when the solute LUMO energy is close to the water HOMO energy the solute can more readily share the water electrons; thus, the solute serves as a Lewis acid. The electrostatic contribution to the HBB is the magnitude of the largest negative formal charge, q_-, on an atom; units are atomic charge units (acu). The corresponding HBA descriptor is the formal charge, q_+, on the most positively charged H atom (in acu).

Earlier papers used ε_b and ε_a; however, increasing ε_b and ε_a mean decreasing basicity and acidity, respectively. Consequently, the transformation, $\varepsilon_{A\ or\ B} = 0.300 - \varepsilon_{a\ or\ b}$, was used so that ε_A and ε_B increase with increasing acidity or basicity; the 0.300 value was chosen to give values of similar size.

So far no adequate descriptor has been found, analogous to δ_{H1}^2, to provide a theoretical model for the solvent cohesive energy. The first thing that comes to mind is to use $1/V_{mc1}$; this amounts to assuming the enthalpy of vaporization does not change much over the set of solvents. Since the Hildebrand solubility parameters are readily available for a large number of solvents it has been suggested that it be used along with the theoretical descriptors.

Equation (3) is the result of applying the TLSER descriptors to the LSER model, equation (1).

$$logSSP = a\ \delta_{H1}^2 V_{mc2} + b\ \pi_{I1}\pi_{I2} + c\ \varepsilon_{B1}\varepsilon_{A2} + d\ \varepsilon_{A1}\ \varepsilon_{B2}$$
$$+ e\ q_{-1}\ q_{+2} + f\ q_{+1}\ q_{-2} + logSSP_0 \qquad (3)$$

This relation can be simplified in the case of multiple solvents for a single solute to give equation (4).

$$logSSP = a_2\ \delta_{H1}^2 + b_2\ \pi_{I1} + c_2\ \varepsilon_{B1} + d_2\ \varepsilon_{A1}\ + e_2\ q_{-1} + f_2\ q_{+1} + logSSP_0 \qquad (4)$$

The TLSER equation, analogous to equation (3) for multiple solutes in a given solvent, is equation (5).

$$logSSP = a_1\ V_{mc2} + b_1\ \pi_{I2} + c_1\ \varepsilon_{B2} + d_1\ q_{-2} + e_1\ \varepsilon_{A2} + f_1\ q_{+2} + logSSP_0 \qquad (5)$$

Within experimental uncertainty, the coefficients in equations (4) and (5) depend on the solute and solvent ,respectively, as well as the appropriate coefficient in equation (2); for example, $b_2 = b\pi_{I2}$ and $b_1 = b\pi_{I1}$. Again, the intercepts should be the same, also within experimental uncertainty. For a given property and set of compounds, the coefficients in these equations are determined using multilinear regression analysis to fit the data. In most cases not all terms are significant.

It is important to note that the theoretical descriptors are from calculations that, strictly speaking, apply to an isolated molecule and, thus the gas phase. However, one might expect that there could be some relationship between the gas phase structure and the structure in a condensed phase.

2. PROCEDURE

Experimental data have been obtained from the literature. PCMODEL (Serena Software, Bloomington, IN) and the in-house-developed molecular modelling package, MMADS, have been used to construct and view all molecular structures [38]. Molecular mechanics is used to optimize the geometry of this original model of the molecule. Optimizing the geometry means that the resulting geometrical structure corresponds to an energy minimum. Viewing the molecular model helps indicate that the structure is reasonable and might be close to a global minimum. Further molecular geometry optimization is done using the MNDO algorithm contained in

MOPAC [39, 40]. The orbital energies, partial charges and polarizability volumes are results of this optimization. The molecular volume for the optimized geometry has been determined using the algorithm of Hopfinger [41]. Multilinear regression analysis using MYSTAT (Systat, Evanston, Il, USA) or MINITAB (Minitab, Inc. State College, PA) has been used to obtain the coefficients in the correlation equation.

The correlation equations were selected based on the following considerations. One: Most important is the need for an adequate sample size. A good rule of thumb is that there be at least three points for each descriptor. Two: The equation coefficients must be significant at the 0.95 level ("large" t-statistic) or higher. The correlation coefficient, R, should be as close to 1 as possible. In the physical sciences it is desirable for the equation to account for 80% or more of the variance which is given by R^2; consequently, a "good" correlation coefficient might have R > 0.90. Depending on the complexity of the system, R values in the range of 0.8 and up can be considered acceptable. Three: There should be small cross correlation amongst the parameters; the variance inflation factor (VIF) is a good measure of the cross correlation. The VIF is defined as $1/(1-R^2)$ where R is the correlation coefficient of one variable against the others; small (closer to one) values imply small cross correlation [42]. Values in the range 1 < VIF < 5 are considered acceptable. Four: Another concern is that there be a minimum number of outliers. Outliers are taken as compounds whose calculated values were three or more standard deviations from the mean. Outlying compounds can be accounted for in several ways. The mechanism or process may be different from the other compounds. Again, the model may be inadequate; TLSER calculations refer to the very dilute (gaseous) phase. The conformation responsible for the reaction may be considerably different from what is calculated for the gaseous ground state. Lastly, it is possible that an experimental value can be in error. Five: The standard deviation, SD, should be small. An SD of 0.3 when the property involves a base ten logarithm corresponds to the uncertainty being a factor of two. Another measure of the fit for the SD is the fraction of the range covered. For example an SD of 0.5 over the interval 0 < logSSP < 8 would imply that the SD is about 6% of the whole range. Another consideration for the SD is that it should be larger than the experimental uncertainties in the measured properties. If it is not it suggests that the results are artifacts. Six: The Fisher index, F, of the reliability of the correlation equation should also be as large as possible.

3. EXAMPLES

We will present correlations for multiple solutes in individual solvents, equation (4); individual solutes in multiple solvents, equation (5); and multiple solutes and solvents combined, equation (3). Furthermore, we will include representative biological and physicochemical properties. The property of interest will be discussed along with the process it represents, the correlation equation presented with its statistical parameters in the pattern illustrated in equation (6) that follows; and then, that is followed by discussion of the physical implications. Table 1 contains a list of some of the systems studied with TLSER.

3.1. A nonspecific toxicity parameter, Könemann's industrial pollutants on Poecilia reticulata. Multiple solutes.

An organic nonelectrolyte is said to exhibit a nonspecific toxicity if it is present in sufficient concentration and if the rate-controlling step in the mechanism involves the transport and partitioning of the compound into the hydrophobic phase (membrane of an organism) from the aqueous phase. No QSAR technique can directly determine the mechanism for the activity, biological property, of a compound. However, some inferences can be made from examining the significant descriptors and looking at outlying compounds. For example, aldehydes can give calculated concentrations significantly larger than experimental concentrations. This means the calculated toxicity is less than experimental toxicity. <u>Recall that larger concentration implies lower toxicity</u>. This behavior could be considered to involve a specific toxicity; an explanation for the aldehydes could be the formation of a Schiff's base between the carbonyl group of the aldehyde and an amine group on the membrane surface.

This heterogenous process for nonspecific toxicity may be represented by this process: solute(aq) $\rightarrow$ solute(membrane). A forward rate constant could be proportional to the concentration, [X], of the aqueous solute and/or the equilibrium constant might be inversely proportional to [X]. Using guppies, Könemann measured the lethal concentration, a measure of [X], at the 50% level, $LC_{50}(K)$, µmol/L, for the compounds in Table 3 [43]. He found a very good correlation for $\log LC_{50}(K)$ with logP, with P being the octanol-water partition coefficient for a set containing most of the compounds studied. When a similar set of data was analyzed with the TLSER descriptors, equation (6) resulted.

$$\log LC_{50}(K)= \quad -1.779\ V_{mc2} \quad -97.87\ \pi_{I2} \quad +39.53\ \varepsilon_{B2} \quad +10.31 \qquad (6)$$

		0.262	6.10	4.25	0.57
±		0.262	6.10	4.25	0.57
t-stat		6.79	16.0	9.31	18.1
P(2-tail)		0.000	0.000	0.000	0.000
VIF		3.60	4.00	1.56	

$$N = 28 \qquad R = 0.992 \qquad SD = 0.21 \qquad F = 477$$

The statistical parameters show the very good fit (R = 0.992 and SD = 0.21) and the strong significance of each term (t-stat > 6.97). The SD = 0.21 value for the logarithm amounts to a factor of 1.6 in the concentration.

The signs on the terms show that the LC_{50} value decreases (toxicity increases) with the volume and polarizability and increases (toxicity decreases) with the HBB. These points are physically reasonable. Increased volume of the solvent can imply increased dispersive interactions; these would tend to favor the interaction with the hydrophobic regions on the membrane compared to the water. Another way of looking at this is that the larger volume would require greater energy to break up the cohesive forces of the water thus favoring the

Table 3
Compounds (Solutes) with TLSER descriptors for Könemann's toxicity study[a]

	V_{mc}	π_I	ε_B	q_-	ε_A	q_+	logLC	calc.	resid
ethanol	0.5420	0.0927	0.1326	0.3240	0.1429	0.1800	5.3800	5.5162	-0.1362
2-propanol	0.7210	0.0955	0.1335	0.3200	0.1448	0.1780	5.0700	4.9592	0.1108
propanone	0.6390	0.0979	0.1381	0.2870	0.1715	0.0232	5.0400	5.0521	-0.0121
2-me-2-propanol	0.8910	0.0976	0.1342	0.3180	0.1441	0.1765	4.6800	4.4788	0.2012
ethoxyethane	0.8972	0.1003	0.1370	0.3423	0.1460	0.0072	4.4600	4.3142	0.1458
3-pentanol	1.0680	0.0996	0.1344	0.3223	0.1469	0.1787	4.0500	3.9760	0.0740
Cl$_2$methane	0.6030	0.1038	0.1207	0.1605	0.1773	0.0555	3.5400	3.8508	-0.3108
1,1-Cl$_2$ethane	0.7920	0.1047	0.1216	0.1633	0.1773	0.0631	3.3100	3.4620	-0.1520
1,2-Cl$_2$ethane	0.7728	0.1064	0.1214	0.1850	0.1789	0.0489	3.0300	3.3219	-0.2919
1-Clbutane	0.9302	0.1076	0.1253	0.2162	0.1693	0.0308	3.0200	3.0785	-0.0585
1,2-Cl$_2$propane	0.9666	0.1045	0.1235	0.1877	0.1744	0.0576	3.0100	3.2460	-0.2360
Cl$_3$methane	0.7528	0.1130	0.1165	0.1123	0.1849	0.0876	2.9300	2.5178	0.4122
benzene	0.8460	0.1204	0.1517	0.0590	0.1413	0.0590	2.9100	3.0192	-0.1092
toluene	0.9995	0.1231	0.1528	0.1007	0.1760	0.0810	2.8700	2.5252	0.3448
1,1,2-Cl$_3$ethane	0.9422	0.1079	0.1202	0.1528	0.1811	0.0683	2.8500	2.8262	0.0238
3-methyltoluene	1.2320	0.1170	0.1532	0.1060	0.1762	0.0814	2.5500	2.7244	-0.1744
2-methyltoluene	1.1890	0.1210	0.1532	0.1060	0.1761	0.0810	2.5200	2.4094	0.1106
1,1,2,2-Cl$_4$ethane	1.0849	0.1133	0.1182	0.1224	0.1790	0.0916	2.3400	1.9647	0.3753
Clbenzene	0.9935	0.1243	0.1493	0.1119	0.1794	0.0781	2.2300	2.2801	-0.0501
1,2-Cl$_2$benzene	1.1480	0.1267	0.1466	0.0821	0.1834	0.0831	1.6000	1.6636	-0.0636
1,3,5-Cl$_3$benzene	1.3040	0.1289	0.1436	0.0857	0.1867	0.0989	1.2600	1.0521	0.2079
1,2,4-Cl$_3$benzene	1.2950	0.1297	0.1451	0.0879	0.1874	0.0988	1.1200	1.0491	0.0709
1,2,3-Cl$_3$benzene	1.2950	0.1288	0.1445	0.0708	0.1862	0.0869	1.1100	1.1134	-0.0034
2,4,5-Cl$_3$toluene	1.4870	0.1276	0.1459	0.0933	0.1877	0.0857	0.9400	0.9446	-0.0046
1,2,3,4-Cl$_4$-benzene	1.4460	0.1312	0.1434	0.6220	0.1899	0.0921	0.5700	0.5664	0.0036
1,2,3,5-Cl$_4$-benzene	1.4510	0.1312	0.1429	0.0768	0.1902	0.1021	0.5700	0.5377	0.0323
1,2,4,5-Cl$_4$-benzene	1.4480	0.1318	0.1435	0.0596	0.1908	0.1027	0.1500	0.5080	-0.3580
Cl$_5$benzene	1.5930	0.1336	0.1417	0.0531	0.1930	0.1059	-0.1500	0.0027	-0.1527

[a]Units on TLSER descriptors in Table 2. Units on LC = LC$_{50}$(K) are µmol/L.

partitioning onto the membrane. The polarizability term also can imply dispersive interactions; the greater these are, the more readily the solute can complex with the membrane. The HBB term implies a lesser tendency to complex to the membrane, thus raising the concentration needed to bring about the effect; this could be through less tendency to interact with a HBA function on the membrane or a greater tendency to interact with the water HBA.

Of the 30 representative compounds used in this study two were outliers, 1,3-bis(chloromethyl)benzene and tetrachloromethane. Calculated values of logLC$_{50}$(K) based on equation (6) are too high for the benzene compound and too low for the chlorinated methane. This suggests that the first compound is more toxic than expected and suggests a different mechanism of interaction with the membrane. Similarly, the second compound is less toxic; a possible explanation for this is that there is a limit set to toxicity by the limited solubility in water. Könemann discussed this "cut-off" in his paper [43]. It is also possible that the compound does not interfere with the membrane functions in the same way the others do.

3.2. Spectral peak position for some ylides. Multiple solvents.

The position of an electronic absorption peak for a compound depends on its environment. The process may be symbolized as follows: e^-(ylide-solvent, ground π) $\rightarrow$ e^-(ylide-solvent, excited π^*) $\Delta E = hcEAS$. Dorohoi and coworkers measured the UV-VIS electronic absorption frequency, EAS in cm^{-1}, for three pyridinium ylides [44] in up to 51 solvents and for seven benzoquinolinium ylides [45] in up to 36 solvents. Pyridinium ylides absorbance peaks were in the range of 22000 to 27000 cm^{-1} while benzoquinolinium ylide peaks were lower and in the range of 19000 to 24000 cm^{-1}. Based on similarities to the spectra to protonated pyridine and pyridinium iodide they made the transition assignment, $\pi \rightarrow \pi^*$. Their purpose was to get information on the solvent effect on the ylide molecules. Experimental uncertainty was 100 cm^{-1}.

The structures of the ylides are given in Table 4. They are zwitterions with the positively charged nitrogen on the cycloimmonium group being covalently bonded to the carbanion. The MNDO calculations indicate a considerable negative partial charge on the carbon portion of the zwitterion but a small partial positive charge on the nitrogen. This data makes it possible to study the effect of the solvent on the frequency of a given solute ylide. Equation (7) is the result of applying the TLSER descriptors to the Δv of the ylide, pyr3 in Table 4, with R1 = R2 = $C(O)OC_2H_5$, group on the carbanion, units are cm^{-1}. This compound had the best correlation of the three pyridinium ylides. Table 5 contains a list of the solvents with their TLSER descriptors. The logarithm is not used since this involves an energy directly. The other two compounds, pyr1 and pyr2 in Table 4, had R values of 0.923 and 0.914 and contained the same set of parameters.

EAS =	$- 29393 \pi_{I1}$	$+ 1456 q_{-1}$	$+ 10532 q_{+1}$	$+ 25430$	(7)
±	5567	426	707	676	
t-stat	5.28	3.43	13.7	37.6	
P(2-tail)	0.000	0.000	0.000	0.000	
VIF	1.6	1.5	1.3		

$$N = 47 \qquad R = 0.957 \qquad SD = 315 \qquad F = 157$$

The most significant parameter, highest t-statistic, was q_{+1} in all of the equations.

Table 4
Ylide parameters and structures

	V_{mc}	π_I	ε_B	q-	ε_A	q+
pyr1	2.4461	0.1303	0.1634	0.4819	0.1054	0.1255
pyr2	2.5539	0.1318	0.1638	0.4980	0.1057	0.1293
pyr3	2.1671	0.1212	0.1629	0.4754	0.1052	0.1284
bnz1	2.9016	0.1746	0.1675	0.3705	0.1065	0.1177
bnz2	3.0799	0.1477	0.1656	0.4928	0.1052	0.1185
bnz3	3.3010	0.1460	0.1646	0.4743	0.1043	0.1218
bnz4	3.6200	0.1472	0.1653	0.4909	0.1049	0.1192
bnz5	3.7959	0.1484	0.1624	0.4883	0.1029	0.1200
bnz6	3.7166	0.1495	0.1657	0.5070	0.1047	0.1759
bnz7	3.5864	0.1612	0.1684	0.4592	0.1030	0.1204

	R_1	R_2
pyr1	$C(O)CH_3$	$C(O)C_6H_5$
pyr2	$C(O)C_2H_5$	$C(O)NHC_6H_5$
pyr3	$C(O)C_2H_5$	$C(O)C_2H_5$
bnz1	$C(O)C_6H_4NO_2(p)$	H
bnz2	$C(O)C_6H_5$	$C(O)CH_3$
bnz3	$C(O)C_6H_5$	$C(O)C_2H_5$
bnz4	$C(O)C_6H_5$	$C(O)C_6H_5$
bnz5	$C(O)C_6H_5$	$C(O)C_6H_4NO_2(m)$
bnz6	$C(O)C_6H_5$	$C(O)NHC_6H_5$
bnz6	$C(O)C_6H_5$	$C(O)C_6H_2(NO_2)_3$

pyridinium ylide

benzoquinolinium ylide

Table 5
TLSER descriptors used with solvents for the ylides[a]

	V_{mc}	π_I	ε_B	q_-	ε_A	q_+	EAS[b]
heptane	1.3517	0.1020	0.1740	0.0220	0.1550	0.0060	22080
Cl$_2$methane	0.6030	0.1040	0.1790	0.1610	0.1200	0.0560	23160
Cl$_3$methane	0.7530	0.1130	0.1840	0.1120	0.1150	0.0880	23280
Cl$_4$methane	0.9160	0.1160	0.1870	0.0700	0.1090	0.0000	21910
Cl$_3$ethene	0.8620	0.1160	0.1610	0.0720	0.1170	0.1090	22500
1,2-Br$_2$ethene	0.9040	0.1180	0.1720	0.1230	0.1190	0.0490	22740
1,2-Cl$_2$ethene	0.7730	0.1060	0.1790	0.1850	0.1210	0.0490	23090
methanol	0.3650	0.0860	0.1690	0.3290	0.1620	0.1930	25230
ethanol	0.5420	0.0930	0.1670	0.3240	0.1570	0.1800	24970
propanol	0.7130	0.0970	0.1670	0.3250	0.1560	0.1800	24950
2-propanol	0.7210	0.0960	0.1660	0.3200	0.1550	0.1780	25000
butanol	0.8980	0.0980	0.1670	0.3250	0.1550	0.1800	24550
2-butanol	0.8970	0.0980	0.1660	0.3220	0.1540	0.1770	24700
pentanol	1.0740	0.1000	0.1670	0.3250	0.1540	0.1800	24350
hexanol	1.2110	0.1040	0.1670	0.3250	0.1540	0.1800	24520
cyclohexanol	1.1220	0.1070	0.1640	0.3220	0.1520	0.1800	24600
octanol	1.5910	0.1030	0.1670	0.3250	0.1530	0.1600	24560
1,2-ethanediol	0.6100	0.0940	0.1670	0.3160	0.1550	0.1820	25560
1,2-propanediol	0.7930	0.0950	0.1650	0.3320	0.1540	0.1890	25290
1,3-propanediol	0.7870	0.0960	0.1670	0.3280	0.1550	0.1820	25140
1,2,3-propanetriol	0.8550	0.0960	0.1670	0.3370	0.1530	0.1940	25490
propanone	0.6390	0.0980	0.1620	0.2870	0.1290	0.0230	23450
2,4-pentanedione	0.9920	0.1020	0.1620	0.2800	0.1260	0.0440	23260
4-CH$_3$-4-OH-2-pentanone	1.2690	0.0990	0.1610	0.3030	0.1280	0.1770	24390
dioxane	0.8600	0.1050	0.1620	0.3270	0.1520	0.0360	22900
propanoic acid	0.7030	0.0980	0.1700	0.3700	0.1310	0.2200	25600
methylethanoate	0.7080	0.1010	0.1690	0.3570	0.1310	0.0270	23400
ethylethanoate	0.8890	0.1020	0.1690	0.3570	0.1310	0.0260	23300
butylethanoate	1.2100	0.1060	0.1680	0.3570	0.1310	0.0270	23020
isopentylethanoate	1.4318	0.1030	0.1680	0.3580	0.1310	0.0260	23210
acetonitrile	0.4510	0.0940	0.1820	0.1150	0.1380	0.0210	23750
benzene	0.8460	0.1200	0.1480	0.0590	0.1590	0.0590	22550
Clbenzene	0.9940	0.1240	0.1510	0.1120	0.1210	0.0780	22950
toluene	1.0000	0.1230	0.1470	0.1010	0.1240	0.0810	22720
2-methyltoluene	1.2320	0.1170	0.1470	0.0950	0.1230	0.0580	22450
4-methyltoluene	1.2160	0.1190	0.1460	0.0950	0.1230	0.0580	22520
1,3,5-(CH$_3$)$_3$benzene	1.3840	0.1190	0.1470	0.1100	0.1240	0.0560	22410
hydroxymethylbenzene	1.0820	0.1200	0.1470	0.3250	0.1240	0.1830	24770
pyridine	0.7910	0.1220	0.1510	0.2300	0.1220	0.0840	23100
benzonitrile	0.9970	0.1280	0.1530	0.0870	0.1170	0.0740	23000
formamide	0.4080	0.0970	0.1630	0.3570	0.1130	0.1570	25190

Table 5 (Continued)
TLSER descriptors used with solvents for the ylides[a]

	V_{mc}	π_I	ε_B	q_-	ε_A	q_+	EAS[b]
N,N-dimethylformamide	0.7700	0.1050	0.1580	0.4510	0.1360	0.0560	23750
dimethylsulfoxide	0.7580	0.1000	0.1520	0.7200	0.1270	0.0530	23720
acrylonitrile	0.5570	0.1080	0.1610	0.0840	0.1220	0.0750	23750
acetophenone	1.1910	0.1200	0.1510	0.2590	0.1210	0.0650	23370
methoxybenzene	1.0900	0.1240	0.1430	0.2860	0.1250	0.0750	23040
2,3-butanediol	0.9680	0.0970	0.1640	0.3300	0.1520	0.1920	24850

[a]symmetrical ylide with $R1 = R2 = C(O)C_2H_5$, [b]cm^{-1}

The correlations show that these TLSER parameters account for more than 80% of the variance. Furthermore, the VIF values show that there is small cross-correlation amongst the variables in this set. There is one outlier, its calculated value is higher than the experimental one. It was retained in the correlation equation.

This equation indicates that the frequency decreases, a red shift, with an increase in solvent polarizability. It also shows that the frequency increases, a blue shift, with increase in the electrostatic HBB and HBA of the solvent. This carries with it the idea that increasing volume and polarizability of the solvent tend to lower the energy of the excited state when the solvent-ylide complex is formed. The opposite is true for the solvent HBB and HBA.

This is physically reasonable since the pyridine ring is highly polarizable so that there can be considerable dispersive interaction there with polarizable solvents. Interpretation of the volume term is not quite as straightforward; increasing volume is associated with greater dispersion forces. It could also suggest less interaction between neighboring ylides because of the greater size of the solvent ylide complex; this amounts to a steric effect or a solvent separated ion pair (SSIP) effect.

The hydrogen-bonding effects can be understood in the context of the ylides interacting with the HBB of the solvent through the positive nitrogen. The HBA of the solvent can help form a solvent-ylide complex near the carbanion in general and the carbonyl oxygens in particular.

Turning to the benzoquinolinium ylides, equation (8) was obtained for the case with $R_1 = C(O)ONO_2$ and $R_2 = H$ and is listed as bnz1 in Table 4.

$$EAS = \quad -14861\, \pi_{I1} \quad + 2076\, q_{-1} \quad + 8444\, q_{+1} \quad + 20252 \quad\quad (8)$$

±	3992	283	499	478
t-stat	3.72	7.34	16.9	42.4
P(2-tail)	0.000	0.000	0.000	0.000
VIF	1.5	1.4	1.1	

N = 33   R = 0.976   SD = 184   F = 194

228

This compound had the best correlation of the seven benzoquinolinium ylides and three pyridinium ylides. The R values for the other benzoquinolinium ylides in numerical order, based on Table 5, were 0.960; 0.950; 0.958; 0.902; 0.961; and 0.878. The parameters, π_{I1} and q_{+1}, were significant in all equations while q_{-1} was not significant in the bnz3, bnz4, bnz6 and bnz7 cases. The ε_{A1} parameter was significant in the bnz6 case; in effect it replaced the q_{-1} parameter while the others, bnz3, bnz4 and bnz7 were only two parameter equations. As in the case of the pyridinium ylides the most significant parameter, highest t-statistic, was q_{+1} in all of the equations.

The interpretation applied to the pyridinium ylide equation applies here since the correlation relations are very similar. The q_{-1} term which is missing in four of the cases is least significant in four of the other six ylides. Pyr1 and bnz2 are the only direct analogues across the two ylide types. The pyridinium analogue has a higher maximum absorbance frequency than the benzoquinolinium compound in a given solvent. One explanation is that the benzoquinolinium structure has a lower lying π^* orbital relative to the π orbital. The greater extent of the conjugated system can account for this. This greater frequency is indicated by the larger constant in the pyridinium equation.

3.3. Spectral peak position for some ylides. Multiple solutes.

Examination of the information in the previous section, 3.2., shows that there is a set of 10 ylide solutes. If all six of the TLSER parameters were to be significant these 10 solutes would not give an adequate correlation equation. However, if three or fewer are significant, and that is how it turns out, it would be satisfactory. The two references of Dorohoi and coworkers [44,45] reveal that there are thirteen solvents in which the spectral peak position have been measured for each of these ten ylides. Thus it is possible to examine how the solute structure effects the frequency in a given solvent.

The correlation equations for these 13 cases were not as good as those for the solvent effect relations discussed in section 3.2., above. The best relation, based on the R value, occurs for the solvent, benzene, and is given in equation (9).

EAS =	$- 80405\, \pi_{I1}$	$-31146\, q_{+1}$	$+ 36763$	(9)
±	14267	12605	2780	
t-stat	5.64	2.47	13.2	
P(2-tail)	0.001	0.043	0.000	
VIF	1.0	1.0		

$N = 10 \qquad R = 0.913 \qquad SD = 655 \qquad F = 17.6$

The relations for the other 12 solvents involve only one descriptor. For trichloromethane π_{I2} is significant and R is 0.853. The other 11 solvents all have ε_{B2} significant and negative. For butyl, ethyl, and methyl ethanoate R is 0.871, 0.860 and 0.881 respectively. Phenylmethanol, pentanol, 2-propanol, propanol, propanone, ethanol, methanol and diethylformamide have R values of 0.870, 0.857, 0.866, 0.856, 0.856, 0.759, 0.762, and 0.853 respectively. With the exception of methanol and ethanol these values are acceptable. Furthermore the descriptors are all significant at the 0.988 level or better.

The negative descriptor coefficients for ε_B imply that increasing solute HBB tends to decrease the frequency. This indicates that associated with increasing basicity of the solute there is a decrease in the separation of the excited and ground pi states. This is in keeping with an increased energy of the ylide HOMO (implying an increase in the basicity) also being closer to its LUMO. Possible also is the idea that the solvent acidity enhances the decrease in the energy separation. In the case of the benzene and trichloromethane solvents, the decreased frequency with increased π_{I2} suggests that there can be some interaction with these highly polarizable solvents to decrease the separation of these pi orbitals. The reason for the poor fit for the methanol and ethanol, solvents with strong hydrogen bonding capabilities, is not readily apparent.

3.4. Spectral peak position for some ylides. Combined solvent-solute sets.

When the solute and solvent data for the ylide studies are combined they provide a large sample space to be used with equation (3). There are 13 solvents (Table 5) for which all ten ylides have spectral data; when another solvent with data for nine of the solutes is added there are 139 sample points. Equation (10) is the resulting correlation equation.

EAS =	238.13 V_{mc1} V_{mc2}	$- 815957\ \pi_{I1}\ \pi_{I2}$	$+ 416418\ \varepsilon_{B1}\ \varepsilon_{A2}$
$\pm$	109.84	54478	81337
t-stat	2.19	15.20	5.21
P(2-tail)	0.000	0.030	0.000
VIF	1.66	2.41	1.17

	$+ 245058\ \varepsilon_{A1}\ \varepsilon_{B2}$	$- 14826\ q_{-1}\ q_{+2}$	$+ 31637\ q_{+1}\ q_{-2}$	$+ 20634$	(10)
	52864	6103	3160	1642	
	4.64	2.43	10.01	12.55	
	0.000	0.016	0.000	0.000	

$$N= 139 \qquad R = 0.887 \qquad SD = 822 \qquad F = 81$$

All six of the terms in equation (3) turn out to be significant at the 0.97 level or better. Two of the parameters, $\pi_{I1}{}^*\pi_{I2}$ and $q_{+1}{}^*q_{-2}$, have a large t-statistic indicating that they are the most significant in contributing to the spectral peak position. The standard deviation of 821 cm^{-1} compares well with the range of about 8000 cm^{-1} and is about eight times larger than the experimental uncertainty of 100 cm^{-1}. Thus, the equation is not very likely to be an artifact.

Since all six terms are significant, the equation is not very selective. However, the polarizability , $\pi_{I1}{}^*\pi_{I2}$, and solvent HBA-solute HBB, $q_{+1}{}^*q_{-2}$, terms are particularly statistically significant. The polarizability term is negative indicating that the energy difference of the two orbitals decreases with increasingly polarizable solutes and solvents (greater dispersive interactions). This is consistent with the individual equations for multiple solvents, equation

(8) and multiple solutes, equation (9), in the case of the polarizable solvents, benzene and trichloromethane. The term involving covalent solvent basicity and solute acidity suggests an increase of the energy difference in the molecular orbitals with increases in solvent HBB and solute HBA; the opposite is true for the electrostatic contributions. Both covalent and electrostatic solvent acidity and solute basicity terms contribute to an increase in the splitting. That was not the case for the solute basicity in the multiple solute relations.

3.5. Decarboxylation rate for 3-carboxybenzisoxazoles. Multiple solvents and combined solvent-solutes sets.

An interesting application of these theoretical descriptors for solvent-solute interactions involves solute and particularly solvent effects on reaction rate. An extensive study of solvent effects on a reaction rate was reported in 1975 by Kemp and Paul [46]. The reaction involved decarboxylation of 3-carboxy-benzisoxazoles to form 2-cyanophenolates in the presence of the base, tetramethylguanidine (TMG), as represented by this chemical equation.

Table 6 contains their rate constant data and illustrates the strong dependence on solvent. The greatest solvent effect occurs for the 6-NO$_2$- compound where the rate constant is 10^8 larger in hexamethylphosphoramide (HMPA) than in water. They suggested three things to explain the solvent effect on the reaction rate: solvent donor hydrogen bond acidity (HBA), ion pair interactions, and dispersive interactions.

The role of solvent donor HBA is supported by these observations. The reaction is slowest in protic solvents; these are capable of intermolecular hydrogen-bonding with the solvent being the donor. Also, the 4-hydroxy substituted compound, capable of intramolecular hydrogen-bonding, decarboxylates more slowly in all solvents [47]. One explanation is that the solvent donor HBA interacts with the carboxyl group to stabilize the reactant relative to the transition state, and, thus, decrease the rate. One can account for the stabilization by considering that in the hydrogen-bonded solvent-solute complex the charge tends to be localized near the hydrogen bond region of the anion rather than delocalized over the whole anion as it would be in the proposed transition state structure.

Ion pairing between the carboxylate and tetramethylguanidinium (TMGH$^+$) ions is supported by the following. Pairing would be expected in aprotic, thus little donor HBA, solvents of low polarity; with little solvent-solute polar attraction, ion-ion effects could be prominent. Grate and co-workers [48] recognized that TMGH+ can be a hydrogen bond donor to the carboxylate. Tight hydrogen-bonded anion-TMGH+ pairs would stabilize the anion relative to the transition state and, thus, decrease the rate. Again, one can account for the stabilization by considering that in the hydrogen-bonded solvent-solute complex the charge tends to be localized near the ion pair contact region rather than delocalized over the whole anion as indicated by the proposed transition state structure. Smid and coworkers [49, 50] provided empirical evidence for the role of ion pairs in the case of the 6-NO_2 compound. They showed that crown ethers and their polymers catalyze the reaction in benzene with potassium ion as the counter ion; the reaction was about a thousand times faster than with TMGH+ in benzene. Apparently the crown ether can form a complex with the cation, thus freeing up the anion to form the transition state. Additionally, they found the rates for the free acid (no base present) in dioxane to be slow and even slower in the presence of a stronger acid. This suggested that the anion was indeed involved. Under these conditions the concentration of the anion is very small and there are few counterions (solvated protons) to pair with it. When the concentration of the anion is taken into account, the rate constant becomes large thus suggesting that the free anion reacts quite rapidly. Consequently they proposed that ion-pairing effects, rather than dispersion interactions (discussed in next paragraph) between solvent and transition state, better explain the reaction rates in aprotic solvents. It should be noted that dispersion effects are always involved in the solvation of ions so that even aprotic but polarizable solvents can effect the separation of ions.

The existence of ion pairs in nonpolar solvents seems reasonable; but, it is not apparent that ion pair concentration can be appreciable in more polar solvents. Grate and coworkers considered the equilibria amongst the various ion pair complexes: tight hydrogen-bonded pairs represented by (cation..anion); solvated, solvent separated pairs represented by (solvated cation)..(solvated anion); and "free" solvated ions represent by (solvated cation) + (solvated anion). They used estimated dissociation constants in acetonitrile, to calculate the concentration of ion pairs for the tetramethylguanidinium benzisoxazole-3-carboxylate solvent separated ion pairs. The equilibrium constant estimation was based on values determined by Kolthoff and coworkers [51], for analogous ion pairs - trimethylammonium 3,5-dinitrobenzoate, tetramethylguanidinium benzoate and 3,5-dinitrobenzoate - in acetonitrile. The equilibrium constants are on the order of 10^{-4} which along with the very low stoichiometric concentrations used by Kemp indicate that there is a significant fraction of ion pairs. Data for more basic, HBB, solvents was not available to make estimations of the fraction of ion pairs; the fraction should decrease with increased solvent basicity.

Kemp and Paul also suggested that transition state stabilization by dispersion relative to the anion is supported by these observations. The reaction is faster in aprotic solvents than in water and is catalyzed by extraction from water into such solvents. This information does not support stabilization of the transition state by dispersion interactions directly. It does imply that the transition state in aprotic solvents is stabilized relative to that in water. That can be explained by less polarity of the non-aqueous solvent compared to water; the higher polarity of the water with its concomitant HBA and HBB would tend to localize the anion charge thus making it more stable than the transition state. An implication is that in solvents less polar than water this difference in stability would be less than in water. Conclusions regarding dispersive interactions must be drawn from data for non-polar, aprotic solvents. Protic solvents, capable of HBA, have already been noted to slow the rate. However, for benzene, the quintessential case of the nonpolar solvent, Kemp's data shows that the rate is faster than in water. The primary solvent-solute effect here is dispersion, benzene is a very polarizable molecule hence capable of considerable dispersive interaction. However, it still does not mean that dispersion strongly influences transition state stabilization in general; other interactions are stronger.

Continuing with the role of dispersive interactions, it should be noted that most of the aprotic solvents with increased rate relative to water, were also polar and their rate increased with polarity. This is contrary to expectation for ion-dipole interactions; these would help localize the charge. One interpretation is that transition stabilization by dispersion is not a major effect at all. The increased rate in polar, aprotic solvents could be explained by ion pairs. These solvents have acceptor HBB and can form a complex with the cation thus separating the ion pairs and freeing the anion.

Evidence for the role of the acceptor HBB mentioned in the previous paragraph comes from the work of Grate and co-workers. Equation (11) shows their correlation equation for 20 solvents. The β term is positive and indicative of the basicity.

$\log k =$	$+ 5.45\, \pi_1^*$	$- 1.46\, \delta 1$	$- 3.03\, \alpha_1$	$+ 1.80\, \beta_1$		
$\pm$	0.89	0.55	0.65	0.66		
t-stat	6.12	2.66	4.65	2.74		
P(2-tail)	0.000	0.019	0.000	0.016		
VIF	2.29	1.89	3.26	2.11		
			$-1.06\, \delta_{H1}^2$		-2.97	(11)
			0.25		0.52	
			2.66		5.73	
			0.001		0.000	
			4.78			

$$N = 20 \qquad R = 0.976 \qquad SD = 0.582 \qquad F = 56.2$$

The statistical parameters indicate the good fit of the data. (The statistical parameters are those commonly used except, perhaps, the VIF which was explained in the procedure section.) However, the most important aspect is the physical interpretation. The α_1 term is negative implying that the rate

Table 6
Logarithms of rate constant, k, for decarboxylation of 6-x-3-carboxybenzisoxazoles.[a]

	NH$_2$	H	CH$_3$O	Cl	NO$_2$	NO$_2$[b]	(NO$_2$)$_2$[c]
water	-5.700	-5.975	-5.611	-5.357	-5.134	-4.025	-2.991
methanol	-4.959	-4.770	-4.553	-4.000	-3.602	-2.410	-1.000
ethanol	-4.602	-3.398	-4.102	-3.509	-3.000	-1.796	-0.456
formamide	-4.377	-4.097	-3.959	-3.456	-3.131	-1.796	
N-methylformamide	-3.745	-3.456	-3.149	-2.553	-2.092	-0.770	
nitromethane	-1.854	-1.509	-1.252	-0.620	-0.237	1.000	2.699
acetonitrile	-1.071	-0.921	-0.638	0.079	0.462	1.800	
dimethylsulfoxide	-0.700	-0.398	0.000	0.612	1.000		
propanone	-0.398	-0.155	0.255	0.944	1.380		
N-dimethylformamide	-0.091	0.176	0.672	1.225	1.568		
tetramethylene sulfone	0.079	0.477	0.653	1.431	1.806		
dimethylacetamide	0.279	0.544	0.903	1.845	2.204		
N-mepyrrolidine	0.602	1.041	1.255	2.000	2.398		
(me)$_6$phosphoramide	0.820	1.255	1.447	2.342	2.845		

[a]From Kemp and Paul, Ref.46. (Pseudo first order rate constant, units = s^{-1}; T=30 °C, Base = tetramethylquanidine.) [b]5-NO$_2$ [c]5,6-(NO$_2$)$_2$

decreases with increasing HBA; this is consistent with experiment. The β_1 term is positive implying that the rate increases with increasing HBB; this is consistent with the explanation in the previous paragraph. The solvent, an HBB, can form a complex with the cation, an HBA, thus freeing the anion to move into the transition state. The π_1^* term is positive indicating that rate increases with increasing solvent polarizability. Furthermore, this is the most statistically significant term. One interpretation is that the polarizability of the solvent can enhance the polarizability of the solutes, highly polarizable themselves, and thus favor delocalization of the charge over the anion. Polarizability is related to the ability to induce a dipole and, hence, related to dispersive interactions which involve induced dipole-induced dipole interactions, primarily. Another interpretation can be that dispersive interactions between the solvent and the anion can help separate the ion pairs; this would increase the rate. The separation can be enhanced by the size increase that goes with increased polarizability. Thus, one can infer a role for dispersion that does not have to involve transition state stability.

Practically concurrent with the work of Grate and colleagues, a study on the carbon kinetic isotope effects on the 5-NO$_2$ compound gave strong evidence that the transition state structure changes very little with changes in solvent [52]. This implied the idea that the transition state structure might not depend strongly on an interaction with the solvent particularly with regard to solvation.

Based on the LSER results, Famini, Grate and Wilson have applied the TLSER parameters to this reaction rate data [53]. While the Kemp and Paul data referred to 30°C and Hildebrand solubility parameters were for 25°C. The 5°C difference represents less than 2% error based on the free energy relation, $\ln k = -E_a/RT$ + constant, model. Table 6 contains their data for seven solutes and 14 solvents while Table 7 contains their data for the 6-NO_2 compound over 24 solvents (which include the 14 in Table 6) as well as the TLSER parameters for the compounds used in this work. The 14 member set compounds are quite polar while the 10 member set compounds have low polarity. The diff(erence) column in Table 7 shows that the calculated values for the 10 solute set are all higher than experiment. Tetramethylene sulfone, an outlier in the set of 14, is quite different in that its calculated value is much less than experiment.

Examination of Table 6 shows that there is barely enough data to examine the multiple solute-single solvent case. A further complication is that the anion descriptors are strongly cross-correlated. There is adequate data for multiple solvent-single solute correlations. This can be done for the 6-NO_2 compound over the 24 solvents in Table 7 as well for the five 6-x compounds (including the 6-NO_2) over the 14 solvents in Table 6. The 24 solvent case did not correlate well; R was 0.857 with ε_{B1}, q_{-1}, and q_{+1} significant (close to it for ε_{B1} since P(2-tail) = 0.055). Equation (12) shows the correlation equation for the 6-NO_2 compound in 13 solvents.

$$\log k = \quad -1.161\ \delta_H^2 \quad +66.34\ \varepsilon_{B1} \quad -26.55\ q_{+1} \quad -6.362 \qquad (12)$$

$\pm$	0.237	10.67	1.69	1.546	
t-stat	5.14	6.22	15.68	4.12	
P(2-tail)	0.003	0.000	0.000	0.003	
VIF	1.73	1.27	1.62		
N = 13	R = 0.994		SD = 0.322	F = 268	

Tetramethylene sulfone was an outlier and was dropped from the set. The other four 6-X compounds had analogous equations with R values of 0.993 or 0.994 and illustrated the very good fit by the 14 solvent data set. When the other 10 solvents (of the 24) are analyzed, the best equation has R = 0.871 with ε_{B1}, q_{-1}, and ε_{A1} significant with tetrahydrofuran an outlier. Also there is strong cross-correlation for the latter two variables. Removing the outlier improves R to 0.952 but cross correlation is still too large. Interestingly, there was strong cross-correlation, also, when the 10 solvent subset was analyzed with the LSER parameters.

Equation (12) is consistent with the LSER results, equation (11). Both indicate that the rate decreases with increasing solvent cohesive interactions, δ_{H1}^2. Also they show that the rate increases with solvent basicity (HBB) and decreases with solvent acidity (HBA). This solvent HBA effect is consistent with experimental data. There is a more general observation to be made in connection with equation (12). As in the LSER case, equation (11), the Hildebrand parameter is statistically significant and not strongly cross

Table 7
TLSER parameters[a] for solvents, solutes and residuals for solvents[b]

	V_{mc}	π_I	ε_B	q_-	ε_A	q_+	δ_H^2	logk[c]	diff[d,e]
water	0.1933	0.0581	0.1237	0.3255	0.1237	0.1628	2.2970	-5.134	0.000
methanol	0.3647	0.0860	0.1314	0.3291	0.1402	0.1803	0.8586	-3.602	0.152
formamide	0.4090	0.0965	0.1371	0.3553	0.1667	0.1572	1.5134	-3.131	-0.068
ethanol	0.5423	0.0927	0.1326	0.3235	0.1429	0.1800	0.6782	-3.000	-0.105
N-meformamide	0.5790	0.1006	0.1381	0.3437	0.1682	0.1557	0.6150	-2.092	0.027
nitromethane	0.4740	0.1093	0.1302	0.3342	0.1817	0.0500	0.6632	-0.237	0.405
acetonitrile	0.4529	0.0937	0.1177	0.1145	0.1622	0.0209	0.5766	0.462	-0.221
dimethylsulfoxide	0.7209	0.1046	0.1475	0.7196	0.1734	0.0525	0.7063	1.000	0.230
propanone	0.6441	0.0972	0.1381	0.2867	0.1715	0.0232	0.3791	1.380	0.362
(me)$_2$formamide	0.7693	0.1042	0.1441	0.4698	0.1649	0.0576	0.5812	1.568	-0.583
(me)$_4$sulfoxide[f]	1.0129	0.1192	0.1368	0.6788	0.1942	0.0814	0.7489	1.806	-2.104
N,N-(me)$_2$acetamide	0.9634	0.1026	0.1452	0.4656	0.1658	0.0285	0.4879	2.204	-0.279
N-methylpyrrolidine	1.0311	0.1053	0.1516	0.4541	0.1495	0.0169	0.5339	2.398	0.250
(me)$_6$phosphoramide	1.8436	0.1107	0.1456	0.6518	0.1874	0.0096	0.3071	2.845	-0.098
trichloromethane	0.7540	0.1114	0.1165	0.1130	0.1849	0.0876	0.3711	-3.090	1.656
tetrachloromethane	0.9058	0.1172	0.1132	0.0704	0.1912	0.0000	0.3088	-2.820	3.596
benzene	0.8463	0.1204	0.1517	0.0594	0.1744	0.0593	0.3506	-2.320	4.068
dimethoxymethane	0.8076	0.0978	0.1369	0.3304	0.1474	0.0190	0.5905	-1.440	3.198
dioxane	0.8598	0.1045	0.1379	0.3275	0.1480	0.0364	0.4184	-1.390	2.742
dichloromethane	0.6045	0.1036	0.1207	0.1605	0.1773	0.0555	0.4088	-1.330	1.033
ethoxyethane	0.9035	0.0996	0.1365	0.3423	0.1455	0.0072	0.2351	-1.050	3.318
benzonitrile	0.9984	0.1274	0.1474	0.0865	0.1833	0.0699	0.5142	0.400	0.535
tetrahydrofuran	0.7889	0.1021	0.1378	0.3270	0.1471	0.0217	0.3615	0.600	1.190
(2-meOet)$_2$ether	1.4171	0.1024	0.1157	0.3572	0.1502	0.0128	0.4000[d]	0.700	1.151

<u>y-x-3-carboxybenzisoxazole anions</u>

	V_{mc}	π_I	ε_B	q_-	ε_A	q_+			
6-NH$_2$	1.3385	0.1356	0.1982	0.5807	0.1531	0.0971			
6-H	1.2249	0.1328	0.1980	0.5709	0.1518	0.0784			
6-CH$_3$O	1.4657	0.1329	0.1978	0.5741	0.1529	0.0840			
6-Cl	1.3791	0.1348	0.1959	0.5688	0.1565	0.0885			
6-NO$_2$	1.3795	0.1430	0.1935	0.5683	0.1648	0.0920			
5-NO$_2$	1.3888	0.1406	0.1940	0.5740	0.1628	0.1129			
5,6-(NO$_2$)$_2$	1.5638	0.1460	0.1900	0.5635	0.1715	0.1172			

[a]δ_H^2 units are Joule/m^3, other units in Table 2. [b]The first 14 solvents were used for the majority of solutes, see Table 6. The next 10 solvents in addition to the first 14 were used for the 6-NO$_2$-compound and N,N-dimethyl formamide. [c]for the 6-NO$_2$ compound. [d]diff = (logk calculated from equation for 6-NO$_2$) − (empirical logk). [e]estimated. [f]tetramethylene sulfoxide.

correlated with the other parameters. This implies that the Hildebrand solubility parameter can serve as the model for solvent cohesive effects along with the TLSER parameters.

Table 6 provides an 81 point combined solvent-solute data set while Table 7 gives 10 more. Equation (13) is the correlation equation for the $N = 81$ case.

$$logk = 319.0 \, \pi_{I1} \pi_{I2} + 412.1 \, \varepsilon_{B1} \varepsilon_{A2} - 40.89 \, q_{+1} q_{-2} - 12.04 \quad (13)$$

±	61.1	70.1	2.82	1.36
t-stat	5.22	5.88	14.49	8.83
P(2-tail)	0.000	0.000	0.000	0.000
VIF	1.84	1.74	1.41	

$$N = 81 \qquad R = 0.945 \qquad SD = 0.809 \qquad F = 214$$

The larger sample space leads to greater statistical reliability. Examination of the results shows that not all of the descriptors are significant. Furthermore, the correlations are good. Correlation coefficient values show that more than 80% of the variance is accounted for with $R > 0.945$ for most cases. Again, SD values indicate that most of the equations provide rate constant, k, values within a factor of 2 to 6 of the empirical values over a range as high as 10^8. As a fraction of the range for logk, the SD values correspond to $0.04 < SD/range < 0.1$; that is, 4 to 10%. Most terms are significant at well over the 0.96 level, the worst cases are for the intercepts of solvent-multiple solute equations which have a very small set of data points, anyway. The VIF values (1.05 to 1.84) indicate that the parameter sets have low cross correlation.

When the 7X10 block in Table 6 is analyzed, removing the 6-NO_2 and 5,6-(NO_2) compounds, R increases to 0.969 and the same terms are significant. When the full $N = 91$ set is examined R decreases to 0.901 and the solvent HBB-solute HBA term, $q_{-1}{}^*q_{+2}$, becomes significant with the same sign as the covalent solvent HBB-solute HBA term. Consequently the three equations are physically consistent.

Equation (13) is consistent with the experimental results discussed earlier. The negative solvent acidity (HBA) term implies that the rate decreases with solvent HBA. Furthermore, it is the statistically most significant term with t-statistic values greater than 10. This decrease in rate with increasing solvent HBA was the most prominent feature of the data of Kemp and Paul. The solvent polarizability term is positive indicating that it increases the rate. Polarizability contributes to dispersive interactions and experimental results suggest that solvent polarizability increases the rate.

It is appropriate to comment on these results with respect to the three mechanisms proposed by Kemp and Paul and discussed earlier. The rate decreases with increasing solvent HBA; the TLSER solvent HBA term is negative. The rate increases with increasing dispersive interactions; the LSER polarizability term is positive. However, it is not clear that this is due to stabilization of the transition state structure as proposed by Kemp and Paul. The apparent increase in rate with solvent polarizability could also imply fewer ion pairs through greater solvent-Bnzx⁻ complex formation. The anion is quite polarizable and a polarizable solvent could tend to separate it to form the counter ion. This separation can be enhanced by the larger size that accompanies increased polarizability. The role of ion pairs in influencing the rate can be inferred from the LSER solvent basicity term being positive. Increased solvent basicity could imply greater interaction of solvent and the tetramethylguanidinium (TMGH⁺) ion. This could diminish the (TMGH⁺..Bnzx⁻) tetramethylguanidinium benzisoxazolate pair concentration; consequently, there would be more unpaired Bnzx⁻ ions to move into the transition state.

3.6. Comparison of TLSER and molecular transform. A multiple solute set.

A classical example of the success of the QSAR approach is the correlation of the minimum blocking concentration for compounds used as anesthetics. The process involved can be modelled as in the first example involving toxicity. Famini and coworkers applied the TLSER descriptors and the molecular transform, the latter was mentioned with the theoretical descriptors in the QSAR section earlier, to the 36 compounds in Table 8. The molecular transform provides a complex description of a molecule condensed to a single number; here it was used in the form of the square root of the area under the Fourier Transform curve (SQRT) since that gave a better fit. Very good correlation equations were obtained with the both the SQRT and TLSER parameter set. The TLSER result was the better of the two and is given in equation (14) with a reduced set of statistical parameters.

$$logMBC = -2.363\ V_{mc2} \quad -1.727\ \pi_{I2} \quad +5.590 \tag{14}$$

t-stat	35.1	4.88
VIF	1.1	1.1

$$N = 36 \qquad R = 0.989 \qquad SD = 0.311 \qquad F = 787$$

This is an excellent fit. The SD value corresponds to a factor of two in the minimum blocking concentration for which the range was $-5 < logMBC < 3$. There were no outliers. The TLSER descriptors also correlated very well ($R = 0.991$) with the molecular transform. The same two parameters were significant with the polarizability having an even greater t-statistic, 8.82.

The very strong dependence on the volume indicates that the bulk/steric effect is most important; larger molecules decrease MBC. One interpretation is that the molecule interferes with what is happening near the membrane surface. The polarizability term also suggests that increased dispersion interactions also contribute to this interference. That is also consistent with greater lipophilicity facilitating interaction with corresponding lipophilic membrane sites. Perhaps the presence of these compounds interferes with process involving electrolytes near their membrane location.

Equation (15) gives the fit using the SQRT; just the minimum set of statistical parameters are included.

Table 8
Significant TLSER parameters for minimum blocking concentrations

| | V_{mc} | π_I | logMBC[a] | | |
			Observed	Calculated	Residual
methanol	0.3669	0.0859	3.09	3.24	-0.15
ethanol	0.5564	0.0909	2.75	2.71	0.04
propanone	0.6385	0.0960	2.60	2.42	0.18
2-propanone	0.7149	0.0955	2.55	2.25	0.30
propanol	0.7201	0.0976	2.40	2.20	0.20
urethane	0.8348	0.1023	2.00	1.85	0.15
ethoxyethane	0.9058	0.0993	1.93	1.73	0.20
butanol	0.8927	0.0976	1.78	1.79	-0.01
pyridine	0.7871	0.1210	1.77	1.64	-0.13
hydroquinone	0.9737	0.1263	1.40	1.11	0.30
aniline	0.9687	0.1246	1.30	1.15	0.15
phenylmethanol	1.0798	0.1201	1.30	0.96	0.34
pentanol	1.0752	0.1003	1.20	1.32	-0.12
phenol	0.9018	0.1238	1.00	1.32	-0.32
toluene	1.0122	0.1201	1.00	1.12	-0.12
benzimidazole	1.0762	0.1350	0.81	0.72	0.09
hexanol	1.3509	0.1002	0.56	0.67	-0.11
nitrobenzene	1.0016	0.1328	0.47	0.93	-0.46
quinoline	1.2285	0.1388	0.30	0.29	0.01
8-(OH)quinoline	1.2904	0.1411	0.30	0.10	0.20
heptanol	1.4332	0.1007	0.20	0.46	-0.26
2-naphthol	1.3431	0.1413	0.00	-0.02	0.02
methylanthranilate	1.3752	0.1245	0.00	0.19	-0.19
octanol	1.5970	0.1022	-0.16	0.05	-0.21
thymol	1.5990	0.1206	-0.52	-0.27	-0.25
o-phenanthroline	1.6218	0.1500	-0.80	-0.83	0.03
ephridine	1.7382	0.1156	-0.80	-0.51	-0.29
procaine	2.3732	0.1203	-1.67	-2.10	0.43
lidocaine	2.5062	0.1157	-1.96	-2.33	0.37
(ø)$_2$hydramine	2.6469	0.1226	-2.80	-2.78	-0.02
tetracaine	2.5779	0.1195	-2.90	-2.57	-0.33
øtoloxamine	2.6612	0.1240	-3.20	-2.84	-0.36
quinine	3.2273	0.1279	-3.60	-4.25	0.65
physostigmine	2.6792	0.1255	-3.66	-2.91	-0.75
caramiphen	3.0572	0.1150	-4.00	-3.62	-0.38
dibucaine	3.4766	0.1238	-4.20	-4.76	0.56

[a]MBC = minimum blocking concentration in mol/L.

$$\log \text{MBC} = -1.328 \text{ SQRT}*10^{-2} + 3.379 \tag{15}$$

$$N = 36 \qquad R = 0.977 \qquad SD = 0.457 \qquad F = 710$$

The two good correlations suggest that the SQRT should correlate well with the TLSER parameters. This is indeed the case as is shown in the next equation, (16), and helps provide some understanding of the molecular transform in terms of the more readily interpreted TLSER parameters.

$$\text{SQRT} = 1.663 \text{ V}_{mc2} + 208.2 \, \pi_{I2} \quad -240.17 \tag{16}$$

t-stat	37.0	8.82	
VIF	1.1	1.1	

$$N = 36 \qquad R = 0.991 \qquad SD = 20.7 \qquad F = 958$$

This indicates that for this set of compounds the molecular transform incorporates both a bulk/steric effect, from the volume and the polarizability of the molecule. This latter feature shows that the transform contains some electronic information about the molecule.

4. CONCLUSIONS

Based on the examples illustrated here and previous investigations the TLSER parameters seem to be applicable to a wide range of properties involving solvent-solute interactions. They have provided statistically significant relations and physically reasonable results for a wide range of such properties. Furthermore, they have shown good agreement with the results from the empirical LSER parameters. Because the TLSER parameters have quite well defined molecular meanings they can serve as a probe to help suggest a picture of what is happening between solvent and solute at the molecular level. Furthermore, these results taken in conjunction with other LSER studies point out that the essential LSER viewpoint, as modelled by equations (1) and (3), seem to provide a very good model for solvent-solute interactions. In addition, the ready availability of the Hildebrand solubility parameter and its significance and low cross correlation in the cases studied here suggests that it can be a convenient parameter to model the solvent cohesive interactions in the TLSER approach.

REFERENCES

1. S. Gupta, *Chem. Rev.*, 87 (1987) 1183.
2. G. N. Burkhardt, *Nature (London)*, 17 (1935) 684.
3. L. P. Hammett, *Chem. Rev.*, 17 (1935) 125.
4. L. P. Hammett, *J. Am. Chem. Soc.*, 59 (1937) 125.
5. O. Exner, *Correlation Anaylsis of Chemical Data*, Plenum Press, New York, 1988, p. 25.
6. C. Hansch, *Acc. Chem. Res.*, B2 (1969) 232.
7. M. J. Kamlet, R. W. Taft and J.-L. M. Abboud, *J. Am. Chem. Soc.*, 91 (1977) 8325.

240

8.   I. A. Koppel and V. A. Palm, *Advances in Linear Free Energy Relationships*, N. B. Chapman and J. Shorter (eds.), (Plenum Press, London, 1972)

9.   M. J. Kamlet, R. M. Doherty, M. H. Abraham and R. W. Taft, *Quant. Struct.-Act. Relat.*, 7 (1988) 71.

10. M. H. Abraham, J. Andonian-Haftan, M. D. Chau, V. Diart, G. S. Whiting, J. W. Grate and R. A. McGill, Submitted.

11. M. H. Abraham, G. S. Whiting, R. Fuchs and E. J. Chambers, *J. Chem. Soc. Perkin Trans. 2*, (1990) 291.

12. M. J. Kamlet, R. W. Taft, G. R. Famini and R. M. Doherty, *Acta Chem. Scand.*, 41 (1987) 589.

13. J. P. Hickey and D. R. Passino-Reader, *Environ. Sci. Technol.*, 25 (1991) 1753.

14. J. J. Gajewski, *J. Org. Chem.*, 57 (1992) 5500.

15. (a) J. G. Kirkwood, *J. Chem. Phys.*, 2 (1934) 351.

     (b) L. Onsager, *J. Am. Chem. Soc.*, 58 (1936) 1486.

16. J. H. Hildebrand, J. M. Prausnitz and R. L. Scott, *Regular and Related Solutions*, Van Nostrand-Reinhold, Princeton, 1970.

17. (a) B. G. Cox, G. R. Hedwig, A. J. Parker and D. W. Watts, *Aust. J. Chem.*, 27 (1974) 477.

     (b) Y. Marcus, *Pure Appl. Chem.*, 55 (1983) 977.

18. M. G. Ford and D. J. Livingstone, *Quant. Struct.-Act. Relat.*, 9 (1990) 107.

19. D. F. V. Lewis, *J. Comp. Chem.*, 8 (1987) 1084.

20. T. Brinck, J. S. Murray and P. Politzer, *Molecular Physics*, 76 (1992) 609.

21. J. S. Murray and P. Politzer, *J. Chem. Research*, 5 (1992) 110.

22. G. H. Loew, M. Poulsen, E. Kirkjian, J. Ferrell, B. S. Sudhindra and M. Rebagliati, *Environ. Health Perspect.*, 61 (1985) 69.

23. L. Pederson, *Environ. Health Perspect.*, 61 (1985) 185.

24. M. Chastrette, M. Rajzmann, M. Chanon and K. F. Purcell, *J. Am. Chem. Soc.*, 107 (1985) 1.

25. D. F. V. Lewis, *Progress in Drug Metabolism*, J. W. Bridges and L. F. Chasseaud (eds.), John Wiley, London, 1990, p. 205.

26. L. B. Kier and L. Hall, *Molecular Connectivity in Structure-Activity Analysis*, Research Studies Press, Letchworth, 1986.

27. J. W. King, R. J. Kassel and B. B. King, *Int. J. of Quantum Chem., Quant. Biol. Symp.*, 17 (1990) 27.

28. (a) M. W. Wong, K. B. Wiberg and M. J. Frisch, *J. Am. Chem. Soc.*, 113 (1991) 4776.

     (b) M. W. Wong, K. B. Wiberg and M. J. Frisch, *J. Am. Chem. Soc.*, 114 (1992) 523.

     (c) M. W. Wong, K. B. Wiberg and M. J. Frisch, *J. Am. Chem. Soc.* 114 (1992) 1645.

29. F. Peradejordi, A. N. Martin and A. Cammarata, *J. Pharm. Sci.*, 60 (1971) 576.

30. G. R. Famini, *Using Theoretical Descriptors in Quantitative Structure Activity Relationships*, V. CRDEC-TR-085, US Army Chemical, Research, Development and Engineering Center, Aberdeen Proving Ground, MD, 1989.

31. L. Y. Wilson and G. R. Famini, *J. Med. Chem.*, 34 (1991) 1668.

32. G. R. Famini, R. J. Kassel, J. W. King and L. Y. Wilson, *Quant. Struct.-Act. Relat.*, 10 (1991) 344.

33. G. R. Famini, W. P. Ashman, A. P. Mickiewicz and L. Y. Wilson, *Quant. Struct.-Act. Relat.*, 11 (1992) 162.

34. G. R. Famini, C. E. Penski and L. Y. Wilson, *J. Phys. Org. Chem.*, 5 (1992) 395.

35. G. R. Famini, B. C. Marquez and L. Y. Wilson, *J. Chem. Soc. Perkin Trans. 2*, (1993) 773.

36. G. R. Famini and L. Y. Wilson, *J. Phys. Org. Chem.*, 6 (1993) 539.

37. S. C. De Vito, G. R. Famini and L. Y. Wilson, *Chemical Res. Toxicol.*, in press.

38. J. M. Leornard and G. R. Famini, *A User's Guide to the Molecular Modeling Analysis and Display System*, CRDEC-R-030, US Army Chemical Research, Development and Engineering Center, Aberdeen Proving Ground, MD, 1989.

39. M. J. K. Dewar and W. Thiel, *J. Am. Chem. Soc.*, 99 (1977) 4899.

40. J. J. P. Stewart, *Mopac Manual*, FJSRL-TR-88-007, Frank J. Seiler Research Laboratory, US Air Force Academy, Colorado Springs, CO, 1988.

41. A. J. Hopfinger, *J. Am. Chem. Soc.*, 102 (1980) 7126.

42. D. A. Belesley, E. Kuh and R. E. Welsh, *Regression Diagnostics*, Wiley, New York, 1980.

43. H. Könemann, *Toxicity*, 19 (1981) 209.

44. D. Dorohoi, D. Iancu and G. Surpataneau, *Analele Stiinifico de Universittii, Al. I Cuza din Iasi 1b*, 27 (1981) 59.

45. D. Dorohoi, D. Iancu and G. Surpataneau, *Analele Stiinifico de Universittii, Al. I Cuza din Iasi 1b*, 29 (1983) 27.

46. D. S. Kemp and K. G. Paul, *J. Am. Chem. Soc.*, 97 (1975) 7305.

47. D. S. Kemp, D. D. Cox and K. G. Paul, *J. Am. Chem. Soc.*, 97 (1975) 7312.

48. J. W. Grate, R. A. McGill and D. Hilvert, *J. Am. Chem. Soc.*, 115 (1993) 8577.

49. J. Smid, A. Varma and S. C. Shah, *J. Am. Chem. Soc.*, 101 (1979) 5764.

50. M. Shirai and J. Smid, *J. Am. Chem. Soc.*, 102 (1980) 2863.

51. (a) I. M. Kolthoff, *Anal. Chem.*, 46 (1974) 1992.
    (b) M. K. Chantooni and I. M. Kolthoff, *J. Phys. Chem.*, 80 (1976) 1307.

52. C. Lewis, P. Paneth, M. H. O'leary and D. Hilvert, *J. Am. Chem. Soc.*, 115 (1993) 1410.

53. G. Famini, J. W. Grate and L. Y. Wilson, submitted.

P. Politzer and J.S. Murray
Quantitative Treatments of Solute/Solvent Interactions
Theoretical and Computational Chemistry, Vol. 1

A General Interaction Properties Function (GIPF): An Approach to Understanding and Predicting Molecular Interactions

Jane S. Murray and Peter Politzer

Department of Chemistry, University of New Orleans, New Orleans, Louisiana, 70148, USA

1. INTRODUCTION

We have developed a general approach that permits the analysis, correlation and prediction of macroscopic properties that are determined by molecular interactions in fluid media; these are most often noncovalent in nature, but can also include charge transfer/polarization. Among the properties that we have been able to treat are boiling points, critical temperatures, pressures and volumes, solution and gas phase acidities, heats of vaporization, solubilities in supercritical fluids, partition coefficients and hydrogen-bonding parameters.

Our methodology can be summarized by the General Interaction Properties Function (GIPF) described by eq. (1) [1]. GIPF is a collective term that we apply to a group of relationships involving one or more of the computed quantities within brackets in eq. (1). The latter will be defined and discussed in detail in section 2; what they represent physically is shown in Figure 1. The molecule's surface area is a measure of its size;

$$\text{Property} = f\left[\text{surface area}, \bar{I}_{S,min}, V_{S,max}, V_{S,min}, \Pi, \sigma^2_{tot}, \nu\right] \qquad (1)$$

$\bar{I}_{S,min}$ reflects the tendency for charge transfer [2]; $V_{S,max}$ and $V_{S,min}$ are indicators of long-range attraction for nucleophiles and electrophiles, respectively [2, 3]; Π and σ^2_{tot} represent local polarity [4] and the variability [5-7] of the surface electrostatic potential, respectively, and ν is an "electrostatic balance" term [7, 8].

Applications of equation (1) involve the use of some subset of these computed molecular quantities in analytical expressions representing the specific physical or chemical property of interest. Examples will be discussed in section 3 of this chapter. A key point that should be stressed is that we are able to use quantities evaluated computationally for an isolated molecule to predict properties of condensed phases, e.g. liquids and solutions. As the range of applications of the GIPF is expanded, it is likely that additional molecular quantities may be needed, besides the seven included in eq. (1). Thus we anticipate that the GIPF approach will be continually evolving.

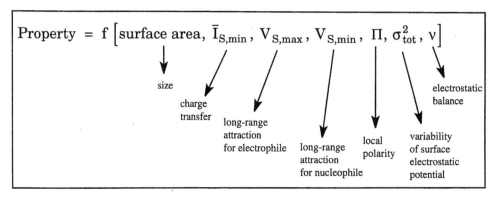

Figure 1. The General Interaction Properties Function (GIPF), with the physical significance of each computed quantity.

2. MOLECULAR QUANTITIES APPEARING IN THE GIPF

2.1. General:

Our General Interaction Properties Function (GIPF), shown in eq. (1) and Figure 1, involves quantities which are evaluated on the molecular surface. This is defined, following Bader et al [9], as the 0.001 au contour of the electronic density $\rho(\mathbf{r})$. Such surfaces have been shown to contain more than 95% of the electron density of a molecule, and to give meaningful molecular shapes and dimensions [10]. They are determined by a molecular property, $\rho(\mathbf{r})$, and consequently show features such as lone pair regions and "bent" bonds that would not appear for surfaces defined by additive atomic volumes, e.g. spheres corresponding to atomic van der Waals radii. The area used in eq. (1) is obtained by establishing a 0.28 bohr grid on the surface and converting the number of points on the grid to units of Å^2 [5-8].

Since most of the quantities in eq. (1) are derived from the electrostatic potential $V(\mathbf{r})$ on the molecular surface, we will begin with a brief discussion of this fundamental property and some examples of its extensive past use in analyzing molecular interactions. We will then proceed to the GIPF quantities that are related to $V(\mathbf{r})$, and also to the one that is not, $\bar{I}_{S,min}$.

2.2. The Electrostatic Potential V(r)

2.2.1. Definition

The electrostatic potential $V(\mathbf{r})$ has emerged over the last two decades as an effective tool for analyzing the reactive behavior of molecules in both electrophilic and nucleophilic processes [11-16]. It is particularly well suited for analyzing intermolecular interactions that do not involve any significant degree of charge transfer or polarization [1, 2, 14-16]. $V(\mathbf{r})$ is a real physical property expressing the net electrical effect of the nuclei and electrons of a molecule at any point $\mathbf{r}$ [13], and is given rigorously by eq. (2).

$$V(\mathbf{r}) = \sum_{A} \frac{Z_A}{|\mathbf{R}_A - \mathbf{r}|} - \int \frac{\rho(\mathbf{r}')\,d\mathbf{r}'}{|\mathbf{r}' - \mathbf{r}|} \tag{2}$$

Z_A is the charge on nucleus A, located at $\mathbf{R}_A$, and $\rho(\mathbf{r})$ is the electronic density function of the molecule. The first term on the right side of eq. (2) gives the effects of the nuclei and is positive; the second term reflects the presence of the electrons and is negative. The sign of $V(\mathbf{r})$ in any particular region therefore depends upon whether the effects of the nuclei or the electrons are dominant there.

In past applications of the electrostatic potential to the interpretation and prediction of molecular reactivity, the focus has been primarily upon (a) the locations and magnitudes of the most negative values of $V(\mathbf{r})$ (i.e. its minima in three-dimensional space, not restricted to any particular surface), and (b) the qualitative nature of the pattern of positive and negative regions. Both types of applications will now be discussed briefly, in sections 2.2.2 and 2.2.3. The GIPF makes use of both the maxima and the minima of $V(\mathbf{r})$ *on the molecular surface*, but includes in addition some quantitative statistically-based measures of the distributions and strengths of the positive and negative potentials over the entire surface. These will be defined and discussed in section 2.2.4.

2.2.2. Spatial Minima and Surface Extrema

Molecular sites reactive toward electrophiles can be identified and ranked by means of either the spatial minima (V_{min}) [3, 11, 12, 14-17] or the minima on the molecular surface ($V_{S,min}$) [2]; however only the maxima of $V(\mathbf{r})$ on the molecular surface ($V_{S,max}$) serve the analogous purpose for nucleophilic attack [3, 18]. The spatial maxima of the electrostatic potential are found only at the positions of the nuclei [19], and reflect the magnitudes of the nuclear charges rather than relative reactivities toward nucleophiles. We have shown, however, that potential maxima on the molecular surface ($V_{S,max}$) do act as good indicators of tendencies for nucleophilic interactions [3, 18, 20]. By definition, V_{min}, $V_{S,min}$ and $V_{S,max}$ are site-specific quantities; they refer to particular points in the space of a molecule, and are of key importance in the analysis of physical and/or chemical properties that can be viewed as being largely site dependent, e.g. hydrogen bonding [3, 16-18], and acidity [2].

While V_{min} and $V_{S,min}$ play similar roles, the former is normally located inside the molecular surface, by which point an approaching electrophile would have significantly polarized the molecule's electronic charge distribution. Since this is not being taken into account, we feel that it is more relatistic to use $V_{S,min}$ in the GIPF.

2.2.3. Patterns of Positive and Negative V(r)

The electrostatic potential surrounding a molecule inherently contains more useful information than merely the locations and magnitudes of the spatial minima and the surface extrema (V_{min}, $V_{S,min}$ and $V_{S,max}$). Both the overall pattern of positive and negative regions and a knowledge of the sites and magnitudes of the extrema are important in analyzing and predicting a molecule's reactive behavior. For example, Figures 2 and 3 show the calculated electrostatic potential of acridine (**1**), plotted first in a plane above the rings and then on the 0.001 au molecular surface. The figures show that both V_{min} and $V_{S,min}$ are in the vicinity of the nitrogen; however this is

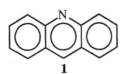

1

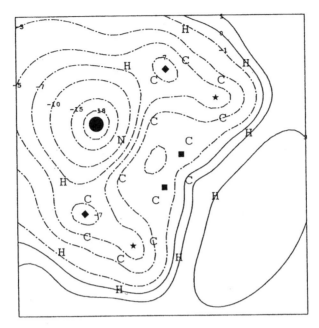

Figure 2. Calculated electrostatic potential of acridine (**1**), in kcal/mole, in a plane 1.75 Å above the molecular plane. The projections of the nuclear positions are shown by their atomic symbols. Dashed contours correspond to negative potentials. The positions of the most negative V(**r**) are indicated. The values are: ●-19.7; ◆-7.7; ★-5.9; ■-1.4.

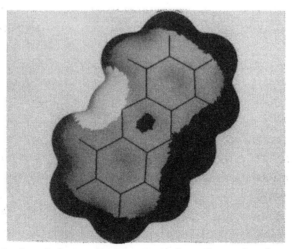

Figure 3. Calculated electrostatic potential on the molecular surface of acridine (**1**). V(**r**), in kcal/mole: black, greater than 0; gray, between 0 and -10; white, more negative than -10.

not the sole factor determining acridine's electrostatic interactions. **1** has negative potentials above and below the planes of its six-membered rings, attributed to the pi electrons; these indicate additional regions that are favorable for electrophilic interactions. $V(\mathbf{r})$ is positive around the edge of the molecule, in the vicinity of its hydrogens (Figures 2 and 3).

Two specific examples of the use of $V(\mathbf{r})$ patterns in understanding intermolecular interactions will be discussed here. These involve (a) some dibenzo-*p*-dioxins and structural analogues, in relation to their biological activities, and (b) some diphenyl ureas, in relation to their observed hydrogen bonding tendencies. (These discussions will be qualitative in nature and not significantly dependent upon knowledge of the $V(\mathbf{r})$ extrema.)

In earlier studies, we computed the electrostatic potentials of dibenzo-*p*-dioxin (**2**), a variety of halogenated dibenzo-*p*-dioxins, including the highly toxic 2,3,7,8-tetrachlorodibenzo-*p*-dioxin (TCDD, **3**), and several structural analogues of TCDD, e.g. **4** and **5** [15, 16, 21]. In the light of evidence indicat-

2

3

4

5

ting that the toxic effects produced by the dibenzo-*p*-dioxins are mediated by a specific receptor [22, 23], we sought to identify key features of their electrostatic potentials that favor the initial recognition process involved in this receptor interaction.

We were able to show that high toxicity is associated with: (a) negative regions over all or most of the lateral portions of the molecules (positions 2, 3, 7 and 8), with intermediate magnitudes being optimum, and (b) positive potentials over the central portions [15, 16, 21]. The latter requires that the negative regions due to the oxygens be small and weak, as is the case for TCDD, **3** [21]. In contrast, the V(**r**) map of dibenzo-*p*-dioxin (**2**), which is non-toxic, shows strongly negative potentials associated with the oxygens overlapping weaker negative ones over the benzene rings, while the lateral regions are positive [21]! The requirement that the oxygen negative regions be small and weak may reflect a need to avoid repulsive interactions with negative potentials of the receptor binding site.

We emphasize that it is this V(**r**) pattern that appears to be linked to high biological activity, whatever may be the particular molecule producing it. For example, **4** has a V(**r**) distribution that closely mimics that of TCDD (**3**), and indeed **4** is nearly as toxic as **3** [22], even though it has no oxygens. On the other hand, even though the structure and the V(**r**) pattern of **5** do resemble **3**, it is much less active [23]. The negative oxygen regions of **5** protrude from the sides of the molecule and are therefore quite pronounced, in marked contrast to those of **3** (described above). We have speculated that the biological inactivity of **5** may be due in part to its oxygen negative potentials being in a position to inhibit an attractive interaction with the receptor.

1,3-bisphenylurea (**6**) is the parent compound of a large family of derivatives, most of which do <u>not</u> cocrystallize with guest molecules [24]. Even

6

6A

7

when put in solution with strong hydrogen bond acceptors, e.g. dimethyl sulfoxide (DMSO), tetrahydrofuran (THF) and triphenylphosphine oxide (TPPO), most diphenyl ureas crystallize with their own kind in a connectivity pattern such as **6A** instead of forming cocrystals (e.g. **7**, which involves THF).

A striking exception to this trend is 1,3-bis(*m*-nitrophenyl)urea (**8**), which does preferentially form cocrystals with hydrogen bond acceptors of at least intermediate strength, such as diethyl ether, acetone, cyclohexanone, DMSO, THF and TPPO, if they are available [24]; in these cocrystal structures, **8** is planar. (**8** is also different in that none of its connectivity patterns for homocrystallization are analogous to **6A**.)

8

We have shown that the tendency for 1,3-bisphenylurea to form homomeric crystals rather than guest-host cocrystals can be attributed largely to a relatively strong electrostatic attraction between extended regions of negative and positive potential along the top and bottom portions, respectively, of molecules of **6** [25]. We view this as a more extensive and global interaction than simply the site-specific hydrogen bonding that is depicted in **6A**, and we believe that it gives added stability to these homomeric crystals that would not be anticipated from hydrogen bonding considerations alone. In the case of **8**, on the other hand, the same type of interaction is hindered by some degree of intramolecular hydrogen bonding between the carbonyl oxygen and the neighboring aryl hydrogens, and also by the steric effects of the NO_2 groups [25]. This may explain its preference for cocrystal hydrogen bonding.

The examples of the dibenzo-*p*-dioxins and the diphenyl ureas demonstrate the importance of considering the overall $V(\mathbf{r})$ pattern in understanding the interactions of a molecule, whether with its own kind or with another chemical species, e.g. a receptor, a hydrogen-bond donor or acceptor, etc. One of the advantages of plotting the potential on the molecular surface rather than in a plane through the molecule, as was commonly done in the past [11-

15], is that it eliminates the uncertainty related to choosing the most appropriate plane(s) in which to compute V(**r**); this can be a very significant point for systems with considerable asymmetry. Surface potentials are particularly important in investigating recognition processes, since they can reveal key steric features, as for example in the activities of **5** and **8** [21, 25].

2.2.4. Statistically-based Interaction Indices Computed From the Entire Surface Electrostatic Potential

Our interpretations of dibenzo-*p*-dioxin toxicities and of diphenyl urea crystal structures are based upon the global natures of their surface electrostatic potentials. While this approach, a qualitative assessment of similarities and differences in V(**r**) patterns, was successful, it does not allow quantitative predictions or the establishment of 'structure-activity relationships. In the course of the past two years, we have accordingly made efforts to develop quantitative measures of key aspects of the *entire* surface potentials of molecules, which would reflect the overall V(**r**) distribution. We have introduced three global statistical quantities defined in terms of a molecule's surface electrostatic potential. These are:

(a) Π, which reflects local polarity [4];

(b) σ_{tot}^2, related to the variability of the electrostatic potential on the surface [5, 6]; and

(c) ν, which indicates the "balance" between a molecule's positive and negative regions of V(**r**) (whether these be strong or weak) [7, 8].

These three global quantities, all of which appear in the GIPF, are defined in eqs. (3) - (5), and shall be discussed in turn. V(**r**$_i$) is the value of V(**r**) at point **r**$_i$

$$\Pi = \frac{1}{n} \sum_{i=1}^{n} \left| V(\mathbf{r}_i) - \overline{V}_S \right| \tag{3}$$

$$\sigma_{tot}^2 = \sigma_+^2 + \sigma_-^2 = \frac{1}{m} \sum_{i=1}^{m} \left[V^+(\mathbf{r}_i) - \overline{V}_S^+ \right]^2 + \frac{1}{n} \sum_{j=1}^{n} \left[V^-(\mathbf{r}_j) - \overline{V}_S^- \right]^2 \tag{4}$$

$$\nu = \frac{\sigma_+^2 \sigma_-^2}{\left[\sigma_{tot}^2 \right]^2} \tag{5}$$

on the surface, and $\bar{V}_S$ is the average value of the potential on the surface: $\bar{V}_S = \frac{1}{n} \sum_{i=1}^{n} V(\mathbf{r}_i)$. In a similar fashion, $V^+(\mathbf{r}_i)$ and $V^-(\mathbf{r}_j)$ are the positive and negative values of $V(\mathbf{r})$ on the surface, and $\bar{V}_{S^+}$ and $\bar{V}_{S^-}$ are their averages: $\bar{V}_{S^+} = \frac{1}{m} \sum_{i=1}^{m} V^+(\mathbf{r}_i)$ and $\bar{V}_{S^-} = \frac{1}{n} \sum_{j=1}^{n} V^-(\mathbf{r}_j)$.

2.2.4.1. Π, a measure of local polarity

Π, as given by eq. (3), is equal to the average deviation of the electrostatic potential on the molecular surface. Eq. (3) is our working version of the more rigorous eq. (6), in which the surface integral represents the total amount by which the electrostatic potential on the molecular surface deviates from

$$\Pi = \frac{1}{A} \int_S \left| V(\mathbf{r}) - \bar{V}_S \right| dS \tag{6}$$

its average value, $\bar{V}_S$ [4]. The integral in eq. (6) is divided by the surface area (and the summation in eq. (3) is divided by the number of surface points) in order to permit comparisons between molecules of different sizes.

Our development of a local polarity scale was prompted by the many examples of molecules with zero dipole moments but nevertheless considerable internal charge separation (e.g. BF_3, *para*-dinitrobenzene and *s*-tetrazine), which would be anticipated to play an important role in determining their interactive behavior. Some insight into internal charge separation can of course be obtained from the magnitudes of the multipole moments. However, Π provides a single very direct quantitative measure of it and is based upon a physical manifestation of the distribution of electronic and nuclear charge, i.e. the electrostatic potential $V(\mathbf{r})$.

It follows directly from eqs. (3) and (6) that any spherically symmetric system, such as a ground state atom [26, 27], has $\Pi = 0$, since $V(\mathbf{r}) = V(r) = \bar{V}_S$ for any surface of radius r. As charge separation increases, so does the average deviation of the surface potential, Π [4]. For most organic molecules, Π ranges from 2 to 20 kcal/mole. To demonstrate the relationship of Π to local polarity, or internal charge separation, Figures 4 and 5 show the surface electrostatic potentials of cyclohexane, benzene, boron trifluoride and water. The first three of these have zero dipole moments, by symmetry. The surface

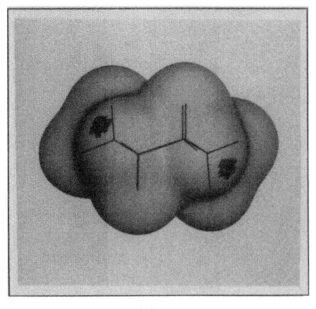

cyclohexane

$\Pi = 2$
(zero dipole
moment)

V(**r**) ranges (kcal/mole):
BLACK: Greater than 5.
GRAY: Between -5 and 5.
WHITE: More negative than -5.

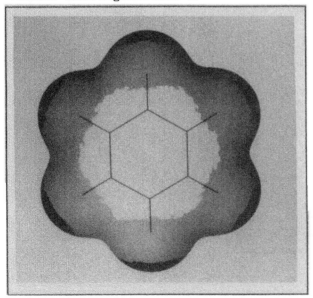

benzene

$\Pi = 5$
(zero dipole
moment)

Figure 4.   Calculated surface electrostatic potentials of cyclohexane and
benzene.

254

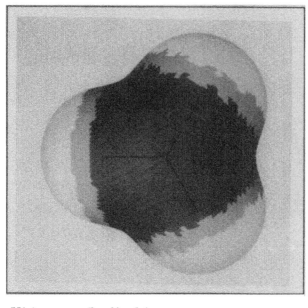

boron trifluoride

F
|
B
F   F

Π = 14
(zero dipole
moment)

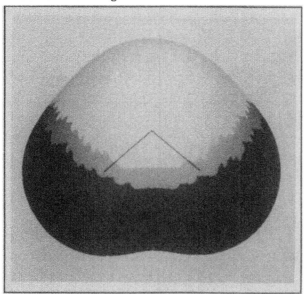

water

O
H   H

Π = 22

Figure 5. Calculated surface electrostatic potentials of brown trifluoride and water.

potential of cyclohexane suggests very little local polarity; it is relatively uniform everywhere. However in proceeding to benzene, boron trifluoride and water, there is clearly an increasing internal separation of charge, indicated by the growing regions of relatively strong positive and negative potentials. This trend is reflected in the Π values: cyclohexane, 2.16; benzene, 4.83; boron trifluoride, 13.99; and water, 21.46.

We have shown that Π correlates with several polarity-polarizability scales and with the dielectric constant [1, 4]. It is also important in repre-senting, via our GIPF equation, octanol/water and acetonitrile/NaCl-saturated-water partition coefficients [1, 28, 29]. These applications will be discussed in a later section.

2.2.4.2 The total variance, σ_{tot}^2, a measure of the variability of the surface potential

The total variance, σ_{tot}^2, given by eq. (4) as a sum of the positive and negative variances, is a measure of the spread, or range of values, of the surface potential. Because the terms are squared, σ_{tot}^2 is particularly sensitive to variations in the magnitude of $V(\mathbf{r})$, emphasizing positive and negative extremes [5-7]. We interpret it as indicative of a molecule's tendency for noncovalent electrostatic interactions. For example, it is effective (in conjunction with molecular volume or surface area) in correlating solubilities in supercritical CO_2, apparently because it reflects solute-solute effects [5, 6]. As will be seen, for some applications it is advantageous to use σ_+^2 or σ_-^2 alone, rather than σ_{tot}^2. When this is found to be the case, it provides additional insight into the natures of the interactions that are involved.

In contrast to Π, the magnitudes of σ_{tot}^2 cover quite a wide range. Figure 6 shows the striking increase in the variability of the surface electrostatic potential in going from naphthalene to indole to indole-3-carboxylic acid. σ_{tot}^2 increases in the same order from 15.9 to 96.6 to 182.9, whereas the corresponding values of Π are 5.12, 8.39 and 11.09. Indeed, Π and σ_{tot}^2 do not necessarily vary in the same direction, as can be seen from a number of examples [1,5].

It was in the course of our initial study of possible relationships between computed solute properties and solubility in several supercritical solvents that

256

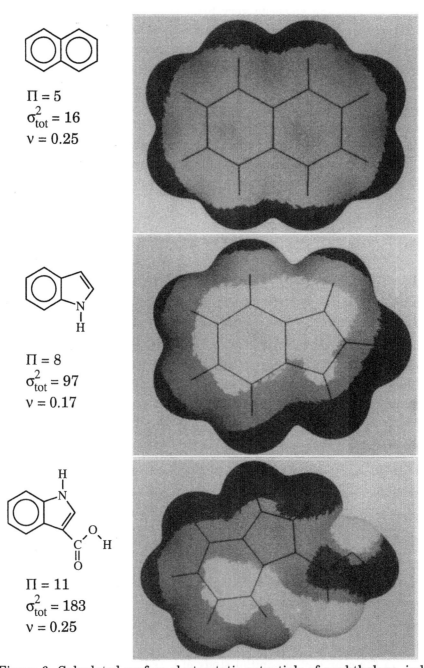

$\Pi = 5$
$\sigma_{tot}^2 = 16$
$\nu = 0.25$

$\Pi = 8$
$\sigma_{tot}^2 = 97$
$\nu = 0.17$

$\Pi = 11$
$\sigma_{tot}^2 = 183$
$\nu = 0.25$

Figure 6. Calculated surface electrostatic potentials of naphthalene, indole and indole-3-carboxylic acid (top to bottom). V(**r**) ranges in kcal/mole: black, greater than 0; gray, between 0 and -10; white, more negative than -10.

we recognized the potential usefulness of a quantitative measure of the variability of V($\mathbf{r}$) on a molecular surface [5]. The compounds for which solubilities in super-critical ethane, ethylene, carbon dioxide and trifluoromethane were available for purposes of correlation analyses were naphthalene (**9**) and a group of indoles (**10** - **17**), shown below. A particularly

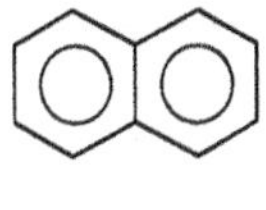

9

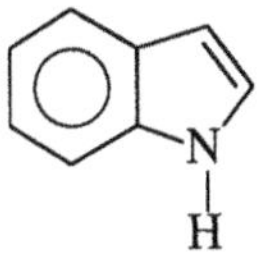

10

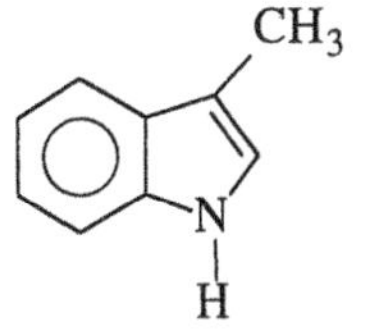

11

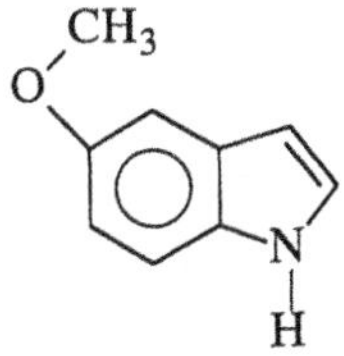

12

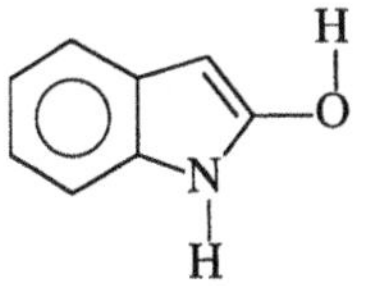

13

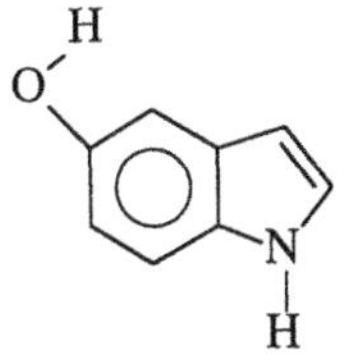

14

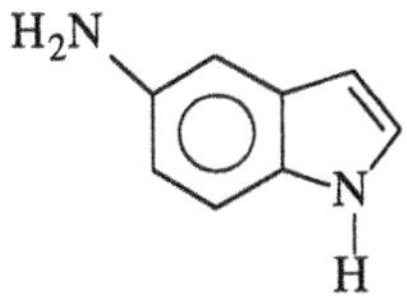

15

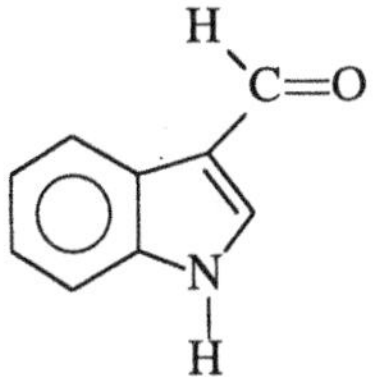

16

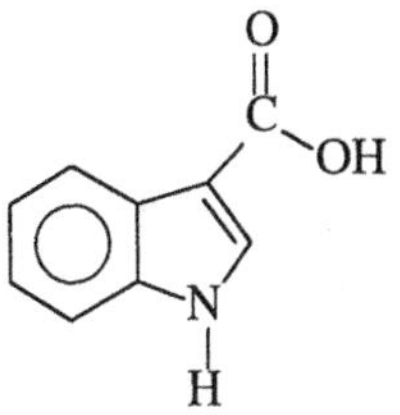

17

interesting feature of the trends in their solubilities is that these are independent of the nature of the solvent; for a given pressure, the solubility decreases from **9** to **17**, in that order, in all four solvents [30-32], despite the significant differences in the latter.

In seeking a relationship between computed quantities and supercritical solubilities, we looked first at our calculated V_{min}, $V_{S,min}$, $V_{S,max}$ and Π values. None of these, alone or in combination, yielded satisfactory corrrelations [5]. We noticed, however, that the solubilities of **9** - **17** decrease as the V($\mathbf{r}$) pattern becomes more variable in both distribution and magnitude. The question then was how to quantify this feature of the surface electrostatic potential, its variability. We investigated several functions, and concluded that σ_{tot}^2 is most effective for this purpose. We found that, to a very good degree of correlation, the solubilities of **9 - 17**, for a given pressure and solvent, decrease as σ_{tot}^2 increases; linear relationships for log(solubility) vs σ_{tot}^2 were found in all four solvents [5].

Since our introduction of σ_{tot}^2 as a useful statistical quantity, we have developed a number of other applications of it [1, 5-8, 28, 29, 33]. These will be discussed in section 3 of this chapter.

2.2.4.3. ν, a measure of "electrostatic" balance

The "balance" parameter ν [eq. (5)] helps to more accurately represent the manner in which σ_{tot}^2 affects electrostatic interactive tendencies [7, 8]. As can be seen from eq. (5), ν attains its maximum value, 0.250, when σ_+^2 and σ_-^2 are equal; thus ν is an indicator of the degree of balance between the positive and negative potentials on the molecular surface. The closer that ν is to 0.250, the more capable is the molecule of interacting to a similar extent (whether strongly or weakly) through both its positive and negative regions. For example, benzene has σ_+^2 and σ_-^2 equal to 7.1 and 9.3 (kcal/mole)^2, respectively, so that ν = 0.246. Benzene is thus expected to interact to a similar degree through both its positive and negative regions, although weakly, as indicated by the magnitudes of σ_+^2, σ_-^2 and σ_{tot}^2. As another example, it is interesting to look at two sets of structural isomers: (a) ethanol and dimethyl ether, and (b) n-butanol and diethyl ether. The ν values are 0.159 and 0.049 for the first pair, and 0.144 and 0.055 for the second. ν is larger for the alcohols, consistent with their being relatively strong as both hydrogen bond donors and acceptors, whereas the ethers act only as hydrogen bond acceptors [34] and thus are less balanced in their interactive behavior. Finally, the three molecules in Figure 6 provide interesting examples of ν. It has nearly the same values for naphthalene and indole-3-carboxylic acid, 0.250 and 0.248, showing that there is essential balance between the positive and negative regions in each molecule, although these are much weaker in naphthalene. For indole, on the other hand, ν = 0.169; the negative potentials dominate over the positive.

While ν sometimes enters our correlations as a separate term [1, 8, 29], often it is as the product $\nu\sigma_{tot}^2$ [1, 7, 8, 33]. We have found the latter to be of key importance for treating properties that depend on how well a molecule interacts electrostatically with other molecules of its own kind; these include boiling points, critical temperatures and heats of vaporization [7, 33].

2.3. The Average Local Ionization Energy, $\bar{I}(r)$
2.3.1. Definition

Most of the molecular quantities in the GIPF are related to the electrostatic potential, as has been discussed in section 2.2. $\bar{I}_{S,min}$, on the other hand, refers to the average local ionization energy $\bar{I}(\mathbf{r})$, defined by eq. (7) [35].

$$\bar{I}(\mathbf{r}) = \sum_i \frac{\rho_i(\mathbf{r})|\varepsilon_i|}{\rho(\mathbf{r})} \tag{7}$$

$\rho_i(\mathbf{r})$ is the electronic density of the ith molecular orbital at the point $\mathbf{r}$, ε_i is its orb-ital energy, and $\rho(\mathbf{r})$ is the total electronic density function.

By Koopmans' theorem, the energy required to remove an electron from a chemical system can be approximated by the absolute value of its orbital energy [36]. Accordingly we interpret $\bar{I}(\mathbf{r})$ as the average energy that would be required to remove an electron from any point $\mathbf{r}$ in the space of an atom or molecule. Thus the positions at which $\bar{I}(\mathbf{r})$ has its lowest values, $\bar{I}_{min}$, are indicative of the least tightly bound electrons and are the sites expected to be the most reactive toward electrophiles. We have found it particularly useful to evaluate $\bar{I}(\mathbf{r})$ on the molecular surface [2, 35, 37-39], and so we normally focus on the sites of lowest surface $\bar{I}(\mathbf{r})$, designated as $\bar{I}_{S,min}$.

The preceding may suggest that the information conveyed by $\bar{I}_{min}$ and $\bar{I}_{S,min}$ is very much the same in its significance as that obtained from V_{min} and $V_{S,min}$. However we believe that the latter are particularly relevant to the approach of an electrophile, and the former to subsequent charge transfer, if any. Thus $\bar{I}(\mathbf{r})$ and $V(\mathbf{r})$ are actually complementary, as shall be demonstrated later in this chapter.

2.3.2. Surface $\bar{I}(r)$ of Aromatic Systems

We have shown for a series of substituted benzenes, **18** [35], some azines and azine N-oxides (e.g. pyridine, **19,** and pyridine N-oxide, **20**) [37, 40], and a group of *para*-substituted anilines, **21** [41], that $\bar{I}_{S,min}$ provides a quantitative measure of the activitating/deactivating and the directing tendencies of the

X = NH$_2$, CH$_3$, H,   Y = NH$_2$, OH,
    F, CHO, NO$_2$ OCH$_3$, CH$_3$,
                  H, SH, CHO,
                  COOH, CF$_3$,
                  CN, N(NO$_2$)$_2$,
                  NF$_2$, NO$_2$

         18   **19**   **20**   **21**

substituents or heteroatoms with regard to electrophilic aromatic substitution. Systems with *ortho/para* directors (e.g. NH$_2$, CH$_3$, F and N$^+$O$^-$) have $\bar{I}_{S,min}$ above the *ortho* and *para* ring positions, while *meta* directors (NO$_2$, CHO and N) produce $\bar{I}_{S,min}$ above the *meta* carbons [35, 37, 40]. In the *para*-substituted anilines (**21**), the strong directing tendencies of the NH$_2$ group dominate, in that the ring $\bar{I}_{S,min}$ are generally *ortho* to the site of the NH$_2$ regardless of the nature of the substituent Y [41]. In addition to the ring $\bar{I}_{S,min}$, others of varying magnitudes are found in the vicinities of atoms that have lone pairs, e.g. O, N, F and S [35, 37, 40, 41].

The magnitudes of the ring carbon $\bar{I}_{S,min}$ are quantitative as well as qualitative indicators of the activating or deactivating abilities of substituents or atoms with respect to electrophilic aromatic substitution [35, 40, 41]; this is shown by excellent correlations that we have found between the ring carbon $\bar{I}_{S,min}$ of benzene derivatives and the corresponding Hammett constants σ_p and σ_m [37, 41]. The latter are well-established measures of the overall electron-withdrawing and -donating tendencies of substituent groups on benzene rings [42-44].

2.3.3. Correlations with Aqueous Acidities

Since $\bar{I}_{S,min}$ values permit site-specific quantitative predictions of reactivity toward electrophiles for a variety of aromatic systems [35, 40, 41], we investigated whether $\bar{I}(\mathbf{r})$ might be related to acidity (e.g. pK$_a$), which reflects the tendency to interact with the electrophile H$^+$. In an initial study of the surface $\bar{I}(\mathbf{r})$ of a series of azines and azoles, we showed that an excellent correlation exists between the ring nitrogen $\bar{I}_{S,min}$ and the pK$_a$ values of the conjugate acids [37]. We then extended this analysis to other classes of acids for which the conjugate bases are anionic species [38]: carboxylic acids, oxoacids, substituted methanes, and other carbon and nitrogen acids. We computed $\bar{I}(\mathbf{r})$ on the molecular surfaces of the anionic conjugate bases and found good linear relationships between aqueous (as well as gas phase) acidity and $\bar{I}_{S,\min}$ for the different groups of carbon, oxygen and nitrogen acids [38]. In fact, a single linear correlation exists between pK_a and $\bar{I}_{S,\min}$ for all of the systems studied. These results provide a predictive capability for estimating pK_a values. For example, we have predicted the pK_a's of **22 - 26** to be 36, 26, 3.1, -2.3 and -5.6, respectively [37-39].

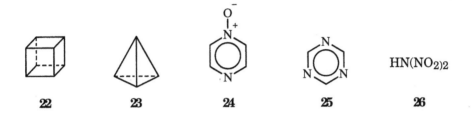

22	**23**	**24**	**25**	**26**

3. APPLICATIONS

3.1. General

In developing a GIPF representation of a particular macroscopic property, it is first necessary to have known values of it for some group of compounds. The next step is to calculate the various quantities in the GIPF, eq. (1), for the corresponding molecules. We then use the SAS statistical analysis program [45] to develop relationships between the known values of the property and various subsets of the computed quantities. We identify which of these are the important ones for that property, and we try to use no more than is necessary for a good correlation.

Most of our applications of the GIPF have involved some portion of the compounds in Table 1, which are listed in order of increasing normal boiling points [46]. Their optimized molecular geometries were computed at the *ab initio* HF/STO-3G* level with the GAUSSIAN 88 code [47], and were used to calculate the HF/STO-5G* electronic densities and electrostatic potentials $V(\mathbf{r})$. In each case, we computed $V(\mathbf{r})$ on a 0.28 bohr grid on the 0.001 electrons/bohr^3 molecular sur-face, and used it to find the various $V_{S,\max}$ and $V_{S,\min}$, and to obtain the global quantities Π, σ_+^2, σ_-^2, σ_{tot}^2 and ν, as given by eqs. (3) - (5). The sites and magnitudes of the $\bar{I}_{S,\min}$ were determined from the molecular orbital energies and their electronic densities on the same surface grid, using eq. (7).

We will first discuss some GIPF relationships that are in terms of only global molecular quantities, and then proceed to others that also include site-specific ones. One important example of the latter type has already been mentioned: the correlation between aqueous acidity (pK_a) and $\bar{I}_{S,min}$, a site-specific quantity.

3.2. Applications Involving Only Global Molecular Quantities
3.2.1. Critical Constants and Normal Boiling Points

A compound's critical constants are of both fundamental and practical interest [48-50], as is its normal boiling point. The latter (T_{bp}) is the temperature at which the liquid and the gaseous phases are in equilibrium at 1 atmosphere of pressure. A compound's critical temperature T_c is always greater than T_{bp}. Above T_c, a compound can only exist in a single fluid phase (a "supercritical fluid"), regardless of the pressure. On a pressure-volume plot, the isotherm corresponding to T_c goes through a saddle-point; the pressure and molar volume at this point are designated as the compound's critical pressure and critical volume, P_c and $\bar{V}_c$. The magnitude of T_c is often interpreted as reflecting the strength of intermolecular interactions [51].

Normal boiling points and critical temperatures, pressures and volumes can be determined experimentally [52] or they can be predicted using theoretical procedures, often empirical or semi-empirical in nature [53-56]. For example, Grigoras has recently introduced a computational technique for estimating normal boiling points and critical constants that is based on molecular surface areas and surface interactions, as computed from atomic radii and atomic charges [56].

As an extension of Grigoras' approach, we recently investigated the possibility of relating the normal boiling points and critical constants to our molecular surface areas and/or our statistical quantities defined in terms of the surface electrostatic potential (Π, σ_{tot}^2 and ν) [7]. For the molecules in Table 1, which comprise a large variety of aliphatic and aromatic hydrocarbons and their derivatives, including alcohols, ethers, ketones, aldehydes, amines and halogenated systems, we have found that T_{bp}, T_c, $\bar{V}_c$ and P_c can be related to $\nu\sigma_{tot}^2$ and surface area by means of eqs. (8) - (11), where α, β and γ are coefficients of positive sign [7].

Table 1

Experimentally-determined normal boiling points[a] and calculated molecular quantities.[b]

Molecule	T_{bp} (K)	surface area	Π	σ_+^2	σ_-^2	σ_{tot}^2	ν	$\nu\sigma_{tot}^2$
CH_4	109.2	55.5	3.15	5.4	3.5	8.9	0.238	2.1
CF_4	144.2	65.5	8.32	66.9	2.9	69.8	0.040	2.8
C_2H_4	169.5	68.9	4.06	7.2	8.3	15.5	0.249	3.9
C_2H_6	184.6	77.5	2.42	3.4	0.6	4.0	0.128	0.5
C_2H_2	189.2	59.8	9.01	36.3	20.5	56.7	0.231	13.1
CHF_3	191.2	63.0	11.74	58.2	11.6	69.8	0.139	9.7
CH_3F	194.8	58.2	9.78	12.3	51.8	64.4	0.154	9.9
C_2F_6	195.0	95.2	8.30	70.9	2.2	73.1	0.029	2.1
CH_2F_2	221.6	60.7	12.55	24.4	22.0	46.4	0.249	11.6
C_3H_6	225.8	90.2	3.83	5.5	9.5	15.1	0.229	3.5
CH_3CF_3	225.9	86.1	10.84	15.4	18.1	33.5	0.248	8.3
C_3H_8	231.1	98.5	2.38	3.1	0.8	3.9	0.163	0.6
C_2H_5F	235.5	80.7	7.80	9.8	71.2	81.1	0.106	8.6
CH_2FCF_3	246.7	89.4	11.48	46.0	12.8	58.8	0.170	10.0
$(CH_3)_2O$	248.3	87.3	9.12	9.0	164.8	173.7	0.049	8.5
CH_3CHF_2	248.5	83.5	10.46	17.2	29.4	46.6	0.232	10.8
$HCCCH_3$	250.0	83.3	8.08	13.9	27.0	40.9	0.224	9.2
$(CH_3)_3CH$	261.6	116.7	2.52	3.1	1.0	4.1	0.184	0.8
$H_2CCHCHCH_2$	268.8	102.6	4.50	7.6	7.5	15.1	0.250	3.8
$CH_3(CH_2)_2CH_3$	272.7	118.8	2.37	2.9	0.8	3.7	0.170	0.6
$(CH_3)_2NH$	280.1	92.8	8.64	17.0	224.2	241.2	0.066	15.9
$C(CH_3)_4$	282.7	132.4	2.76	3.1	1.1	4.2	0.193	0.8
CH_3CH_2Cl	285.5	94.2	9.00	14.3	28.4	42.7	0.223	9.5
CH_2OCH_2	286.4	75.3	10.33	11.6	130.0	141.5	0.075	10.6
$CH_3CH_2NH_2$	289.8	92.2	9.84	27.5	264.3	291.9	0.085	24.8
$(CH_3CH_2)_2O$	307.7	131.4	6.68	8.0	129.8	137.8	0.055	7.6
$CH_3(CH_2)_3CH_3$	309.3	139.7	2.35	2.8	0.9	3.6	0.194	0.7
$(CH_3)_2S$	310.5	98.5	4.41	7.2	25.4	32.6	0.172	5.6
$(CH_3)_3CCl$	324.1	130.4	8.06	13.2	42.5	55.7	0.181	10.1
CH_2CHCHO	325.7	90.0	9.36	18.0	143.7	161.7	0.099	16.0
CH_3COCH_3	329.4	99.4	9.40	15.9	159.8	175.7	0.082	14.4

(continued)

Table 1

Experimentally-determined normal boiling points[a] and calculated molecular quantities (continued).[b]

Molecule	T_{bp} (K)	surface area	Π	σ_+^2	σ_-^2	σ_{tot}^2	ν	$\nu\sigma_{tot}^2$
CH_3COOCH_3	330.2	109.3	10.03	9.7	129.2	138.9	0.065	9.0
CH_3OH	338.2	64.7	12.79	49.6	181.5	231.0	0.169	39.0
tetrahydrofuran	340.2	112.2	8.17	6.2	184.0	190.2	0.032	6.1
$CH_3(CH_2)_4CH_3$	342.0	159.6	2.33	2.7	0.9	3.6	0.188	0.7
CF_3COOH	345.6	95.6	13.95	141.8	38.7	180.5	0.168	30.3
CF_3CH_2OH	347.2	96.2	15.20	85.0	50.2	135.2	0.233	31.5
CCl_4	349.7	120.3	5.22	28.8	2.5	31.3	0.073	2.3
CH_3CH_2OH	351.7	87.1	10.05	45.1	182.4	227.5	0.159	36.2
C_6H_6	353.3	115.3	4.83	7.1	9.2	16.3	0.246	4.0
C_6F_6	353.7	132.0	10.35	39.1	6.1	45.3	0.116	5.3
$c\text{-}C_6H_{12}$	353.9	136.8	2.16	2.5	0.7	3.2	0.171	0.5
CH_3CN	354.8	75.9	17.12	23.6	167.8	191.4	0.108	20.7
$(CH_3)_3COH$	355.5	123.5	7.69	31.1	182.7	213.8	0.124	26.5
$CH_3CHOHCH_3$	355.6	107.0	8.70	35.5	184.2	219.7	0.135	29.7
C_6H_5F	358.3	117.7	5.56	12.0	32.9	45.0	0.195	8.8
$(CH_3CH_2)_2S$	365.3	140.9	3.55	4.2	22.3	26.6	0.132	3.5
CH_3NO_2	374.0	81.2	19.90	34.4	81.7	116.0	0.209	24.2
$C_6H_5CH_3$	383.8	136.0	4.63	6.8	11.1	17.9	0.236	4.2
H_2NCHO	384.2	68.9	17.31	85.5	233.6	319.1	0.196	62.5
pyridine	388.7	110.4	8.55	18.5	212.3	230.8	0.074	17.1
$CH_3(CH_2)_3OH$	390.4	127.9	7.54	35.0	165.9	201.0	0.144	28.9
$CH_2(NH_2)_2$	390.4	106.8	14.82	39.0	234.6	273.7	0.122	33.4
CH_3COOH	391.1	86.4	12.89	41.2	112.1	153.3	0.197	30.2
$CH_3(CH_2)_6CH_3$	398.9	200.6	2.32	2.6	1.0	3.6	0.201	0.7
C_6H_5Cl	405.2	132.2	6.25	14.4	22.9	37.4	0.236	8.8
$(CH_3)_2NCHO$	426.2	112.2	11.07	18.6	158.8	177.4	0.094	16.7
$C_6H_5OCH_3$	428.2	144.2	7.43	15.9	61.3	77.2	0.164	12.7
C_6H_5Br	429.2	137.0	5.94	13.4	18.8	32.2	0.243	7.8
$CH_3CH_2S\text{-}(CH_2)_2Cl$	429.2	163.8	6.66	11.3	28.9	40.2	0.202	8.1
$(CH_3)_2NCOCH_3$	438.2	130.8	10.08	17.3	169.6	186.9	0.084	15.7

(continued)

Table 1

Experimentally-determined normal boiling points[a] and calculated molecular quantities (continued).[b]

Molecule	T_{bp} (K)	surface area	Π	σ_+^2	σ_-^2	σ_{tot}^2	ν	$\nu\sigma_{tot}^2$
m-$C_6H_4Cl_2$	446.2	148.3	6.31	19.7	10.5	30.2	0.227	6.9
p-$C_6H_4Cl_2$	447.2	148.5	6.24	18.1	10.1	28.3	0.228	6.5
o-C_6H_4ClOH	448.1	137.7	6.75	23.9	69.7	93.6	0.190	17.8
$Cl(CH_2)_5Cl$	453.2	170.3	9.51	17.9	23.3	41.5	0.242	10.0
o-$C_6H_4Cl_2$	453.7	146.3	7.62	22.4	23.2	45.6	0.250	11.4
C_6H_5OH	454.9	124.7	8.63	63.8	73.7	137.4	0.249	34.2
$C_6H_5NH_2$	457.2	129.5	9.28	50.4	95.5	145.8	0.226	33.0
C_2Cl_6	459.2	161.8	5.42	28.0	3.4	31.4	0.097	3.0
$(CH_3)_2SO$	462.2	107.8	15.39	24.3	271.7	296.0	0.075	22.2
C_6H_5CN	463.9	135.6	9.98	18.4	176.9	195.3	0.085	16.6
m-$C_6H_4ClOCH_3$	466.7	160.8	8.52	26.0	47.4	73.4	0.229	16.8
m-C_6H_4BrCl	469.2	152.0	6.12	18.9	9.8	28.7	0.225	6.5
p-C_6H_4BrCl	469.2	152.6	6.06	17.5	9.3	26.8	0.227	6.1
$HOCH_2CH_2OH$	470.6	96.4	13.52	68.5	157.2	225.7	0.211	47.6
p-$C_6H_4ClOCH_3$	470.7	160.8	8.32	26.0	47.8	73.8	0.228	16.8
o-$C_6H_4ClOCH_3$	471.7	159.9	9.60	23.7	89.9	113.6	0.165	18.7
o-C_6H_4BrCl	477.2	150.1	7.34	21.7	21.6	43.3	0.250	10.8
$1,3,5$-$C_6H_3Cl_3$	481.2	164.6	6.00	11.9	5.4	17.3	0.215	3.7
$2,4$-$Cl_2C_6H_3OH$	483.2	153.9	7.18	29.2	46.5	75.7	0.237	17.9
$1,2,4$-$C_6H_3Cl_3$	486.7	160.5	7.14	18.0	12.5	30.5	0.242	7.4
m-C_6H_4ClOH	487.2	139.7	8.61	80.1	53.6	133.7	0.240	32.1
$(ClCH_2CH_2)_2S$	490.2	172.4	9.55	21.1	22.4	43.5	0.250	10.9
naphthalene	491.2	159.9	5.12	8.1	7.8	15.9	0.250	4.0
$1,2,3$-$C_6H_3Cl_3$	491.7	160.9	8.27	22.5	18.5	40.9	0.249	10.2
p-C_6H_4ClCN	496.2	152.1	10.33	18.9	157.7	176.5	0.096	16.9
o-C_6H_4ClCN	505.2	150.3	11.69	21.8	154.8	176.5	0.108	19.1
m-$C_6H_4ClNO_2$	508.7	154.1	11.85	23.0	119.0	142.0	0.136	19.3
p-$C_6H_4ClNO_2$	515.2	154.2	11.57	20.1	126.0	146.1	0.119	17.4
5-OCH_3-indole	519.2	179.3	9.56	58.4	59.2	117.6	0.250	29.4
o-$C_6H_4ClNO_2$	519.2	152.8	13.44	22.9	125.9	148.9	0.130	19.4
C_6H_5COOH	522.2	143.3	8.24	41.0	106.8	147.9	0.200	29.6

(continued)

Table 1

Experimentally-determined normal boiling points[a] and calculated molecular quantities (continued).[b]

Molecule	T_{bp} (K)	surface area	Π	σ_+^2	σ_-^2	σ_{tot}^2	ν	$\nu\sigma_{tot}^2$
indole	526.2	149.1	8.39	76.0	20.7	96.6	0.169	16.3
$C_6(CH_3)_6$	538.2	221.9	3.89	3.8	15.9	19.7	0.156	3.1
3-CH_3-indole	538.2	169.1	7.55	63.9	22.5	86.4	0.193	16.7
m-$C_6H_4(NO_2)_2$	564.2	160.4	17.08	35.3	67.9	103.2	0.225	23.2
2-napthol	568.2	169.8	8.14	56.5	57.4	113.9	0.250	28.5
anthracene	613.2	207.1	5.30	8.8	6.8	15.6	0.246	3.8
phenanthrene	613.2	203.0	5.28	9.7	7.1	16.8	0.244	4.1
acridine	618.2	204.3	6.49	16.1	82.7	98.8	0.136	13.4

[a]Experimentally-determined boiling points are taken from reference 48.
[b]Calculated quantities are taken from reference 3. Units are: surface area, Å^2; Π, kcal/mole; σ_-^2, σ_+^2 and σ_{tot}^2, $(\text{kcal/mole})^2$.

$$T_{bp} = \alpha(\text{area}) + \beta(\nu\sigma_{tot}^2)^{0.5} - \gamma \tag{8}$$

$$T_c = \alpha\sqrt{\text{area}} + \beta\sqrt{(\nu\sigma_{tot}^2)^{0.5}} - \gamma \tag{9}$$

$$\overline{V}_c = \alpha(\text{area})^{1.5} + \beta \tag{10}$$

$$P_c = -\alpha(\text{area}) + \beta(\nu\sigma_{tot}^2)/\text{area} + \gamma \tag{11}$$

Eqs. (8) - (11) and Table 2 summarize our best single- or dual-parameter relationships for normal boiling point, critical temperature, critical volume and critical pressure. In each case, many possible variables were tested in our statistical analyses, including powers of area and $\nu\sigma_{tot}^2$, as well as Π, $\nu\sigma_{tot}^2(\text{area})$, $\nu\sigma_{tot}^2/\text{area}$, σ_+^2, σ_-^2, σ_{tot}^2, and ν, taken separately [7]. For $\overline{V}_c$, the inclusion of $\nu\sigma_{tot}^2$ in addition to $(\text{area})^{1.5}$ resulted in an improvement of only 0.002 in R and was not viewed as significant. For T_{bp}, T_c and P_c, only marginal improvements were obtained with the addition of a third and/or fourth variable.

Table 2.
Summary of correlations between computed properties and V_c, T_{bp}, T_c and P_c.[a]

correlation	N	values of constants			R	standard deviation
$T_{bp} = \alpha(\text{area}) + \beta(v\sigma_{tot}^2)^{0.5} + \gamma$		α	β	γ		
	100	2.714	33.57	-69.50	0.948	37.0
	22[b]	2.759	44.80	-97.38	0.959	41.0
$T_c = \alpha\sqrt{\text{area}} + \beta\sqrt{(v\sigma_{tot}^2)^{0.5}} + \gamma$		α	β	γ		
	66	74.68	152.5	-530.1	0.909	60.7
	20[b]	74.17	163.8	-536.1	0.941	53.3
$\bar{V}_c = \alpha(\text{area})^n + \beta$		n	α	β		
	58	1.5	0.168	43.4	0.986	15.2
	16[b]	1.5	0.168	39.8	0.993	13.8
$P_c = \alpha(\text{area}) + \beta(v\sigma_{tot}^2)/\text{area} + \gamma$		α	β	γ		
	57[c]	-0.1764	48.69	61.73	0.910	4.8
	20[b]	-0.1512	64.42	57.07	0.934	3.7

[a]These correlations include all molecules in Table 1 for which experimental data were available. They are the same or very similar to those reported in reference 7, the difference being that CH_3CN

has been included in the correlations for T_{bp}, T_c and $\bar{V}_c$. N is the number of molecules in the data

set and R is the correlation coefficient. The units for T_{bp}, T_c, $\bar{V}_c$ and P_c are K, K, cm^3/mole and bar, respectively.
[b]Molecules containing atoms other than carbon and hydrogen are omitted from data set.
[c]Molecules with more than one fluorine are omitted from the data set.

Looking at eqs. (8) - (11), it is seen that T_{bp}, T_c and $\bar{V}_c$ increase with surface area, while P_c decreases; however T_{bp}, T_c and P_c all increase with $v\sigma_{tot}^2$, the tendency for electrostatic interactions. It is interesting to note that the two independent variables in eq. (9) are the square roots of those in eq. (8). This may be viewed as an attenuation of the effects of size and electrostatic interaction, and may reflect the much lower density of a fluid at T_c compared to the liquid at its normal boiling point [48].

The critical temperature can be regarded as the point beyond which the molecular velocities are so great that no amount of pressure suffices to effect coalescence. As intermolecular attractive forces increase, the higher are the velocities needed to prevent coalescence, and hence T_c increases. Strong intermolecular attractions also increase P_c, since higher velocities mean that the greatest pressure that can still produce coalescence must likewise be higher. That P_c increases with diminishing molecular size is presumably because smaller molecules have greater velocities, at any given temperature.

The dependence of T_{bp} and T_c on surface area (a measure of size) is not surprising, since boiling points are known to increase with molecular weight within families of similar compounds (e.g. chlorinated alkanes) [57-59]. (Fluoro derivatives are an exception to this [60].) However there is not a *general* correlation between boiling point and molecular weight that cuts across families. [For example, Table 1 shows that T_{bp} is much higher for CH_3OH than for the heavier $(CH_3)_2NH$.] Therefore it is particularly pleasing that by including $v\sigma_{tot}^2$ (a measure of the tendency for electrostatic interactions) together with the surface area, we are able to represent T_{bp}, T_c, $\bar{V}_c$ and P_c for the wide variety of compounds in Table 1.

3.2.2. Heats of Vaporization

Typically, normal boiling points and the enthalpies of vaporization at those temperatures vary in the same order [57-59]. For instance, for the group of 41 compounds listed in Table 3, which are primarily halogenated hydrocarbons, the linear correlation coefficient between T_{bp} and $\Delta\bar{H}_{vap}(T_{bp})$ is 0.979. It is therefore to be anticipated that there will be a GIPF representation for $\Delta\bar{H}_{vap}(T_{bp})$ similar to that for T_{bp}. Indeed, for the same 41 compounds, we found,

$$\Delta\bar{H}_{vap}(T_{bp}) = 5.666(\text{area})^{0.5} + 4.136(v\sigma_{tot}^2)^{0.5} - 41.30 \tag{12}$$

The correlation coefficient is 0.965 and the standard deviation is 2.42 kJ/mole.

3.2.3. Solubilities in Low Density Supercritical Fluids

The solubility of a solid in a supercritical fluid reflects primarily two factors: the vapor pressure of the solute and solute-solvent interactions. At low fluid densities, the former has been observed to be the more important

Table 3.
Experimentally-determined normal boiling points (T_{bp}),[a] enthalpies of vaporization (ΔH_{vap})[a], and calculated molecular properties[b].

Molecule	T_{bp} (K)	ΔH_{vap} (T_{bp}) [kJ/mole]	surface area	σ_+^2	σ_-^2	σ_{tot}^2	ν	$\nu\sigma_{tot}^2$
CH_4	109.2	8.2	55.5	5.4	3.5	8.9	0.238	2.12
CF_4	145.2	11.96	65.5	66.9	2.9	69.8	0.040	2.78
C_2H_4	169.5	13.5	68.9	7.2	8.3	15.5	0.249	3.86
C_2H_6	184.6	14.7	77.5	3.4	0.6	4.0	0.128	0.51
C_2H_2	189.2	17.0	59.8	36.3	20.5	56.7	0.231	13.12
CHF_3	191.2	16.76	63.0	58.2	11.6	69.8	0.139	9.67
CF_3Cl[c]	191.7	15.50	82.7	50.6	2.3	52.9	0.042	2.22
CH_3F	194.8	16.7	58.2	12.3	51.8	64.4	0.154	9.89
C_2F_6	195.0	16.2	98.6	71.5	2.2	73.7	0.029	2.13
CF_3Br[c]	215.3	17.68	87.3	44.5	3.0	47.5	0.059	2.80
C_3H_6	225.8	18.4	90.2	5.5	9.5	15.1	0.229	3.46
CH_3CF_3	225.9	19.2	88.5	16.0	18.5	34.5	0.249	8.58
C_3H_8	231.1	19.04	98.5	3.1	0.8	3.9	0.163	0.64
CHF_2Cl[c]	232.4	18.79	80.4	69.4	8.6	78.0	0.098	7.64
CF_2Cl_2[c]	243.4	19.97	97.6	33.2	2.1	35.3	0.056	1.98
CH_3Cl[c]	249.0	21.5	73.6	18.1	24.0	42.1	0.245	10.31
$HCCCH_3$	250.0	22.1	83.3	13.9	27.0	40.9	0.224	9.18
$H_2CCHCHCH_2$	268.8	22.47	102.6	7.6	7.5	15.1	0.250	3.77
$CH_3(CH_2)_2CH_3$	272.7	22.44	118.8	2.9	0.8	3.7	0.170	0.63
CH_3Br[c]	276.7	23.91	77.6	18.0	18.9	36.9	0.250	9.23
$CHFCl_2$[c]	282.2	24.92	93.7	63.8	7.7	71.5	0.096	6.86
CH_3CH_2Cl	285.3	24.7	94.2	14.3	28.4	42.7	0.223	9.51
$CH_3(CH_2)_3CH_3$	309.3	25.8	139.7	2.8	0.9	3.6	0.194	0.70
CH_2Cl_2[c]	313.2	28.06	91.1	46.3	13.8	60.1	0.177	10.78
CH_3I[c]	315.6	27.3	88.9	20.3	24.0	44.3	0.248	10.99
$CHCl_3$[c]	334.9	29.24	107.6	53.5	7.4	60.9	0.107	6.52
$CH_3(CH_2)_4CH_3$	342.0	28.85	159.6	2.7	0.9	3.6	0.188	0.68
CF_3COOH	345.6	33.3	95.6	141.8	38.7	180.5	0.168	30.40
CF_3CH_2OH	347.2	40.0	96.2	85.0	50.2	135.2	0.233	31.56
CCl_4	349.7	29.82	120.3	28.8	2.5	31.3	0.073	2.30

(continued)

Table 3.
Experimentally-determined normal boiling points (T_{bp}),[a] enthalpies of vaporization (ΔH_{vap})[a], and calculated molecular properties[b].

Molecule	T_{bp} (K)	ΔH_{vap} (T_{bp}) [kJ/mole]	surface area	σ_+^2	σ_-^2	σ_{tot}^2	ν	$\nu\sigma_{tot}^2$
C_6H_6	353.3	30.72	115.3	7.1	9.2	16.3	0.246	4.01
C_6F_6	353.7	31.7	132.0	39.1	6.1	45.3	0.116	5.27
c-C_6H_{12}	353.9	29.97	136.8	2.5	0.7	3.2	0.171	0.55
C_6H_5F	358.3	31.19	117.7	12.0	32.9	45.0	0.195	8.77
$C_6H_5CH_3$	383.8	33.2	136.0	6.8	11.1	17.9	0.236	4.22
C_6H_5Cl	405.2	35.19	132.2	14.4	22.9	37.4	0.236	8.82
C_6H_5Br	429.2	37.9	137.0	13.4	18.8	32.2	0.243	7.82
m-$C_6H_4Cl_2$	446.2	38.6	148.3	19.7	10.5	30.2	0.227	6.9
p-$C_6H_4Cl_2$	447.2	38.6	148.5	18.1	10.1	28.3	0.228	6.46
o-$C_6H_4Cl_2$	453.7	40.6	146.3	22.4	23.2	45.6	0.30	11.40
C_2Cl_6	459.2	45.9	161.8	28.0	3.4	31.4	0.097	3.03

[a]References 46 and N. A. Lange, *Handbook of Chemistry*, Handbook Publishers, New York, 1956.

[b]Calculated properties for many of the molecules have been reported earlier in references 1 and 7. Those designated with a superscript c are from reference 60. Units are: surface area, Å^2; σ_-^2, σ_+^2 and σ_{tot}^2, (kcal/mole)^2; ν is unitless.

[61-63]. Our finding that the solubilities of **9 - 17** in supercritical ethane, ethylene, carbon dioxide and trifluoromethane increase as σ_{tot}^2 of the solute decreases [5] is fully consistent with this experimentally-established relationship between supercritical solubility and solute vapor pressure, since a small value of σ_{tot}^2 is indicative of weak solute-solute interactions and thus favors high vapor pressure. For each of the four supercritical solvents, we obtained an equation of the form of eq. (13), in which α and β depend upon the solvent. However, while σ_{tot}^2 effectively reflects electronic factors influencing solubility

$$\log (\text{solubility}) = -\alpha \sigma_{tot}^2 + \beta \tag{13}$$

for **9 - 17** under the conditions being considered, in general the size of the solute molecule is also expected to be important [64-66]. (**9 - 17** are very similar in size, which is presumably why the inclusion of either area or volume does not significantly improve the solubility correlations represented by eq. (13) [5].)

The extension of our investigation to a larger and more heterogenous group of twenty-one solutes (**1, 9 - 17** and **27 - 33**) confirmed that solute size does need to be taken into account [6]. Our best dual-variable relationship for

27	**28**	**29**	**30**

31	**32**	**33**

solubility in supercritical CO_2 at 14 MPa is of the form of eq. (14), where α, β and γ are positive coefficients. The correlation coefficient is 0.948 [6]; it decreases slightly, by 0.01, when volume is replaced by area. Eq. (14) shows that solubility in low-density supercritical CO_2 varies inversely with solute size and σ_{tot}^2. Increases in size and σ_{tot}^2 favor low vapor pressure and hence impede solubility in low-density supercritical CO_2.

$$\ln(\text{solubility}) = \alpha\,(\text{vol})^{-1.5} - \beta\left(\sigma_{tot}^2\right)^2 - \gamma \tag{14}$$

In order to focus on the secondary factor in low-density supercritical solubility, solute-solvent interactions, Dobbs and Johnston [61] have defined an enhancement factor, E [eq. (15)],

$$E = y_2\,P/P_2^{sat} \tag{15}$$

Table 4.
Experimental and calculated data.[a]

solute molecule	enhancement factor E, in CO_2, 20 MPa	sol in CO_2, 20 MPa (mole fraction x 10^4)	surface area	σ^2_{tot}	ν	$\nu\sigma^2_{tot}$
(acridine structure)	743,000.	7.8[b]	204.3	98.8	0.136	13.48
(anthracene structure)	531,000.	0.69[b]	207.1	15.6	0.246	3.84
(phenanthrene/chrysene structure)	385,000.	12.5[c]	203.0	16.8	0.244	4.10
(phthalic anhydride structure)	250,000.	22.5[b]	151.7	119.6	0.155	18.58
(anthranilic acid, COOH / NH$_2$ structure)	188,000.	1.13[b]	154.0	157.7	0.226	35.63
(naphthol, OH structure)	164,000.	5.57[b]	169.8	113.9	0.250	28.47
(benzoic acid, COOH structure)	134,000.	23.5[b]	143.3	147.9	0.201	29.61
(hexamethylbenzene structure)	71,200.	17.8[c]	221.9	19.7	0.156	3.07

(continued)

Table 4.
Experimental and calculated data (continued).[a]

solute molecule	enhancement factor E, in CO_2, 20 MPa	sol in CO_2, 20 MPa (mole fraction x 10^4)	surface area	σ^2_{tot}	v	$v\sigma^2_{tot}$
(3-methylindole, with CH_3; N–H)	16,300.	32.5[d,e]	169.1	86.4	0.193	16.64
(indole, N–H)	13,600.	55.[f]	149.1	96.6	0.169	16.29
(naphthalene)	11,600.	170.[e]	159.9	15.9	0.250	3.97
(5-methoxyindole, H_3CO; N–H)	5,810.	18.0[e]	179.3	117.6	0.250	29.40

[a]All experimental data are at 308°K. The calculated quantities are taken from reference [8]. Units are: volume: Å^3; surface area: Å^2; σ^2_{tot}, σ^2_+ and σ^2_-: (kcal/mole)^2.

[b]Taken from reference [71].
[c]Taken from reference [72].
[d]Taken from reference [31].
[e]Taken from reference [73].
[f]Taken from reference [30].

where y_2 and P^{sat}_2 are the solubility and vapor pressure of the solute and P is the pressure of the system. P^{sat}_2/P can be viewed as the ideal gas solubility (i.e., in the absence of interactions); E therefore indicates how much greater is the actual solubility, y_2, due to solute-solvent interactions. For the group of twelve organic solutes shown in Table 4, we have found the following relationship between E, in supercritical CO_2 at 20 MPa, and computed solute quantities [8]:

$$E = -\alpha \, (\text{area})^{-1.5} + \beta \sigma_{tot}^2 + \gamma v - \varepsilon (v \sigma_{tot}^2) - \eta \qquad (16)$$

α, β, γ, ε and η are positive coefficients.

Eq. (16) indicates that solute-solvent interactions increase with the size of the solute, and with the variance and balance of its surface electrostatic potential, as measured by σ_{tot}^2 and v; on the other hand, increasing size and σ_{tot}^2 lead to a lower vapor pressure and thus diminish solubility in low density supercritical fluids [eq. (14)]. In other words, the area and σ_{tot}^2 of a solute each have simultaneously opposing effects. Indeed, it can be seen in Table 4 that some of the molecules with the largest enhancement factors have very low solubilities. This may explain the need for the cross-term $v\sigma_{tot}^2$ in eq. (16), which comes in with a negative sign (at least in supercritical CO_2); it may be a damping term which reflects the fact that high σ_{tot}^2 and v values also promote solute-solute interactions, which in turn impede solubility.

3.2.4. Critical Temperatures of Mixtures

With the expanding use of supercritical fluids for a variety of laboratory and industrial applications [50, 67-69], it has become important to have practical methods for obtaining reliable estimates of the critical temperatures of mixtures. As was discussed in section 3.2.1., we have demonstrated that the critical properties and normal boiling points of large groups of compounds of various types can be correlated and predicted in terms of computed molecular properties which reflect size and interaction tendencies: surface area, σ_{tot}^2 and v [7]. We have subsequently extended this analysis to the critical temperatures of X/CO_2 mixtures, where X is a solute with mole fractions of 0.10 and 0.25 in mixtures with CO_2 [70].

For the twelve solutes X listed in Table 5, we obtained critical temperatures at these mole fractions by interpolation of experimental data [74]. Our calculated HF/STO-5G*//HF/STO-3G* surface areas and Π and v values are also listed in Table 5.

We have found that relationships of the form given in eq. (17) account for the trend in T_c for each mole fraction separately [70]. α, β, γ and ε are coefficients of positive sign, that differ in magnitudes for the two mole

Table 5.
Interpolated critical temperatures of X/CO_2 mixtures[a] and computed properties of solutes X.[b]

Solute, X	Critical temperature, °K		surface area (Å^2)	Π (kcal/mole)	ν
	0.10 mole fraction X	0.25 mole fraction X			
methane, CH_4	291.5	275.6	55.5	3.15	0.239
ethane, C_2H_6	298.8	293.1	77.5	2.42	0.128
hydrogen chloride, HCL	305.0	306.7	48.3	10.70	0.088
diethyl ether, $C_2H_5-O-C_2H_5$	304.6	330.5	131.4	6.68	0.055
propane, C_3H_8	306.4	314.1	98.5	2.38	0.163
hydrogen sulfide, H_2S	306.8	312.8	53.9	6.27	0.220
propene, $CH_3-CH=CH_2$	308.7	316.1	90.2	3.83	0.229
dimethyl ether, CH_3-O-CH_3	310.2	322.9	87.3	9.12	0.049
chloromethane, CH_3Cl	316.1	334.2	72.5	10.40	0.246
n-butane, C_4H_{10}	317.1	338.3	118.8	2.37	0.169
sulfur dioxide, SO_2	318.1	340.5	63.3	12.60	0.242
n-pentane, C_5H_{12}	330.2	377.8	139.7	2.35	0.194

[a]Obtained by interpolation of experimental data in reference [74].
[b]Reference [73].

$$T_c \, (X / CO_2 \text{ mixture}) = \alpha(\text{area})^2 - \beta(16.1 - \Pi)^2 - \gamma\left(\frac{\text{area}^3}{\nu}\right) + \varepsilon \qquad (17)$$

fractions. In both cases, increasing solute size gives a higher T_c for the X/CO_2 mixture. The second term on the right side of eq. (17) reflects the difference in local polarity between CO_2 (Π = 16.1) and the solute; as this difference increases, the effect is to decrease T_c. The third term is greatest in magnitude for large, electrostatically unbalanced solute molecules, which accordingly produce much smaller increases in the T_c of the X/CO_2 mixture than would be predicted from the size of X alone. A good example of this is diethyl ether, which is larger than n-butane, but for which the mixtures with CO_2 have lower T_c values due to its small ν (Table 5).

3.2.5. Octanol/Water Partition Coefficients

Octanol/water partition coefficients are among the properties most frequently used to test Quantitative Structure Activity Relationships (QSAR) developed for the purposes of drug design and other biological and toxicological applications. However pharmacological and environmental research often concerns poorly characterized or not yet synthesized compounds for which partition coefficients are not known. Several methods for estimating them have therefore been introduced. A widely used approach is to add up empirically determined fragment contributions [75-77]. While this generally leads to very reliable results, it is limited by the availability of data and by the requirement that the parameterization have been conducted for the same fragment types as are in the molecules of interest.

Another approach has been to correlate partition coefficients with computed quantities, such as molecular surface area or volume and semi-empirically derived atomic charges [78-81]. A significant number of these parameters may be needed, since the charges have to be weighted differently depending upon the atom type and because additional empirical corrections are often necessary. An exception is the work of Famini et al, in which only two parameters are employed [81].

Kamlet and coworkers [82] have successfully used their linear solvation energy relationship (LSER) to correlate octanol/water partition coefficients P_{ow}:

$$\log P_{ow} = mV_I + s(\pi^* + d\delta) + a\alpha_m + b\beta_m + C \tag{18}$$

P_{ow} is the ratio of the equilibrium concentration of a solute in octanol to that in water. V_I is the van der Waals volume of the solute molecule, π^* is a polarity/polarizability term, δ is a polarizability correction, α_m and β_m represent hydrogen-bond-donating and -accepting tendencies, respectively, and C is a constant. For 245 organic molecules of different types, eq. (18) gave a correlation coefficient of 0.996 and a standard deviation of 0.131. It should be noted that Kamlet et al eliminated certain groups of compounds from the correlation, e.g. pyridine and its derivatives, primary and secondary amines, and nitroalkanes. More recent work by Abraham has shown that these deficiencies can be corrected if a newly defined set of LSER parameters is used [83]. While a relationship such as that of Kamlet et al may not always be practical for studies of novel drug molecules, since π^*, α_m and β_m are estimated through elaborate experimental procedures, the results are still very encouraging.

The excellent correlation obtained with eq. (18) suggests that it may be possible to describe the partitioning of a solute between water and octanol by means of our GIPF approach. β_m was found to be one of the dominant terms in eq. (18), along with V_I, and since hydrogen bonding has been shown to be mainly electrostatic in nature [84, 85], it seemed reasonable to use a solute's electrostatic potential together with an estimate of its size as a starting point in trying to predict its partition coefficient. Accordingly we investigated possible relationships between log P_{ow} and the quantities Π, σ_+^2, σ_-^2, σ_{tot}^2, v and surface area, using the data set given in Table 1 and in reference 28.

As shown in Table 6, a fair correlation is obtained with the simple relationship,

$$\log P_{ow} = \alpha(\text{area}) - \beta\sigma_{tot}^2 - \gamma \tag{19}$$

The correlation coefficient is 0.936, but the standard deviation is rather large, 0.552. Both are improved, to 0.946 and 0.513, by introducing Π:

$$\log P_{ow} = \alpha(\text{area}) - \beta\sigma_{tot}^2 - \gamma\Pi + \delta \tag{20}$$

From the signs of the coefficients α, β, and γ in eqs. (19) and (20), we can conclude that partitioning into octanol is favored by large surface area, while high σ_{tot}^2 and Π values favor partitioning into water. It was further found advantageous to multiply Π by the area, giving eq. (21). The effect of this is to make the last term size dependent, whereas Π is defined to be size independent. The correlation coefficient becomes 0.950 and the standard deviation is 0.493.

$$\log P_{ow} = \alpha(\text{area}) - \beta\sigma_{tot}^2 - \gamma(\text{area})\Pi - \delta \tag{21}$$

Finally, we tested the possibility that eq. (21) could be improved by using σ_-^2 in place of σ_{tot}^2 [eq. (22)]; this would give greater emphasis to the negative portions of the molecular surface, consistent with the conclusions of Famini *et al* [81] and Kamlet *et al* [82] that the dominant factors determining P_{ow} are size and hydrogen-bond-accepting ability.

Table 6.
Relationships between experimental octanol/water partition coefficients and calculated quantities for 70 molecules from Table 1.

Equation	Correlation Coefficient	Standard Deviation
$\log P_{ow} = 0.0243(\text{area}) - 0.0109\sigma_{tot}^2 - 0.2611$	0.936	0.552
$\log P_{ow} = 0.0233(\text{area}) - 0.00869\sigma_{tot}^2 - 0.0644\Pi + 0.188$	0.946	0.513
$\log P_{ow} = 0.0285(\text{area}) - 0.00848\sigma_{tot}^2 - 0.000603(\text{area})\Pi - 0.415$	0.950	0.493
$\log P_{ow} = 0.0298(\text{area}) - 0.00912\sigma_-^2 - 0.000849(\text{area})\Pi - 0.529$	0.961	0.437

$$\log P_{ow} = \alpha(\text{area}) - \beta\sigma_-^2 - \gamma(\text{area})\Pi - \delta \tag{22}$$

Eq. (22) does have a somewhat better correlation coefficient and standard deviation, 0.961 and 0.437, than does eq. (21) (Table 6).

The log P_{ow} values obtained with eq. (22) are in generally good agreement with the experimental data, but there are a few notable discrepancies. By far the worst are ethylene glycol, $HO-CH_2-CH_2-OH$, N,N-dimethylformamide, $(CH_3)_2NCHO$, and N,N-dimethylacetamide, $(CH_3)_2NCOCH_3$. (If these three molecules were omitted, the correlation coefficient for eq. (22) would be 0.976 and the standard deviation would be 0.329 [28].) For all three molecules, our results underestimate the tendency for partitioning into water. In the case of ethylene glycol, a possible reason is suggested by noting that although it has twice as many hydroxyl oxygens as methanol, its σ_-^2 is actually smaller (Table 1), reflecting the fact that the electrostatic potential is more negative around the oxygen in methanol than around those in ethylene glycol.

3.2.6. Octanol/Water and Acetonitrile/NaCl-Saturated-Water Partition Coefficients of Some Nitroaromatics

For both fundamental and applied purposes it is interesting to explore the liquid-liquid partitioning behavior of solutes in other solvent pairs besides the traditional octanol/water system. For a group of molecules including benzene, toluene and a number of their mono-, di- and trinitro derivatives, shown in Table 7, we have investigated the use of our GIPF [eq. (1)] in correlating their partitioning between (a) octanol and water (P_{ow}), and (b) acetonitrile and NaCl-saturated-water (P_{aw}) [29]. Our calculated properties for these molecules are given in Table 6, along with experimentally-determined values of log P_{ow} and log P_{aw} [86].

Our best dual-parameter relationships for log P_{ow} and log P_{aw} for the molecules in Table 7 are of the forms given in eqs. (23) and (24) [29] (Table 8).

$$\log P_{ow} = \alpha(\text{area}) - \beta(\text{area})\Pi - \gamma \tag{23}$$

$$\log P_{aw} = \alpha(\text{area}) - \beta\Pi + \gamma \tag{24}$$

The linear correlation coefficients are 0.980 and 0.971, respectively. These equations indicate that increasing size favors partitioning into the solvent composed of larger molecules, octanol in the case of eq. (23) and aceto-nitrile in eq. (24), while increasing (area)Π in eq. (23) or Π in eq. (24) favors partitioning into water, which is the more polar solvent in each case. As can be seen in Table 8, eqs. (23) and (24) can be improved by the introduction of a third variable, σ_+^2 in the case of log P_{ow} and ν for log P_{aw}.

It is interesting to look at the log P_{ow} expressions in Table 8 in the context of the relationships found for the larger, more heterogenous set of seventy organic molecules [28] (Table 6). Our best dual-parameter equation for the eleven molecules in Table 7 gives log P_{ow} as a function of area and (area)Π, two of the terms in our best three-parameter relationship for the larger set of molecules (Table 6). However, unlike the latter, the inclusion of σ_-^2 in eq. (23) does not significantly improve our correlation. Instead, it is now the introduction of σ_+^2, to give eq. (25), that markedly improves both the correlation coefficient, from 0.980 to 0.993, and the standard deviation, which decreases from 0.099 to 0.065 [29]. Thus our best representation of log P_{ow} for the nitroaromatics, eq. (25), can be viewed as being analogous in form to our best

$$\log P_{ow} = \alpha(\text{area}) - \beta(\text{area})\Pi - \gamma\sigma_+^2 - \delta \tag{25}$$

general log P_{ow} relationship [eq. (22)], except that they emphasize different aspects of the electrostatic interaction: σ_-^2 in eq. (22) and σ_+^2 in eq. (25). The trend in the σ_+^2 values for the nitroaromatics in Table 7 reflects the increase in the positive potential above the aromatic ring that occurs in going from one to three nitro groups [29]; these indicate an increasing tendency for interaction

Table 7.

Experimentally-determined partition coefficients (P_{ow} and P_{aw}) and computed surface quantities for some nitroaromatics.[a]

Solute	log P_{ow}	log P_{aw}	surface area (Å^2)	Π (kcal/mole)	σ_+^2	σ_-^2 (kcal/mole)^2	σ_{tot}^2
benzene	2.13	2.56	115.3	4.83	7.1	9.2	16.3
toluene	2.73	2.91	136.0	4.63	6.8	11.1	17.9
nitrobenzene	1.85	2.49	142.8	12.13	16.9	105.2	121.9
2-nitrotoluene	2.30	2.87	160.8	10.65	17.3	128.3	146.6
3-nitrotoluene	2.45	2.90	164.2	11.07	14.5	113.9	128.4
4-nitrotoluene	2.42	2.86	163.9	11.24	15.4	114.7	130.1
1,3-dinitrobenzene	1.49	2.61	160.4	17.08	35.3	67.9	103.2
2,4-dinitrotoluene	1.98	2.92	183.4	15.83	35.4	77.4	112.8
2,6-dinitrotoluene	2.02	3.03	180.6	14.77	45.5	80.2	125.7
1,3,5-trinitrobenzene	1.18	2.78	183.0	18.70	105.3	47.4	152.7
2,4,6-trinitrotoluene	1.86	3.30	204.1	17.13	104.8	53.4	158.2

[a]Log P_{ow} and log P_{aw} are partition coefficients in octanol/water and acetonitrile/NaCl-saturated-water, respectively, taken from reference [86].

Table 8.

Summary of relationships between computed quantities and octanol/water and acetonitrile/NaCl-saturated-water partition coefficients (P_{ow} and P_{aw}).

Relationship	Correlation coefficient	Standard deviation
Octanol/water (P_{ow})		
log P_{ow} = 0.0163(area) − 0.1456Π + 1.2079	0.926	0.188
log P_{ow} = 0.0298(area) − 0.00100(area)Π − 0.6878	0.980	0.099
log P_{ow} = 0.0304(area) − 0.00100(area)Π − 0.00049σ_{tot}^2 − 0.7373	0.980	0.105
log P_{ow} = 0.0289(area) − 0.00099(area)Π − 0.00067σ_-^2 − 0.6272	0.982	0.101
log P_{ow} = 0.0296(area) − 0.00088(area)Π − 0.00369σ_+^2 − 0.7691	0.993	0.065
Acetonitrile/NaCl-saturated-water (P_{aw})		
log P_{aw} = 0.0189(area) − 0.00033(area)Π + 0.4706	0.933	0.092
log P_{aw} = 0.0158(area) − 0.0567Π + 0.9792	0.971	0.061
log P_{aw} = 0.0160(area) − 0.0585Π + 0.7033v + 0.8365	0.988	0.042

with nucleophiles, such as the oxygens of water and to a lesser extent of octanol, where the oxygen is sterically more hindered.

The experimental partition coefficients in Table 7 reveal an interesting contrast. Within each series of molecules, the nitrobenzenes and the nitrotoluenes, log P_{ow} shows a general decrease with increasing nitration, whereas P_{aw} tends to increase with the number of NO_2 groups. In seeking to understand the reasons for these opposing trends, it is relevant to note in Table 8 that the areas all enter the P_{ow} and P_{aw} equations with approximately the same weights; on the other hand, the coefficients of the Π and (area)Π terms are 2.5 to 3 times larger in the P_{ow} equations. (The local polarity of the solute is more important for the two solvents with considerably differing Π values, octanol (<7.7) and water (21.6), than for the two which are more similar, acetonitrile (17.1) and water [4].) The log P_{aw} values within each series follow approximately the same trends as the surface areas, which increase with the number of NO_2 groups. But for log P_{ow}, the contribution of the (area)Π term is sufficiently large (and opposite in sign to the area term) that it reverses the trend that would be predicted from the areas alone.

3.3. Applications Which Include Site-Specific Quantities
3.3.1. Relationships for Gas-Phase Acidities:
Group V-VII Hydrides and Their Anions

The complementarity between the electrostatic potential $V(\mathbf{r})$ and the average local ionization energy $\bar{I}(\mathbf{r})$ was brought out in developing a GIPF representation of the gas phase enthalpies of protonation of the first-, second- and third-row hydrides, and their anions, of Groups V - VII of the periodic table [2]. Since the anions require larger basis sets, all optimized structures were computed at the HF/6-31+G* level; however this has not yet been extended to the third row, so for the atoms As, Se and Br, we used a double-zeta valence basis complemented by one set of d polarization functions and one set of diffuse s and p [2]. Table 9 lists our computed $V_{S,min}$ and $\bar{I}_{S,min}$ [2], along with experimentally-determined gas phase protonation enthalpies [87, 88].

The trends in Table 9 are quite striking. When looking at either the hydrides or the anions separately, $V_{S,min}$ shows a vertical decrease in magni-tudes, while $\bar{I}_{S,min}$ increases horizontally [2]. The magnitudes of ΔH°_{pr} decrease both vertically and horizontally. These trends show that

Table 9.

Calculated and Experimental Properties.[a]

Property	Group V		Group VI		Group VII	
	NH_3	NH_2^-	H_2O	OH^-	HF	F^-
$V_{S,min}$	-44.9	-173.1	-40.8	-181.9	-25.6	-178.2
$\bar{I}_{S,min}$	11.99	1.73	15.14	3.66	19.26	5.96
ΔH°_{pr}	-204.0	-403.7	-166.5	-390.8	-117.0	-371.4
	PH_3	PH_2^-	H_2S	SH^-	HCl	Cl^-
$V_{S,min}$	-22.3	-132.7	-21.0	-137.9	-12.5	-142.3
$\bar{I}_{S,min}$	10.76	1.63	11.33	3.31	14.18	5.13
ΔH°_{pr}	-188.6	-370.9	-170.2	-351.1	-135.0[b]	-333.7
	AsH_3	AsH_2^-	H_2Se	SeH^-	HBr	Br^-
$V_{S,min}$	-16.3	-128.7	-18.9	-131.8	-10.6	-133.0
$\bar{I}_{S,min}$	10.67	1.61	10.50	3.12	12.82	4.71
ΔH°_{pr}	-179.2	-362.1	-171.3	-342.7	-139.0	-323.6

[a]The units of $V_{s,min}$, $\bar{I}_{S,min}$ and ΔH°_{pr} are kcal/mole, eV and kcal/mole, respectively. The calculated $V_{S,min}$ and $\bar{I}_{S,min}$ are taken from reference [2]. The ΔH°_{pr} values are from reference [87] unless otherwise indicated. [b]Reference [88].

neither $V_{S,min}$ nor $\bar{I}_{S,min}$ can correlate with all of the protonation enthalpies on the same plot [2], but they suggest a complementarity between $V_{S,min}$ and $\bar{I}_{S,min}$ in relation to ΔH°_{pr}.

Indeed, we have found that the experimental enthalpies of protonation can be expressed as linear combinations of $V_{S,min}$ and $\bar{I}_{S,min}$, as given by eq. (26), where α, β and γ are coefficients of positive sign [2].

$$\Delta H^\circ_{pr} = \alpha V_{S,min} + \beta \bar{I}_{S,min} - \gamma \qquad (26)$$

Relationships of this form hold for the anions and hydrides taken separately, and also for the entire group together. In that order, the correlation coefficients are 0.984, 0.980, and 0.997.

In section 2.3.3, we described a good correlation between aqueous acidity, as measured by pK_a, and $\bar{I}_{S,min}$ of the conjugate base. This encompasses a variety of acids in which the hydrogen is bonded to a first-row atom (carbon, nitrogen or oxygen). The results that have just been presented suggest that the earlier correlation would also require a $V_{S,min}$ term if it included acids in which the hydrogen is bonded to a second- or third-row atom. We did in fact find that pK_a for the hydrides in Table 9 obeys eq. (27) [2],

$$pK_a = -\alpha V_{S,min} - \beta \bar{I}_{S,min} - \gamma \qquad (27)$$

with a correlation coefficient of 0.962 and a standard deviation of 4.2, for values ranging from -9.5 to 34.

The $V_{S,min}$ terms in eqs. (26) and (27) can be viewed as representing the electrostatic contribution to the interaction energy, whereas $\bar{I}_{S,min}$ reflects polarization/charge transfer. Such a partitioning of an interaction energy into an electrostatic portion plus one or more additional contributions is not a new concept and has especially been used in the area of Lewis acid-base interactions [89-91]. We are currently exploring the approach represented by eq. (26), perhaps augmented by one or more of our other computed properties (e.g. area, Π, σ_{tot}^2, ν), as a means of analyzing other types of reactions. Clearly, the relative magnitudes of the $V_{S,min}$ and $\bar{I}_{S,min}$ contributions are expected to depend upon the species involved and their environment.

For example, the influence of $V_{S,min}$ (as reflected by the α/β ratio) is likely to decrease in aqueous solution since water, with its high dielectric constant, will presumably diminish the electrostatic interaction between the reacting species. Indeed, the influence of $V_{S,min}$ is much less in eq. (27) than in eq. (26) [2].

3.3.2. Correlations with Hydrogen Bond Basicity

We have shown for several families of molecules (azines, alkyl ethers, primary amines and molecules containing double-bonded oxygens, taken separately) that the V_{min} of the hydrogen-bond-accepting heteroatoms (O or N) correlate well with solvatochromic parameters that are established measures of hydrogen-bond-accepting tendency [3]. These parameters, β_2^H for solutes and β for solvents, are used in linear solvation energy relationships (LSER) as indicators of hydrogen bond basicity in solute-solvent interactions [82, 83]. Our success, discussed in section 3.3.1, in developing a general expression for hydride gas phase acidities in terms of both $V(\mathbf{r})$ and $\bar{I}(\mathbf{r})$ [2] prompted us to explore whether V_{min} or $V_{S,min}$ could be combined with $\bar{I}_{S,min}$ and perhaps other terms in a multi-variable relationship to describe β and β_2^H without the necessity for grouping molecules into families.

For a general assortment of twenty-four molecules, we have found that β_2^H and β can be described well by relationships of the form of eq. (28). α, β, γ and ε are coefficients of positive sign. The linear correlation coefficents are 0.977 for β_2^H and 0.973 for β. This has been examined in more detail in chapter 3 of this book.

$$\beta_2^H \text{ (or } \beta) = -\alpha V_{S,min} - \beta \bar{I}_{S,min} - \gamma\Pi + \varepsilon \tag{28}$$

3.3.3. Correlations with Hydrogen Bond Acidity

As was mentioned in section 2.2.2, we have demonstrated that surface electrostatic potential maxima ($V_{S,max}$) provide a means for ranking the sites on molecules in regard to propensities for nucleophilic interactions [3, 18, 20], includ-ing the tendency of a hydrogen to be donated in a hydrogen bond. We showed that good correlations exist, within families of compounds, between $V_{S,max}$ and the solvatochromic parameters that measure hydrogen-bond-donating ability, α for solvents and α_2^H for solutes [3, 18].

We have now suceeded in eliminating the family dependence of these correlations by means of our GIPF. Our best relationship for twenty hydrogen-bond donors of various types, including molecules with –OH, –CH, –NH and –SH groups, is represented by eq. (29), where α, β, γ and ε are coefficients of positive sign; the linear correlation coefficient is 0.983.

$$\alpha_2^H = \alpha(\sigma_+^2 V_{S,max}) + \beta(\nu\sigma_{tot}^2) + \gamma(area)^2 + \varepsilon \tag{29}$$

This relationship has been examined in more detail in chapter 3 of this book.

4. DISCUSSION AND SUMMARY

We have presented a general procedure for correlating and predicting macroscopic properties that depend primarily upon molecular interactions. A gratifying and somewhat remarkable feature is that we are able to represent solution and liquid phase properties solely in terms of quantities evaluated computationally for isolated molecules. Thus we avoid the problems associated with taking into account the effects of the medium.

There are several other points that we would also like to emphasize:

(1)   The GIPF approach is based upon well-defined quantities which are computed specifically for each molecule. Accordingly it is able to reflect any unusual electronic or structural features of the molecule. Furthermore, since no experimental data are needed as input for the GIPF, it can be used to predict properties for compounds that have not yet been prepared or isolated.

(2)   We represent each property in terms of the smallest number of quantities, typically two or three, that produces a good correlation. This makes it possible to isolate the key factors that determine a particular property, and facilitates efforts to design molecules with improved performance in that respect.

(3)   One of our primary objectives is to make the relationships very general, covering the widest possible ranges of compounds, even though it is clear that better correlation coefficients and standard deviations could be obtained if different classes (e.g. hydrocarbons, alcohols, acids, etc.) were treated separately.

The GIPF procedure can be viewed as an *ab initio* QSAR (Quantitative Structure-Activity Relationship) approach. Our calculations have been, for the most part, at the minimum-basis-set self-consistent-field level, with polarization functions included for atoms in the second and third rows of the periodic table. Minimum-basis SCF calculations are known to give generally satisfactory results for structures [92] and for one-electron properties, such as the electronic density and the electrostatic potential [14, 16, 93-95].

We continue to expand the scope of the GIPF approach, in terms of both the number and the types of properties being represented. This may require that additional molecular quantities be introduced to augment the eight presently included in eq. (1). (There are actually ten, since σ_+^2 or σ_-^2 is sometimes used alone.) In the past, we have found it convenient to use surface area as a measure of molecular size, although volume may well be more appropriate. It seems likely that the size term may often be representing the contribution of the molecular polarizability, which is known to be directly related to molecular volume [96-98]. A continuing emphasis in this work shall be upon achieving an understanding of the role of each term in the GIPF representation of a property.

ACKNOWLEDGEMENT

We greatly appreciate the support provided by ARPA/ONR Contract No. N00014-91-J-1897, administered by ONR.

REFERENCES:

1. J. S. Murray, T. Brinck, P. Lane, K. Paulsen and P. Politzer, J. Mol Struct. (Theochem), 307 (1994) 55.
2. T. Brinck, J. S. Murray and P. Politzer, Int. J. Quant. Chem., 48 (1993) 73.
3. J. S. Murray and P. Politzer, J. Chem. Res., S (1992) 110.
4. T. Brinck, J. S. Murray and P. Politzer, Mol. Phys., 76 (1992) 609.
5. P. Politzer, P. Lane, J. S. Murray and T. Brinck, J. Phys. Chem., 96 (1992) 7938.
6. P. Politzer, J. S. Murray, P. Lane and T. Brinck, J. Phys. Chem., 97 (1993) 729.
7. J. S. Murray, P. Lane, T. Brinck, K. Paulsen, M. E. Grice and P. Politzer, J. Phys. Chem., 97 (1993) 9369.
8. J. S. Murray, P. Lane, T. Brinck and P. Politzer, J. Phys. Chem., 97 (1993) 5144.
9. R. F. W. Bader, M. T. Carroll, J. R. Cheeseman and C. Chang, J. Am. Chem. Soc., 109 (1987) 7968.
10. R. F. W. Bader, W. H. Henneker and P. E. Cade, J. Chem. Phys., 46 (1967) 3341.
11. E. Scrocco and J. Tomasi, in *Topics in Current Chemistry*, vol. 42, (Springer-Verlag, Berlin, 1973).
12. E. Scrocco and J. Tomasi, Advances Quantum Chemistry, 11 (1978) 115.
13. P. Politzer and D. G. Truhlar, eds., *Chemical Applications of Atomic and Molecular Electrostatic Potentials,* (Plenum Press, New York, 1981).
14. P. Politzer and K. C. Daiker, in *The Force Concept in Chemistry*, B. M. Deb, (Van Nostrand Reinhold Company, New York, 1981) ch. 6.
15. P. Politzer and J. S. Murray, in *Theoretical Biochemistry and Molecular Biophysics: Vol. 2, Proteins*, (Adenine Press, Schenectady, NY, 1991).
16. P. Politzer and J. S. Murray, in *Reviews in Computational Chemistry*, vol. 2, K. B. Lipkowitz and D. B. Boyd, eds., (VCH Publishers, New York, 1991) ch 7.
17. J. S. Murray, S. Ranganathan and P. Politzer, J. Org. Chem., 56 (1991) 3734.
18. J. S. Murray and P. Politzer, J. Org. Chem., 56 (1991) 6715.
19. R. K. Rathak and S. R. Gadre, J. Chem. Phys., 93 (1990) 1770.
20. J. S. Murray, P. Lane, T. Brinck and P. Politzer, J. Phys. Chem., 95 (1990) 844.
21. P. Sjoberg, J. S. Murray, T. Brinck, P. Evans and P. Politzer, J. Mol. Graphics, 8 (1990) 81.

22. A. Poland and J. C. Knutson, Ann. Rev. Pharmacol. Toxicol., 22 (1982) 517.
23. A. Poland, W. F. Greenlee and A. S. Kende, Ann. N. Y. Acad. Sci., 330 (1979) 214.
24. M. C. Etter, Z. Urbanczyk-Lipowska, M. Zia-Ebrahimi and T. W. Pananto, J. Am. Chem. Soc., 112 (1990) 8415.
25. J. S. Murray, M. E. Grice, P. Politzer and M. C. Etter, Mol. Eng., 1 (1991) 95.
26. G. Delgado-Barrio and R. F. Prat, Phys. Rev. A, 12 (1975) 2288.
27. J. K. Nagle, J. Am. Chem. Soc., 112 (1990) 4741.
28. T. Brinck, J. S. Murray and P. Politzer, J. Org. Chem., 58 (1993) 7070.
29. J. S. Murray, T. Brinck and P. Politzer, J. Phys. Chem., 97 (1994) 13807.
30. S. Sako, K. Ohgaki and T. Katayama, J. Supercrit. Fluids, 1 (1989) 1.
31. S. Sako, K. Shibata, K. Ohgaki and T. Katayama, J. Supercrit. Fluids, 2 (1989) 3.
32. T. Nakatani, K. Ohgaki and T. Katayama, J. Supercrit. Fluids, 2 (1989) 9.
33. T. Brinck, H. Hagelin, J. S. Murray and P. Politzer, in preparation.
34. P. Politzer and J. S. Murray, in *Supplement E: The Chemistry of Hydroxyl, Ether and Peroxide Groups*, vol. 2, S. Patai, ed., (John Wiley & Sons, Chichester, England, 1993) ch. 1.
35. P. Sjoberg, J. S. Murray, T. Brinck and P. Politzer, Can. J. Chem., 68 (1990) 1440.
36. T. A. Koopmans, Physica, 1 (1933) 104.
37. T. Brinck, J. S. Murray, P. Politzer and R. E. Carter, J. Org. Chem., 56 (1991) 2934.
38. T. Brinck, J. S. Murray and P. Politzer, J. Org. Chem., 56 (1991) 5012.
39. J. S. Murray, T. Brinck and P. Politzer, J. Mol Struct. (Theochem), 255 (1992) 271.
40. P. Lane, J. S. Murray and P. Politzer, J. Mol Struct. (Theochem), 236 (1991) 283.
41. M. Haeberlein, J. S. Murray, T. Brinck and P. Politzer, Can. J. Chem., 70 (1992) 2209.
42. L. P. Hammett, *Physical Organic Chemistry*. (McGraw-Hill, New York, 1940).
43. O. Exner, *Correlation Analysis of Chemical Data*. (Plenum Press, New York, 1988).
44. C. Hansch, A. Leo and R. W. Taft, Chem. Rev., 91 (1991) 165.
45. SAS, SAS Institute Inc., Cary, NC 27511.
46. D. R. Lide, *Handbook of Chemistry and Physics*. 71st ed. (CRC Press, Boca Raton, FL, 1990).
47. M. J. Frisch, M. Head-Gordon, H. B. Schlegel, K. Raghavachari, J. S. Binkley, C. Gonzalez, D. J. Defrees, D. J. Fox, R. A. Whiteside, R. Seeger, C. F. Melius, J. Baker, R. Martin, L. R. Kahn, J. J. P. Stewart, E. M. Fluder, S. Topiol and J. A. Pople, GAUSSIAN 88. (Gaussian Inc., Pittsburgh, PA, 1988).

48. I. N. Levine, *Physical Chemistry*. 3rd. ed. (McGraw-Hill Book Co., New York, 1988).
49. M. D. Palmieri, J. Chem. Ed., 10 (1988) A254.
50. K. P. Johnston and J. M. L. Penninger, eds., *Supercritical Fluid Science Technology*, ACS Symp. Series 406, (American Chemical Society, Washington, DC, 1989).
51. R. J. Sadus, Fluid Phase Equil., 77 (1992) 269.
52. C. P. Hicks and C. L. Young, Chem. Rev., 75 (1975) 121.
53. D. Ambrose and R. Townsend, Trans. Faraday Soc., 64 (1968) 2622.
54. G. R. Somayajulu, Chem. Eng. Data, 34 (1989) 106.
55. A. S. Teja, R. J. Lee, D. Rosenthal and M. Anselme, Fluid Phase Equil., 56 (1990) 153.
56. S. Grigoras, J. Comp. Chem., 11 (1990) 493.
57. N. L. Allinger, M. P. Cava, D. C. DeJongh, C. R. Johnson, N. A. Lebel and C. L. Stevens, *Organic Chemistry*. (Worth Publishers, New York, 1971).
58. R. T. Morrison and R. N. Boyd, *Organic Chemistry*. 3rd. ed. (Allyn and Bacon, Boston, 1973).
59. D. S. Kemp and F. Vellaccio, *Organic Chemistry*. (Worth, New York, 1980).
60. J. S. Murray and P. Politzer, submitted for publication.
61. J. M. Dobbs and K. P. Johnston, Ind. Eng. Chem. Res., 26 (1987) 1476.
62. C. T. Lira, in *Supercritical Fluid Extraction and Chromatography*, B. A. Charpentier and M. R. Sevenants, eds., ACS Symposium Series 366, (American Chemical Society, Washington, 1988) ch. 1.
63. K. P. Johnston, D. G. Peck and S. Kim, Ind. Eng. Chem. Res., 28 (1989) 1115.
64. R. B. Hermann, J. Phys. Chem., 76 (1972) 2754.
65. G. G. Hall and C. M. Smith, J. Mol Struct. (Theochem), 179 (1988) 293.
66. C. M. Smith, J. Mol Struct. (Theochem), 184 (1989) 103,343.
67. T. G. Squires and M. E. Paulaitis, eds., *Supercritical Fluids*, ACS Symposium Series 329, (American Chemical Society, Washington, 1987).
68. F. V. Bright and M. E. P. McNally, eds., *Supercritical Fluids Technology*, ACS Symposium Series 488, (American Chemical Society, Washington, 1992).
69. E. Kiran and J. F. Brennecke, eds., *Supercritical Fluid Engineering Science*, ACS Symposium Series 514, (American Chemical Society, Washington, 1993).
70. P. Lane, unpublished work.
71. J. M. Dobbs, J. M. Wong, R. J. Lahiere and K. P. Johnston, Ind. Eng. Chem. Res., 26 (1987) 56.
72. J. M. Dobbs, J. M. Wong and K. P. Johnston, J. Chem. Eng. Data, 31 (1986) 304.
73. K. D. Bartle, A. A. Clifford, S. A. Jafa and G. F. Shilstone, J. Phys. Chem. Ref. Data, 20 (1991) 713-755.

74. J. D'Ans, J. Bartels, P. Ten Bruggencate, A. Eucken, G. Joos and W. A. Roth, eds., *Landolt-Börnstein, Zahlenwerte und Funktionen aus Physik, Chemie, Astronomie, Geophysik, Technik,* (Springer-Verlag, Berlin, 1969).

75. R. F. Recker, *The Hydrophobic Fragmental Constants.* (Elsevier, New York, 1977).

76. C. Hansch and A. J. Leo, *Substituent Constants for Correlation Analysis in Chemistry and Biology.* (John Wiley & Sons, New York, 1979).

77. A. J. Leo, Chem. Rev., 93 (1993) 1281.

78. G. Klopman and L. D. Iroff, J. Comp. Chem., 2 (1981) 157.

79. N. Bodor, Z. Gabanyi and C. Wong, J. Am. Chem. Soc., 111 (1989) 3783.

80. A. Kantola, H. O. Villar and G. H. Loew, J. Comp. Chem., 12 (1991) 681.

81. G. R. Famini, C. A. Penski and L. Y. Wilson, J. Phys. Org. Chem., 5 (1992) 395.

82. M. J. Kamlet, R. M. Doherty, M. H. Abraham, Y. Marcus and R. W. Taft, J. Phys. Chem., 92 (1988) 5244.

83. M. H. Abraham, Chem. Soc. Rev., 22 (1993) 73.

84. U. Dinur and A. T. Hagler, in *Reviews in Computational Chemistry*, vol 2, K. B. Lipkowitz and D. B. Boyd, eds., (VCH Publishers, New York, 1991) ch. 4.

85. S. Scheiner, in *Reviews in Computational Chemistry*, vol. 2, K. B. Lipkowitz and D. B. Boyd, eds., (VCH Publishers, New York, 1991), ch. 5.

86. D. C. Leggett, P. H. Miyares and T. F. Jenkins, J. Sol. Chem., 21 (1992) 105.

87. S. G. Lias, J. E. Bartmess, J. F. Liebman, J. L. Holmes, R. D. Levin and W. G. Mallard, J. Phys. Chem. Ref. Data, 17 (1988) 1.

88. P. W. Tiedeman, S. L. Anderson, S. T. Ceyer, T. Hiroka, C. Y. Ng, B. H. Mahan and Y. T. Lee, J. Chem. Phys., 71 (1979) 605.

89. G. Klopman, J. Am. Chem. Soc., 90 (1968) 223.

90. R. S. Drago, G. C. Vogel and T. E. Needham, J. Am. Chem. Soc., 93 (1971) 6014.

91. H. Umeyama, K. Morokuma and S. Yamabe, J. Am. Chem. Soc., 99 (1977) 330.

92. W. J. Hehre, L. Radom, P. v. R. Schleyer and J. A. Pople, *Ab Initio Molecular Orbital Theory.* (Wiley-Interscience, New York, 1986).

93. C. Gatti, P. J. MacDougall and R. F. W. Bader, J. Chem. Phys., 88 (1988) 3792.

94. R. J. Boyd and L.-C. Wang, J. Comp. Chem., 1 (1989) 367.

95. J. M. Seminario, J. S. Murray and P. Politzer, in *The Application of Charge Density Research to Chemistry and Drug Design*, (Plenum Press, New York, 1991).

96. K. M. Gough, J. Chem. Phys., 91 (1989) 2424.

97. K. E. Laidig and R. F. W. Bader, J. Chem. Phys., 93 (1990) 7213.

98. T. Brinck, J. S. Murray and P. Politzer, J. Chem. Phys., 98 (1993) 4305.

P. Politzer and J.S. Murray
Quantitative Treatments of Solute/Solvent Interactions
Theoretical and Computational Chemistry, Vol. 1
© 1994 Elsevier Science B.V. All rights reserved.

Estimation of Chemical Reactivity Parameters and Physical Properties of Organic Molecules Using SPARC

S. H. Hilal[a], L. A. Carreira[b] and S. W. Karickhoff[a]

[a]Environmental Research Laboratory, U.S. Environmental Protection Agency, Athens, GA 30605, U.S.A.

[b]Department of Chemistry, University of Georgia, Athens, GA 30602, U.S.A.

1. GENERAL INTRODUCTION

The major differences between behavior profiles of molecules in the environment are attributable to their physicochemical properties. The need for physical and chemical constants of chemical compounds has greatly accelerated both in industry and government as assessments are made of pollutant exposure and risk. Although considerable progress has been made in process elucidation and modeling for chemical and physical processes, values for the fundamental thermodynamic and physicochemical properties (i.e., rate/equilibrium constants, Henry's law constant, distribution coefficients between immiscible solvents, solubility in water, etc.) have been achieved for only a small number of molecular structures.

For most chemicals, only fragmentary knowledge exists about those properties that determine each compounds's fate in the environment. Chemical-by-chemical measurements of the required properties is not practical because of expense and because trained technicians and adequate facilities are not available for measurement efforts involving thousands of chemicals. In fact physical and chemical properties have actually been measured for, perhaps, 1 percent of the approximately 70,000 industrial chemicals listed by the U.S. Environmental Protection Agency's Office of Toxic Substances (OTS) [1].

Fortunately, estimation techniques that employ the judgment of expert chemists are available to provide the required data in a cost-effective manner. These techniques include the application of linear free energy relationships (LFER) [2,3], structure activity relationships (SAR) [4,5], and other estimation methods particularly in the drug and environmental fields. Even so, methodologies and values often are not available for the parameters needed in the sophisticated mathematical models used for environmental exposure assessment.

Recently [6-10], we described our approach for predicting numerous physical properties and chemical reactivity parameters of organic compounds strictly from molecular structure using the new prototype computer program SPARC (SPARC Performs Automated Reasoning in Chemistry). The goal of this new computer program is to apply the reasoning process that an organic chemist might undertake in reactivity analysis. The approach primarily involves deductive reasoning and is theory/mechanism oriented. The computational approach is based on existing mathematical models of chemistry.

Our new computer program costs the user only a few minutes of computer time and provides greater accuracy and a broader scope than is possible with conventional estimation techniques. The user needs to know only the molecular structure for the compound to predict a property of interest. The user provides the program with the molecular structure either by direct entry as SMILES (Simplified Molecular Input Line Entry System) notation or via the molecular editor that will generate the structure and translate it to SMILES notation. SPARC is programmed with the ALS (Applied Logic Systems) version of Prolog (PROgramming in LOGic). It is executable on machines with the UNIX operating system or 386/486 MS DOS machines with 16 MB of extended memory.

SPARC presently predicts ionization pK_a, electron affinity, and numerous physical properties such as vapor pressure (at any temperature), boiling point (at any pressure), activity coefficient/distribution coefficient (for any solvent), retention times for gas and liquid chromatography, etc. In this chapter we report the calculation of ionization equilibrium constants, electron affinity, and gas chromatography retention indices for a squalane liquid phase strictly from molecular structure for a large number of organic compounds using SPARC.

1.1. SPARC Computational Approach

SPARC does not do "first principles" computation; rather, it analyzes chemical structure relative to a specific reactivity query much as an expert chemist might. SPARC utilizes directly the extensive knowledge base of organic chemistry. For physical properties, intermolecular interactions are expressed as a summation over all the interaction forces between molecules (i.e., dispersion, induction, dipole and H-bonding). Each of these interaction forces is expressed in terms of a limited set of molecular-level descriptors (density-based volume, molecular polarizability, molecular dipole, and H-bonding parameters) that, in turn, are calculated from molecular structure. For chemical reactivity, molecular structure is broken into functional units. Reaction centers with known intrinsic reactivity are identified and the impact on reactivity of appended molecular structure is quantified using mechanistic perturbation models.

A "toolbox" of mechanistic perturbation models has been developed that can be implemented where needed for a specific reactivity query. Resonance models were developed and calibrated on light absorption spectra [6]; electrostatic models were developed on ionization equilibrium constants [7,8]. Solvation models (i.e., dispersion, induction, H-bonding, dipole, etc.) have been developed on physical properties (i.e., vapor pressure, solubilities, distribution coefficient, gas chromatographic retention times, etc.) [9,10]. Ultimately these mechanistic components will be fully implemented for the aforementioned chemical and physical property models and will be extended to additional properties such as hydrolytic and redox processes.

The computational approaches in SPARC are a blend of conventional LFER, SAR, and Perturbed Molecular Orbital (PMO) methods [11,12]. In general, SPARC utilizes LFER to compute thermodynamic or thermal properties and PMO theory to describe quantum effects such as delocalization energies or polarizabilities of π electrons. In reality, every chemical property involves both quantum and thermal contributions and necessarily requires the use of both perturbation methods for prediction.

Any predictive method should be understood in terms of the purpose for which it is conceived and should be structured by appropriate operational constraints. SPARC's predictive methods can be characterized as engineering applications in environmental assessments. More specifically these methods provide :
(a) an *a priori* estimate of physicochemical parameters for physical and chemical process models when measured data are not available, (b) guidelines for ranking a large number of chemical parameters and processes in terms of relevance to the question at hand, thus establishing priorities for measurements or study, (c) an evaluation or screening mechanism for existing data based on "expected" behavior, and (d) guidelines for interpreting or understanding existing data and observed phenomena.

2. CHEMICAL REACTIVITY PARAMETERS: ESTIMATION OF IONIZATION pK_a

INTRODUCTION

A knowledge of the acid-base ionization properties of organic molecules is essential to describing chemical transport, transformation or potential environmental effects. For ionizable compounds, solubility, partitioning phenomena, and chemical reactivity are all highly dependent upon the state of ionization in the solution phase. The ionization pK_a of an organic compound is vital to environmental exposure assessment because it can be used to define the degree of ionization and the propensity for sorption to soil and sediment by cation exchange. These, in turn, can determine mobility, reaction kinetics, bioavailability, complexation etc. In addition to being highly significant in evaluating environmental fate and effects, acid-base ionization equilibria provide an excellent development arena for electrostatic effects models. Because the gain or loss of protons results in a change in molecular charge, these processes are extremely sensitive to electric field effects within the molecule.

Unfortunately, up to now no reliable method has been available for predicting pK_a over a wide range of molecular structures either for simple compounds or for complicated molecules such as dyes. The object of this study was to demonstrate the application of SPARC to the prediction of pK_a for a wide range of molecular structures.

2.1. SPARC's Chemical Modeling

Chemical properties describe molecules in transition, that is, the conversion of a reactant molecule to a different state or structure. For a given chemical property, the transition of interest may involve electron redistribution within a single molecule or bimolecular union to form a transition state or distinct product. The behavior of chemicals depends on the differences in electronic properties of the initial state of the system and the state of interest. For example, a light absorption spectrum reflects the differences in energy between the ground and excited electronic states of a given molecule. Chemical equilibrium constants depend on the energy differences between the reactants and products. Electron affinity depends on the energy differences between the LUMO (Lowest Unoccupied Molecular Orbital) state and the HOMO (Highest Unoccupied Molecular Orbital) state.

For any chemical property addressed in SPARC, the energy differences between the initial state and the final state are small compared to the total binding energy of the reactant involved. Calculating these small energy differences by *ab initio* computational methods is difficult, if not impossible. On the other hand, perturbation methods provide these energy differences with more accuracy and with more computational simplicity and flexibility than *ab initio* methods. These methods treat the final state as a perturbed initial state and the energy differences between these two energy states are determined by quantifying the perturbation. For pK_a, the perturbation of the initial state, assumed to be the protonated form, versus the unprotonated final form, is factored into the mechanistic contributions of resonance and electrostatic effects and any other additional perturbations such as hydrogen bonding, steric contributions or solvation.

2.2. pK_a Computational Procedure

Molecular structures are broken into functional units called the reaction center and the perturber. The reaction center ,C, is the smallest subunit that has the potential to ionize and lose a proton to a solvent. The perturber ,P, is the molecular structure appended to the reaction center, C. The perturber structure is assumed to be unchanged in the reaction. The pK_a of the reaction center is known either from direct measurement or inferred indirectly from pK_a measurements. The pK_a of the reaction center is adjusted for the molecule in question using the mechanistic perturbation models described below.

Like all chemical reactivity parameters addressed in SPARC, pK_a is analyzed in terms of some critical equilibrium component:

$$P-C_i \xrightarrow{\ pK_a\ } P-C_f$$

where C_i denotes the initial protonated state. C_f is the final unprotonated state of the reaction center, C. P is the "perturber". The pK_a for a molecule of interest is expressed in terms of the contributions of both P and C.

$$pK_a = (pK_a)_c + \delta_p (pK_a)_c \tag{1}$$

where $(pK_a)_c$ describes the ionization behavior of the reaction center, and $\delta_p(pK_a)_c$ is the change in ionization behavior brought about by the perturber structure. SPARC computes reactivity perturbations, $\delta_p(pK_a)_c$, that are then used to "correct" the ionization behavior of the reaction center for the compound in question in terms of potential "mechanisms" for interaction of P and C as

$$\delta_p (pK_a)_c = \delta_{ele}\, pK_a + \delta_{res}\, pK_a + \delta_{sol}\, pK_a + ... \tag{2}$$

where $\delta_{res}pK_a$, $\delta_{ele}pK_a$ and $\delta_{sol}pK_a$ describe the differential resonance, electrostatic and solvation effects of P with the protonated and unprotonated states of C, respectively. Electrostatic interactions are derived from local dipoles or charges in P interacting with charges or dipoles in C. $\delta_{ele}pK_a$ represents the difference in the electrostatic interactions of the P with the two states. $\delta_{res}pK_a$ describes the change in the delocalization of π electrons of the two states due to P. This delocalization of π electrons is assumed to be into or out of the reaction center. Additional perturbations include direct interactions of the structural elements of P that are contiguous to the reaction center such as H-bonding or steric blockage of solvent access to C. For example, in the ionization of aniline, $-NR_2$ is the reaction center (denoted C) and the phenyl group is the perturber (denoted P).

The ionization equilibrium constant can be expressed as

$$pK_a = (pK_a)_c + \delta_{res}\, pK_a \tag{3}$$

where $(pK_a)_c$ is the pK_a for the reaction center NR_2 and is equal to 8.93, and $\delta_{res}pK_a$ is the resonance contribution to pK_a.

2.2.1. pK$_a$ Modeling Approach.

The modeling of the perturber effects for chemical reactivity relates to the structural representation S--$_iR_j$--C, where S--$_iR_j$ is the perturber structure, P, appended to the reaction center, C. S denotes substituent groups that "instigate" perturbation. For electrostatic effects, S contains (or can induce) electric fields; for resonance, S donates/receives electrons to/from the reaction center. R links the substituent and reaction center and serves as a conductor of the perturbation ("conducts" resonant π electrons or electric fields). A given substituent, however, may be a part of the structure, R, connecting another substituent to C, and thus functions as a "conductor" for the second substituent. The i and j denote anchor atoms in R for S and C, respectively.

For each reaction center and substituent SPARC catalogs appropriate characteristic parameters. Substituents include all non-carbon atoms and aliphatic carbon atoms contiguous to either the reaction center or a π-unit. Some heteroatom substituents containing π groups are treated collectively as substituents ($-NO_2$, $-C\equiv N$, $-C=O$, $-CO_2H$, etc). The specification of these collective units as substituents is strictly facilitative. The only requisites are that they be structurally and electronically well defined (charge and/or dipolar properties relatively insensitive to the remainder of the perturber structure). Also, these units must be terminal with regard to resonance interactions (no pass-through conjugation). All hydrogen atoms are dropped and "bookkept" only through atom valence. Heteroatom substituents in these π units are replaced by an isoelectronic carbon equivalent plus an appended atom Q. For example $-C=O-$ becomes C=C-Q, which is now treated in SPARC as perturbed ethylene.

In computing the contribution of any given substituent to $\delta_p(pK_a)_c$, the effect is factored into three independent components for the structural components C, S, and R:
1. substituent strength, which describes the potential of a particular substituent to "exert" a given effect,
2. molecular network conduction, which describes the "conduction" properties of the molecular structure R, connecting S to C with regard to a given effect, and
3. reaction center susceptibility, which rates the response of the reaction center to the effect in question.

The contributions of each structural component are quantified (i.e., parameterized independently). For example, the strength of the substituent's electrostatic field effect depends only on the substituent; likewise, the conduction of R is modeled to be independent of the specific identities of both the substituent and the reaction center. The susceptibility of the reaction center to field effects quantifies the differential interaction of the initial state versus the final state with the electric field, but again, this susceptibility gauges only the initial state versus the final state of the reaction center and is independent of both R and S. *The rationale for the factoring is to remove, to the extent possible, both structural and reaction specificity from effects parameterization.* This provides parameter "portability" and, hence, effects-model portability to other structures and to other types of reactivity.

2.2.2. Electrostatic Effects Models

Electrostatic effects on reactivity derive from charges or electric dipoles in the appended perturber structure, P, interacting through space with charges or dipoles in the reaction center, C. Direct electrostatic interaction effects (field effects) are manifested by a fixed charge or dipole in a substituent interacting through the intervening molecular cavity with a charge or dipole in the reaction center. The substituent can also "induce" electric fields in R that can interact electrostatically with the reaction center. This indirect interaction is called the "mesomeric field effect." In addition, electrostatic effects derived from electronegativity differences between the reaction center and the substituent are termed sigma induction. These effects are transmitted progressively through a chain of σ-bonds between atoms. For compounds containing multiple substituents, electrostatic perturbations are computed for each singly and summed to produce the total effect.

With regard to electrostatic effects, reaction centers are classified according to the electrostatic change accompanying the reaction. For example, monopolar reactions proceed with a change in net charge ($\delta q_c \# 0$) at the reaction center and are denoted C_m; dipolar reactions, C_d, produce no net change in charge but involve a change in the dipole moment ($\delta \mu_c \# 0$, $\delta q_c = 0$, etc.,). The nature and the magnitude of this electrostatic change accompanying the reaction determine the "susceptibility" of a given reaction to electric fields existing in structure, P.

2.2.2.1 Field Effects Model.

For a given dipolar or charged substituent interacting with the change in the charge at the reaction center, the direct field effect may be expressed as a multipole expansion

$$\delta(\Delta E)_{field} = \frac{\delta q_c\, q_s}{r'_{cs} D_e} + \frac{\delta q_c\, \mu_s \cos\theta_{cs}}{r_{cs}^{2} D_e} + \frac{\delta \mu_c\, q_s \cos\theta_{cs}}{r'^{\,2}_{cs} D_e} .. + .. \tag{4}$$

where q_s is the charge on the substituent, approximated as a point charge located at point, s'; μ_s is the substituent dipole located at point s (this dipole includes any polarization of the anchor atom i effected by S); q_c ($\delta \mu_c$) is the *change* in charge (dipole moment) of the reaction center accompanying the reaction, both presumed to be located at point c; θ_{cs} is the angle the dipole subtends to the reaction center; D_e is the effective dielectric constant for the medium; and r_{cs} (r_{cs}') is the distance from the substituent dipole (charge) center to the reaction center.

This electrostatic interaction between S and C depends on the magnitude and relative orientation of the local fields of S and C and the dielectric properties and distances through the conducting medium. All uncharged dipole substituents and positively charged substituents will increase the acidity of any acid, no matter what the charge and hence, exert a +F. For a negatively charged substituent, the dipole field component tends to lower the pK_a, whereas the negative charge field component tends to raise the pK_a.

In modeling electrostatic effects, only those terms in Equation 4 containing the "leading" nonzero electric field change in the reaction center are retained. For example, acid-base ionization is a monopole reaction that is described by the first two terms of equation 4; electron affinity is described by only the second term, whereas a dipole transition in light absorption will be described by the three terms.

Once again in order to provide parameter "portability" and, hence, effects-model portability to other structures and to other types of chemical reactivity, the contribution of each structural component is quantified independently:

$$\delta_{field}(pK_a)_c = \rho_{ele}\,\sigma_p = \rho_{ele}\,\sigma_{cs}F_S \tag{5}$$

σ_p characterizes the field strength that the perturber exerts on the reaction center. ρ_{ele} is the susceptibility of a given reaction center to electric field effects that describe the electrostatic charge accompanying the reaction. ρ_{ele} is presumed to be independent of the perturber. The perturber potential, σ_p, is further factored into a field strength parameter, F (characterizing the magnitude of the field component, charge or dipole, on the substituent), and a conduction descriptor, σ_{cs} of the intervening molecular network for electrostatic interactions. This structure-function specification and subsequent parameterization of individual component contributions enables one to analyze a given molecular structure (containing some arbitrary assemblage of functional elements) and "piece together" the appropriate component contributions to give the resultant reactivity effect. For molecules containing multiple substituents, the substituent field effects are computed for each substituent and summed to produce the total effect as

$$\delta_{field}(pK_a)_c = \rho_{ele}\sum \sigma_{cs}F_s \tag{6}$$

The electrostatic susceptibility, ρ_{ele}, is a data-fitted parameter inferred directly from measured pK_as. This parameter is determined once for each reaction center and stored in the SPARC database. In parameterizing the electrostatic field effects models, the ionization of the carboxylic acid group is chosen to be the reference reaction center with an assigned ρ_{ele} of 1. For all the reaction centers addressed in SPARC, electrostatic interactions are calculated relative to a fixed geometric reference point that is chosen to approximate the center of charge for the carboxylate anion, $r_{cj} = 1.3$ unit, where the length unit is the aromatic carbon-carbon length (1.40Å). The ρ_{ele} for other reaction centers reflects electric field changes for these reactions gauged relative to the carboxylic acid reference, but also subsumes any difference in charge distribution relative to the reference point, c.

With regard to the substituent parameters, each uncharged substituent has one field-strength parameter, F_μ, characterizing the dipole field strength; whereas, a charged substituent has two, F_q and F_μ. F_q characterizes the effective charge on the substituent and F_μ describes the effective substituent dipole inclusive of the anchor atom i, which is assumed to be a carbon atom. If the anchor atom i is a noncarbon atom, then F_μ is adjusted based on the electronegativity of the anchor atom relative to carbon. The effective dielectric constant, D_e, for the molecular cavity, any polarization of the anchor atom i affected by S, and any unit conversion factors for charges, angles, distances, etc., are included in the F's. Tables 1 and 2 list the characteristic reaction center and substituent parameters, respectively.

Table 1
Reaction Center Parameters

Reaction Center	ρ_{elc}	ρ_{res}	χ_c	$(pK_a)_c$
CO_2H	1.000	-1.118	2.60	3.75
AsO_2H	0.653	-0.817	2.22	6.63
PO_2H	0.489	-0.394	2.72	2.23
PSOH	0.291	-0.402	2.69	1.55
PS_2H	0.101	-0.802	2.63	1.96
BO_2H_2	0.355	-0.050	3.04	8.32
SeO_3H	1.207	-0.400	2.30	4.64
SO_3H	0.451	-4.104	2.09	-0.09
OH	2.706	18.44	2.49	14.3
SH	2.195	4.348	2.76	7.40
NR_2	3.571	19.36	2.40	9.83
in-ring N	5.726	-11.27	2.31	2.28
=N	5.390	-4.631	2.47	5.33

The distance between the reaction center and the substituent, r_{cs}, for both charges and dipoles is computed as a summation of the respective distance contributions of C, R and S as

$$r_{cs}^o = r_{cj} + r_{ij} + r_{is} \qquad (7)$$

For ring systems this zero-order distance is adjusted for direct through space interactions of S and C as opposed to through the molecular cavity. These adjustments are significant only when C and S are ortho or perri to each other.

Table 2
Substituent Characteristic Parameters[a]

Substituent	F_μ	F_q	M_F	E_r	r_{is}	χ_s
CO_2H	2.233	0.000	0.687	0.072	0.80	3.43
$CO2^-$	1.639	-0.603	0.560	2.978	1.00	2.68
AsO_3H^-	0.300	-0.500	0.500	0.190	1.20	2.60
AsO_3^{-2}	0.600	-1.000	0.300	0.150	1.20	2.60
PO_3H^-	0.600	-0.786	0.400	0.220	1.20	3.32
PO_3^{-2}	0.600	-2.500	0.400	0.840	1.20	2.90
BO_2H2	1.078	0.000	1.010	1.484	0.80	2.40
SO_3^-	6.315	-1.224	2.491	1.407	0.80	2.82
OH	1.506	0.000	-3.116	7.240	0.80	2.76
SH	2.931	0.000	-1.871	3.000	0.80	2.76
O^-	1.913	-1.566	-3.546	11.00	0.50	3.01
S^-	1.727	-1.537	-1.437	9.368	0.50	3.34
NR_2	1.190	0.000	-4.939	17.42	0.70	2.58
NR_2H^+	3.978	0.779	-2.505	21.70	0.50	3.23
CH_3	-1.10	0.000	-2.065	0.129	0.63	2.30
NO_2	7.746	0.000	2.515	3.677	1.00	3.79
NO	6.714	0.000	4.127	1.691	1.00	3.80
$C{\equiv}N$	5.649	0.000	3.141	3.196	0.80	3.71
OR	2.138	0.000	-4.767	1.987	0.80	2.90
SR	2.323	0.000	-1.234	1.952	0.80	2.80
I	4.270	0.000	0.000	4.928	0.75	2.95
Br	3.756	0.000	-0.031	3.012	0.70	3.19
Cl	3.622	0.000	-0.066	1.498	0.70	3.37
F	3.164	0.000	-1.718	0.800	0.65	3.67
in-ring N	5.310	0.000	0.929	2.055	0.00	3.30
in-ring NH^+	1.379	3.785	6.995	8.708	0.00	3.80
SO_2	6.451	0.000	2.038	4.176	0.80	3.60
$=N$	1.533	0.000	0.544	4.918	0.00	3.80
$P=O$	3.004	0.000	2.043	2.300	0.80	3.08
$=NH^+$	2.000	1.000	2.800	2.600	0.00	3.80
$C=O$	3.195	0.000	1.584	2.281	0.00	3.60

[a] F_μ = dipole field parameter; F_q = charge field parameter; M_F = mesomeric effect parameter; E_r = resonance parameter; r_{is} length parameter; χ_s = electronegativity parameter.

$$r_{cs} = A\, r_{cs}^{\,o} \tag{8}$$

where A is an adjustment constant and is assumed to depend only on bond connectivity into and out of the R-π, unit (e.g., points i and j).   For R-π units recognized by SPARC, "A factors" for each pair (i,j) are empirically determined from data (or inferred from structural similarity to other R-π units) as shown in Table 3. The distance through R (r_{ij}) is calculated by summation over delineated units in the shortest molecular path from i to j. All aliphatic bonds contribute 1.0 unit; double and triple bonds contribute 0.9 and 0.8 units, respectively. For ring systems SPARC contains a template listing distances between each constituent atom pair as illustrated in Table 3.

The dipole orientation factors, $\cos\Theta_{ij}$, are presently ignored (set to 1.0) except in those cases where S and C are attached to the same rigid R-π unit. In these situations, they are assumed to depend solely on the point(s) of attachment, (i,j), and are pre-calculated and stored in SPARC databases.

Table 3
Position on Ring and Geometry Parameters

Molecule	Position on ring		Geometry parameters		
	Reaction Center	Substituent	r_{ij}	A_{ij}	$\cos\theta_{ij}$
benzene	1	2	1.0	0.25	0.53
	1	3	1.7	0.87	0.88
	1	4	2.0	1.0	1.00
naphthalene	1	2	1.0	0.25	0.53
	1	3	1.7	0.87	0.88
	1	4	2.0	1.00	1.00
	1	5	2.6	0.73	0.81
	1	6	3.0	0.63	0.83
	1	7	2.7	0.64	0.81
	1	8	1.7	0.47	0.77
	2	1	1.0	0.25	0.53
	2	3	1.0	0.25	0.53
	2	4	1.7	0.81	0.91
	2	5	3.0	0.63	0.83
	2	6	3.6	0.98	0.96
	2	7	3.4	0.80	0.84

2.2.2.2 Mesomeric Field Effects.

Structure 1

The substituent can "induce" electric fields in the R that can interact electrostatically with C. This indirect interaction is called the "mesomeric field effect". The amino group in structure 1 should exert a +F effect and lower the pK_a. However, the observed effect is exactly the opposite of that predicted. The pK_a of m-aminopyridine is 6.1 and is greater than the pK_a of pyridine (5.2). In this case, the NH_2-induced charges are ortho and para to the in-ring N. These charges will interact indirectly with the dipole of the nitrogen in the ring and raise the pK_a.

The contribution of the mesomeric field can be estimated as a collection of discrete charges, q_R, with the contribution of each described by appropiate substituent-monopole terms of Equation 4. As is the case in modeling the direct field effects, the mesomeric effect components are resolved into three independent components S, R, and C and as

$$\delta_{M_F}(pK_a)_c = \rho_{ele}\, q_R\, M_F \qquad (9)$$

where, M_F is the mesomeric field effect constant that is characteristic of the substituent S. It describes the ability or strength of a given substituent to induce a field in R_π. q_R describes the location and relative charge distributions in R, and ρ_{ele} describes the susceptibility of a particular reaction center to electrostatic effects. Since the reaction center does not discriminate the sources of electric fields, ρ_{ele} is the same as that described previously in discussions of the field effects.

In modeling the mesomeric field effect, the intensity and the location of charges in R depend on both the substituent and the R_π network involved. The contributions of S and R_π are resolved by replacing the substituent with a reference probe or NBMO (NonBonded Molecular Orbital) charge source. This NBMO reference source was chosen to be the methylene anion, $-CH_2^{\cdot}$, for which the charge distribution in any arbitrary R_π network can be calculated as described in Appendix I. In calculating NBMO charge distribution across essential single bonds in R_π, substituent-specific electrometric and steric factors are considered as described in Appendix II.

The mesomeric substituent strength parameter describes the π-induction ability of a particular substituent relative to the CH_2^-. The magnitude of a given M_F parameter describes the relative field strength, whereas the sign of the parameter specifies the positive or negative character of the induced charge in R_π. The mesomeric field effect for a given substituent is given by:

$$\delta_{M_F}(pK_a)_c = \rho_{ele} M_F \sum_k \frac{q_{ik}}{r_{kc}} \qquad (10)$$

where q_{ik} is the charge induced at atom k, with the reference probe attached at atom i calculated using PMO [11,12] theory (see Appendix I for calculation procedure). r_{kc} is the through-cavity distance to the reaction center as described previously for direct fields effect. Because induction does not change total molecular charge, the sum of all induced charges must be zero. This is achieved by placing, at the location of the substituent, a compensating charge, q_s, equal to but opposite to the total charge distributed within the R_π network.

2.2.2.3 Sigma Induction Effects Model
Sigma induction derives from electronegativity differences between two atoms. The electron cloud that bonds any two atoms is not symmetrical except when the two atoms are the same and have the same substituents; hence, the higher electronegativity atom will polarize the other. The effect is believed to be transmitted progressively between atoms. The substituent electronegativity effect is important only at the atom to which the substituent is attached and any effect beyond the second atom is negligible.

The interaction energy of this effect depends on the difference in electronegativity between the reaction center and the substituent and on the number of substituents bonded to the reaction center. Sigma induction effects are resolved into two independent structural component contributions of S and C.

$$\delta_{sig}(pK_a)_c = \rho_{ele} \sum [\chi_c - \chi_s] \qquad (11)$$

where ρ_{ele} is the susceptibility of a given reaction center to electric field effects. Once again, because the reaction center does not discriminate the source of the electric fields, ρ_{ele} is the same as described for the direct field effect. χ_c is the effective electronegativity of the reaction center. χ_s is the effective electronegativity of the substituent. The electronegativity of the reaction centers and the substituents are estimated based on the electronegativity of the methyl group that was chosen to be the reference group.

2.2.3. Resonance Effects Model
Resonance involves variations in charge transfer between the π system and a suitable orbital of the substituent. The interaction of the substituent orbital with a π-orbital of a reaction center can lead to charge transfer either to or from

that reaction center. Electron withdrawing reaction centers will localize the charge over itself. As a result, the acidic state will be stabilized more than the basic state making these compounds less acidic. For electron donating reaction centers, resonance will stabilize the basic state more than the acidic state and lower the pK_a.

Resonance stabilization energy in SPARC is a differential quantity, related directly to the extent of π electron delocalization in the neutral state versus the ionized state of the reaction center. The source or sink in P may be the substituents or R-π units contiguous to the reaction center. As with the case of electrostatic perturbations, structural units are classified according to function. Substituents that withdraw electrons are designated S+ while electron-donating groups are designated S-. The R-π units withdraw or donate electrons, or serve as "conductors" of π electrons between resonance units. Reaction centers are likewise classified as C+ or C-, denoting withdrawal or donation of electrons, respectively.

In SPARC, the resonance interactions describe the delocalization of an NBMO out of C_i or C_f into a contiguous R-π or a conjugated substituent. To model this effect, the reaction center is replaced by a surrogate electron donor, CH_2^-. The distribution of NBMO charge from this surrogate donor is used to quantify the acceptor potential for the substituent and the molecular conductor. The resonance perturbation of the initial state versus the final state for an electron-donating reaction center is given by:

$$\delta_{res} (pK_a)_c = \rho_{res} (\Delta q)_c \qquad (12)$$

where $(\Delta q)_c$ is the fraction loss of NBMO charge from the surrogate reaction center calculated based on PMO theory (see Appendix I). ρ_{res} is the susceptibility of a given reaction center to resonance interactions. ρ_{res} quantifies the differential "donor" ability of the two states of the reaction center relative to the reference donor CH_2^-. In parameterization of resonance effects, resonance strength is defined for all the substituents (i.e., the ability to donate or receive electrons); resonance susceptibility is defined for all the reaction centers; and resonance "conduction" in R_π networks is modeled so as to be portable to any array of R_π units or to linking any resonance source or sink group.

2.2.4. Solvation Effects Model

If a base is more solvated than its conjugate acid, its stability increases relative to the conjugate acid. For example methylamine is a stronger base than ammonia, and diethylamine is stronger still. These results are easily explainable due to the sigma induction effect. However, trimethylamine is a weaker base than dimethylamine or methylamine. This behavior can be explained by differential hydration of the reaction center of interest and the reference reaction center.

The initial state and the final state of the reaction center frequently differ substantially in degree of solvation, with the more highly charged moiety solvating more strongly. Thus, steric blockage of the reaction center is distinguished from the steric-induced twisting of the reaction center incorporated in electron delocalization interactions. Differential solvation is a significant effect in the protonation of organic bases (e.g., $-NH_2$, in-ring N, =N) but is less important for acidic compounds except for highly branched aliphatic alcohols.

In SPARC's reactivity models, differential solvation of the reaction center is incorporated in $(pK_a)_c$, ρ_{res} and ρ_{ele}. If the reaction center is bonded directly to more than one hydrophobic group or if the reaction center is ortho to an aromatic bridge (perri), then $\delta_{sol}(pK_a)$ must be calculated. The $\delta_{sol}(pK_a)$ contributions for each reaction center bonded directly to more than one hydrophobic group are quantified based on the sizes and the numbers of hydrophobic groups attached to the reaction center and\or to the number of the aromatic bridges that are ortho to the reaction center.

2.2.5. Intramolecular H-bonding Effects Model

Intramolecular hydrogen bonding is a direct site coupling of a proton-donating (α) site with a proton accepting (β) site within the molecule. Reaction centers might interact with substituents through intramolecular hydrogen bonding and thus impact the pK_a. The C_i and C_f frequently differ substantially in degree of hydrogen bonding strength with a substituent.

In aromatic, π-ring or π-aliphatic systems where the reaction center is contiguous to the substituent and where a stable 5- or 6-member ring may be formed, $\delta_{H-B}(pK_a)_c$ must be estimated. $\delta_{H-B}(pK_a)_c$ is a differential quantity that describes the H-bonding differences of the initial state versus the final state of a reaction center with a substituent. SPARC estimates the hydrogen bonding contributions for each reaction center with each substituent, $\delta_{H-B}(pK_a)_c$, taking into consideration steric effects. For reaction centers that might hydrogen bond with more than one substituent, the hydrogen bonding contribution for each substituent is calculated and the stronger contributor is selected.

2.2.6. Statistical Effects Model

All the SPARC perturbation models presented thus far describe the ionization of an acid at a single site. If a molecule contains multiple equivalent sites, a statistical correction $\delta_{stat}(pK_a)_c$ is required. For example, if a first ionization equilibrium constant, K, is computed for a single site, and if the molecule has n such sites, then

$$\delta_{stat}(pK_a)_c = \log \frac{n_a}{n_b} \tag{13}$$

where a and b refer to the acid and conjugate base sites, respectively.

306

2.3. Results and Discussion

Figure 1 shows sample calculations of 4-hydroxybenzoic acid pK$_a$'s. It is important to reemphasize that the reaction parameters describing a given reaction center (Table 1) are the same regardless of the appended molecular structure. Likewise, substituents (Table 2) are independent of the rest of the molecule. Once again, this structure factoring and function specification enables one to construct, for a given reaction center of interest, essentially any molecular array of appended units, and to compute the resultant reactivity.

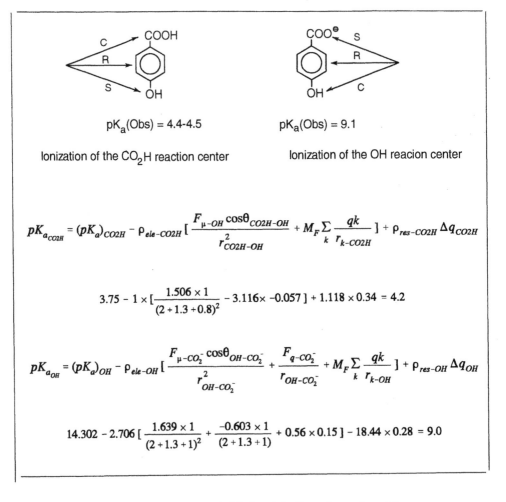

Figure 1. Sample calculation of pK$_a$ for 4-hydroxybenzoic acid.

The SPARC ionization pK_a calculator was first parameterized using measured ionization constants for more than 775 compounds from the International Union of Pure and Applied Chemistry (IUPAC) [13-15]. The RMS (Root Mean Squared) deviation for the set was found to be equal to 0.22 pK_a units. The pK_a calculator was then tested on more than 4000 pK_a's for 3500 compounds including multiple pK_as up to the sixth pK_a *spanning a range of over 30 pK_a units* [16]; the result of this rigorous test is shown in Figure 2. The reaction centers in this study were $-CO_2H$, $-OH$, $-SH$, $-PO_2H$, $-AsO_2H$, $-BO_2H_2$, $-SeO_3H$, $-PSOH$ and $-PS_2H$ as acids, $-NR_2$, in-ring N and $-N=$ as bases. The $(pK_a)_c$ for CO_2H, OH, SH and NR_2 were measured values, whereas the rest of the reaction centers were trained values inferred directly from pK_a measurements (see Table 1). The RMS deviation for this large set of compounds was found to be less than 0.35 pK_a units, which is comparable to experimental error. The RMS deviation for the acid set and the NR_2 set were less than 0.31. The $=N$ and the in ring N set have an RMS equal to 0.4 pK_a units. We believe that the measurements are not better than 0.4 pK_a units for in-ring N and even higher for azo dye compounds where the experimental error in the measured pK_a's for some azo dyes can be as high as 2 pK_a units (for more details see reference 7).

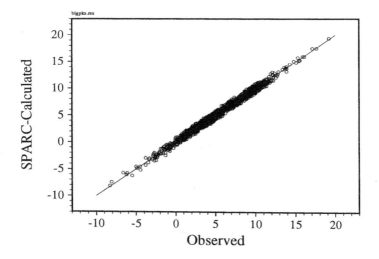

Figure 2.   Observed versus SPARC-calculated pK_a's for more than 4000 pK_a's for 3500 Organic Compounds. The RMS deviation was found to be less than 0.35 pK_a units.

SPARC's pK_a calculator was also used to estimate 358 pK_a's for 214 azo dyes and a number of related aromatic amines. This set includes some azo dyes that have up to eight multiple ionization sites. The RMS deviation was found to be 0.62 pK_a units as shown in Figure 3. The reported RMS interlaboratory deviations between the different observed values for azo dyes and related aromatic amines where more than one measurement was reported is 0.64 [7]. We believe, therefore, that the errors in our calculated values are comparable to experimental error for these complicated azo dyes.

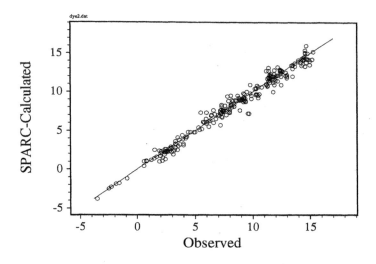

Figure 3.   Observed vs. SPARC calculated values for 358 pK_a's for 214 azo dyes and a number of related aromatic amines. The RMS deviation was found to be 0.62 pK_a unit.

The perturbations of some reaction centers such as CO_2H or any other oxy acid are small [16], whereas OH, NR_2, in-ring N and =N reaction centers have large perturbations. Table 4 shows the perturbations of the =N reaction center and how the SPARC pK_a calculator predicts these large perturbations. The $(pK_a)_c$ for =N is 5.39 (see Table 1). Resonance and mesomeric effects raise the pK_a substantially as in the case of guanidine. Electrostatic effects (field and mesomeric) of the nitro group lower the pK_a by more than 14 pK_a units. The electrostatic and hydrogen bonding effects between the two =N groups lower the two pK_a's in diguanide. Solvation and electrostatic effects will also lower the pK_a as is the case for azobenzene, which is shown in Table 4. For test performance of the SPARC pK_a calculator see references 6,7 and 16.

Table 4
Observed vs. SPARC-Calculated pK$_a$ of =N Reaction Center for Selected Molecules

Molecule	Observed	Calculated
H$_2$N–C(=NH)–NH$_2$	13.9	13.7
H$_3$C–C(=NH)–NH–C(=NH)–NH$_2$	12.5 , 3.4	12.3 , 3.4
H$_3$C–C(=NH)–NH$_2$	12.25	12.0
HO–C$_6$H$_4$–C(=NH)–NH$_2$	12.5	11.9
CH$_3$O–C(=NH)–NH$_2$	10.4	11.0
imidazole (=NH)	6.9	6.5
Ph–C(=NH)–Ph	6.9	6.9
o-CH$_3$-C$_6$H$_4$–C(=NOH)–NH$_2$	3.99	3.5
Ph–N=C–C$_6$H$_4$–Cl	2.8	2.8
H$_2$N–C(=NNO$_2$)–NH$_2$	-0.93	-1.0
Ph–N=N–Ph	-2.78	-2.1
Ph–N=N–C$_6$H$_4$–NO$_2$	-3.5	-3.6

3.   ESTIMATION OF ELECTRON AFFINITY

INTRODUCTION

One of the fundamental properties of gaseous negative ions is the lowest energy required to remove an electron. This energy is called the electron affinity (EA). The EA of a molecule plays an important role not only in gas-phase ions, but also in condensed-phase and charge-transfer complexes in chemistry, biology and physics. Although gaseous negative ions in molecules were first observed around 1900 and have been studied extensively since then, the first reliable EA was not obtained until the early 1960's for O_2 [17]. Since that time, numerous methods [18] have been utilized to predict and measure electron affinity. Despite the availability of a large number of methods and the fundamental importance of electron affinity values, the complexity of molecular negative ions and the inherent difficulties in determining EA have prevented the determination of this important property for many molecules of interest. Wide disagreement also exists among reported values [18]. This study seeks to apply the SPARC chemical reactivity models to the calculation of electron affinity strictly from the molecular structures of a wide range of nonpolymeric organic compounds.

3.1. SPARC Computational Methods

The EA property of a molecule describes the conversion of the neutral molecule to a molecular negative ion when both the neutral molecule, E, and the negative ion, E⁻, are in their most stable state. Electron affinity is defined as the difference in energy between a neutral molecule plus an electron at rest at infinity and the molecular negative ion when both the neutral molecule and the negative ion are in their ground electronic, vibrational and rotational states, i.e. the energy change for the reaction at $0°$ K.

$$E + e^- \underset{K_{-1}}{\overset{K_1}{\rightleftharpoons}} E^-$$

The added electron enters the LUMO (Lowest Unoccupied Molecular Orbital), which in the negative ion becomes the HOMO (Highest Occupied Molecular Orbital). The lower the energy of the LUMO the higher will be the electron affinity and vice versa. The energy differences between the LUMO state and the HOMO state are small compared to the total binding energy of the reactant involved; hence, perturbation theories can be used to calculate the energy differences between the two states. These theories treat the final state as a perturbed initial state and the energy differences between these two energy states can be determined by quantifying the perturbation. This perturbation of the HOMO state versus the LUMO state is factored into mechanistic components of resonance, field and sigma induction contributions. Similar to SPARC's approach to ionization pK_a, molecular structures are broken into functional units having known intrinsic electron affinity $(EA)_c$. This intrinsic behavior is adjusted for the molecule in question using the mechanistic perturbation models described in Section 2.

3.2. Electron Affinity Models

As was the case for ionization pK_a, the SPARC computational procedure starts by locating the potential sites within the molecule at which a particular reaction of interest could occur. In the case of electron affinity these reaction centers, C, are the smallest subunit(s) that could form a molecular negative ion. Any molecular structure appended to C is viewed as a "perturber" (P). All reactions to be addressed in SPARC are analyzed in terms of critical equilibrium components:

$$P\!-\!C_i \xrightarrow{\quad EA \quad} P\!-\!C_f$$

where C_i and C_f denote the initial state and the final state or the LUMO state and the HOMO state of the reaction center, respectively; P is the structure that is presumed unchanged by the reaction; and EA denotes the electron affinity reaction. EA is expressed as a function of the energy required to add an electron to the LUMO state. It represents the energy difference between the neutral molecular state and the molecular negative ion state. To model this energy difference, EA is expressed in terms of the summation of the contributions of all the components, perturber(s) and reaction center(s), in the molecule:

$$EA = \sum_{c=1}^{n} [\, (EA)_c + \delta_p \, (\Delta EA)_c \,] \tag{14}$$

where the summation is over n, which is defined as the number of subunits that could form a molecular negative ion or simply as the number of reaction centers in the molecule. $(EA)_c$ is the electron affinity for the reaction center. The electron affinity of the reaction center is assumed to be unperturbed and independent of P. $\delta_p(\Delta EA)_c$ is a differential quantity that describes the change in the EA behavior affected by the perturber structure. $\delta_p(\Delta EA)_c$ is factored into mechanistic contributions as

$$\delta_p \, (\Delta EA)_c = \delta \, (\Delta EA)_{field} + \delta \, (\Delta EA)_{res} + \delta \, (\Delta EA)_{sig} \tag{15}$$

where $\delta(\Delta EA)_{field}$ describes the difference in field interactions of P with the two states, $\delta(\Delta EA)_{res}$ describes the change in the delocalization of π electrons of the two states due to P (this delocalization of π electrons is assumed to be into or out of the reaction center), and $\delta(\Delta EA)_{sig}$ describes the change in sigma induction of P with the two states.

3.2.1. The Field Effects Model

Field effects derive from charges or electrical dipoles in the appended structure P, interacting with the charges or the dipoles of the reaction center, C, through space. The molecular conductor, R, acts as a low dielectric conductor. This effect follows from the fact that the bonds between most atoms are not completely covalent, but possess a partial ionic character that imposes electrical asymmetry either in the substituent or the reaction center bonds. The field interaction between the substituent and the reaction center depends on the magnitude and the relative orientation of the local fields of S and C, the dielectric properties of the conduction medium D_e, and the distance through the molecular cavity.

The field effect of a given S is given as

$$\delta(\Delta EA)_{field} = \frac{\delta q_c \; \mu_s \cos\theta_{cs}}{r_{cs}^2 \; D_e} \tag{16}$$

where μ_s is the substituent dipole located at point S; δq_c is the change in charge of the reaction center accompanying the reaction, presumed to be located at point C; Θ_{sc} gives the orientation of the substituent dipole relative to the reaction center; r's are the appropriate distances of separation; and D_e gives the effective dielectric constant for the intervening conduction medium.

Field effects are resolved into three independent structural component contributions representing the change in dipole field strength of S, a conduction factor of R, and the change in the charge at C as

$$\delta(\Delta EA)_{field} = \rho_{ele} \, \sigma_p = \rho_{ele} \, \sigma_R F_S \tag{17}$$

where σ_p describes the potential of P to "create" an electric field irrespective of C, and ρ_{ele} is the susceptibility of a given reaction center to electric field effects that describes the change in the electric field of the reaction center accompanying the reaction. The perturber potential, σ_p, is further factored into the field strength parameter, F_S, (describing the magnitude of the dipole field component on the substituent) and a conduction descriptor, σ_R, of the intervening molecular network for the electrostatic interactions.

The ρ_{ele} for the electron affinity of the reaction centers are data-fitted parameters that are inferred directly from electron affinity measurements data as shown in Table 5. The field strength parameter, F_S, for each substituent is inferred from measurements of ionization pK_a (Table 2). The distances among the various components and the orientation angle are calculated from geometry models and stored in the SPARC database as explained in the ionization pK_a section (Table 3).

Table 5
Reaction Center Characteristic Parameters

Reaction Center	ρ_{ele}	ρ_{res}	χ_{cs}	$(EA)_c$
-NO$_2$	-0.05	2.0	3.28	0.57
-C≡N	-0.03	1.2	----	0.29
-C=O	-0.02	3.2	----	-0.39
in-ring N	-0.16	0.76	----	0.20
-NR$_2$	0.02	-1.01	----	0.17
-OH	0.11	-1.80	----	0.42
-OR	0.06	-0.67	----	0.24
-CH$_3$	0.01	-0.14	2.30	0.01
-F	-0.01	-0.11	4.35	0.16
-Cl	-0.02	-0.09	4.15	0.25
-Br	-0.05	-0.07	3.95	0.26

3.2.2. Sigma Induction Model

Sigma induction derives from electronegativity differences between two atoms. This effect is transmitted progressively through a chain of σ-bonds among atoms. This is a short-range interaction that is strong when the two atoms are bonded to each other and any effect beyond the second atom is negligible. As is the case in modeling field effects, sigma induction effects are resolved into the three independent structural component contributions of S, R and C, characterizing the change in the difference of the electronegativity between the substituent and the reaction center, a conduction factor of R, and the change in the electrostatic effects of the reaction center.

$$\delta(\Delta EA)_{sig} = \rho_{ele}\, d\chi_{cs} \tag{18}$$

where ρ_{ele} is the susceptibility of a given reaction center to electric field effects and $d\chi_{cs}$ is the difference in the effective electronegativity of the reaction center and the substituent.

3.2.3 Resonance Model

Resonance involves the delocalization of π electrons into or out of the reaction center. This long range interaction is transmitted through the π-bond network and results in a different distribution of electron density than would be the case if there were no resonance. The resonance reactivity perturbation, $\delta_{res}(\Delta EA)$, is the differential resonance stabilization of the initial versus final state of the reaction center. It is a differential quantity, related directly to the extent of electron delocalization in the neutral state versus the molecular ion state of the reaction center. The source or sink in P may be the substituents or R-π units contiguous to the reaction center.

As explained in the estimation of ionization pK_a, the reaction center is replaced by a surrogate electron donor, CH_2^-. The distribution of NBMO charge from this surrogate donor is used to quantify the acceptor potential for the perturber structure, P. The reactivity perturbation is given by:

$$\delta(\Delta EA)_{res} = \rho_{res} (\Delta q)_c \qquad (19)$$

where $(\Delta q)_c$ is the fraction loss of NBMO charge from the surrogate reaction center, and the susceptibility, ρ_{res}, of a given reaction center to resonance quantifies the differential "donor" ability of the two states of the reaction center relative to the reference donor CH_2^-.

3.3   Results and Discussion

In modeling any property in SPARC, the contributions of the structural components C, S, and R are quantified (parameterized) independently. For example, the strength of a substituent in creating an electrostatic field effect depends only on the substituent regardless of the C, R, or property interest. Likewise, the molecular network conductor R is modeled so as to be independent of the identities of S, C, or the property being estimated. Hence, S and R parameters for electron affinity are the same as those for pK_a (Table 2). The susceptibility of a reaction center to an electrostatic effect quantifies only the differential interaction of the initial state versus the final state with the electrostatic fields. The susceptibility gauges only the reaction $C_{initial} - C_{final}$ and is completely independent of both R or S. For instance, for electron affinity the electrostatic susceptibility reflects the electrostatic perturbations of the LUMO state versus the HOMO state, which once again, is totally independent of C or R. Thus, no modifications in pK_a models or extra parameterization for either S or R are needed to calculate electron affinity from pK_a models other than inferring the electronegativity and electron affinity susceptibility of the reaction centers to electrostatic and resonance effects.

Figure 4 shows a sample calculation of electron affinity for 4-chloronitrobenzene. SPARC first computes the resonance and electrostatic perturbations of the appended substituent Cl para to the reaction center NO_2 through the molecular conductor of the benzene ring. Next SPARC computes the perturbation of the appended substituent NO_2 para to the reaction center Cl through the benzene ring. Finally, SPARC sums these perturbations with the base electron affinities for NO_2 and Cl. The susceptibilities of the NO_2 and Cl for resonance and electrostatic effects are shown in Table 5. The substituent parameters and the distance between NO_2 and Cl are obtained as described in Section 2.

Figure 5 shows the SPARC-calculated values versus the observed values of electron affinity. The RMS deviation was found to be 0.14 eV, which is about

$$E.A_{(4\text{-chloronitrobenzene})} \quad = \quad \delta(EA)_{NO2} \quad + \quad \delta(EA)_{Cl}$$

Reaction center	NO_2	Cl
Substituent	Cl	NO_2
Molecular network	benzene	benzene

$$\delta(EA)_{no2} = \delta_{no2}\, EA_{bas} + \delta_{no2}\, EA_{field} + \delta_{no2}\, EA_{res}$$

$$\delta(EA)_{no2} = EA_{no2} + \frac{\rho_{ele} \times F_s \times \cos\theta}{(r_{cj} + r_{ij} + r_{is})^2} + \rho_{res} \times \Delta q$$

$$\delta(EA)_{no2} = 0.51 + \frac{-0.05 \times 3.622 \times 1.0}{(1.3 + 2.0 + 0.65)^2} + 2.0 \times 0.236 = 0.97$$

$$\delta(EA)_{cl} = \delta_{cl}\, EA_{bas} + \delta_{cl}\, EA_{field} + \delta_{cl} EA_{res}$$

$$\delta(EA)_{cl} = EA_{cl} + \frac{\rho_{elec} \times F_s \times \cos\theta}{(r_{cj} + r_{ij} + r_{is})^2} + \rho_{res} \times \Delta q$$

$$\delta(EA)_{cl} = 0.25 + \frac{-0.02 \times 7.46 \times 1.0}{(1.3 + 2.0 + 1)^2} + (-0.09 \times 0.273) = 0.217$$

$$EA_{4\text{-chloronitrobenzene}} \quad = \quad 0.97 \quad + \quad 0.217 \quad = 1.187 \text{ eV}$$

Figure 4. Sample calculation of 4-chloronitrobenzene EA.

equal to measurement error in charge transfer experiments. The electrostatic contributions to EA for the molecules studied were found to be much smaller than the resonance contributions; electrostatic perturbations were almost the same for the LUMO state and the HOMO state. Figure 6 shows the calculated and observed electron affinity for multiple cyano groups attached to a benzene ring where the electrostatic contributions are included and neglected. Although electrostatic contributions are generally small when many large dipole substituents are present, the effects are significant.

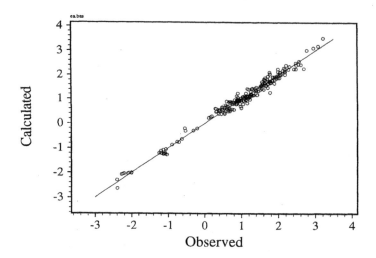

Figure 5.   Observed vs. SPARC calculated electron affinity. The RMS was found to be 0.14 eV.

Benzene is the progenitor of the other aromatic compounds. The added electron enters the LUMO state, which in the negative ion becomes the SOMO (Singly Occupied Molecular Orbital). Since the LUMO energy is still high, a stable negative ion is not formed in the gas phase. The LUMO energy can be lowered and stabilized either by expansion of the π conjugation or by introduction of electron-withdrawing substituents. For unsubstituted benzene, naphthalene and anthracene, the electron affinity is seen to increase significantly due to increase in the resonance contributions going from benzene to anthracene. A positive electron affinity (a stable negative ion) is observed only for anthracene.

Introducing π-electron withdrawing substituents such as cyano, aldehyde and ketone strongly perturb the benzene and raise the electron affinity significantly. The LUMO states for these substituents are close to the degenerate π^* benzene LUMOS. This leads to a strong interaction between the LUMO states of these substituents and one of the degenerate LUMO states of benzene [32], which in turn, lowers and stabilizes the energy of the LUMO states of these molecules. The order of the resonance stabilizing effect of the LUMO state of these substituents is -C=O > -NO$_2$ > -C≡N > in ring N. For benzene substituted by any electron-donating groups like F, Cl, Br, NR$_2$, and OR, the perturbations of the LUMO state versus the HOMO state are extremely small. Hence, the EA for these molecules will be close to the EA of benzene (-1.2) as shown in Table 6. The same trend was found for all of these groups when attached to any other aromatic or ethylenic compound.

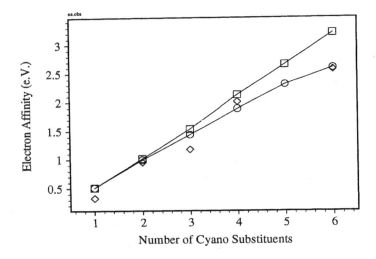

Figure 6.   The effects of multi-cyano groups attached to a benzene on EA. The circles are the EAs including the electrostatic contributions, the squares are the EAs assuming the electrostatic is zero, and diamonds are the observed EAs from ref. 25.

Table 6
Observed vs. SPARC-Calculated of Substituted Aromatics EA

#	Molecule	Observed	Calculated	Ref.
1	benzene	-1.2	-1.2	(20)
2	aniline	-1.1	-1.1	(22)
3	phenol	-1.0	-0.9	(22)
4	benzonitrile	0.3	0.5	(23)
5	nitrobenzene	1.0	1.0	(20)
6	bromobenzene	-0.7	-0.8	(22)
7	chlorobenzene	-0.8	-0.8	(22)
8	fluorobenzene	-0.9	-0.9	(20)
9	benzaldehyde	0.4	0.5	(25)
10	benzophenone	0.6	0.7	(20)
11	methylbenzene	-1.1	-1.2	(22)
12	ethylbenzene	-1.2	-1.2	(22)
13	isopropylbenzene	-1.1	-1.2	(22)
14	tert-butylbenzene	-1.1	-1.2	(22)
15	naphthalene	-0.2	-0.2	(20)

318

Table 6 (cont'd)
Observed vs SPARC-Calculated of Substituted Aromatics EA

#	Molecule	Observed	Calculated	Ref.
16	1-nitronaphthalene	1.2	1.1	(20)
17	2-nitironaphthalene	1.2	1.0	(20)
18	1-cyanonaphthalene	0.7	0.6	(20)
19	2-cyanonaphthalene	0.6	0.5	(20)
20	4-methoxynitronaphthalene	1.1	1.0	(20)
21	2-methylnitronaphthalene	1.0	1.2	(20)
22	1-naphthaldehyde	0.7	0.6	(20)
23	2-naphthaldehyde	0.6	0.5	(20)
24	anthracene	0.6	0.6	(20)
25	9-anthraldehyde	1.3	1.2	(20)
26	9-acetylanthracene	1.0	0.9	(20)
27	1-chloroanthracene	0.8	0.8	(23)
28	2-chloroanthracene	0.8	0.8	(20)
29	9-chloroanthracene	0.9	0:8	(20)
30	9-nitro-anthracene	1.4	1.1	(20)
31	1-cyanoanthracene	1.3	1.0	(20)

Poly Substituted Aromatics

#	Molecule	Observed	Calculated	Ref.
32	p-xylene	-1.1	-1.2	(22)
33	o-xylene	-1.1	-1.2	(22)
34	m-xylene	-1.1	-1.2	(22)
35	1,3,5-trimethylbenzene	-1.2	-1.3	(22)
36	1,2,4-trimethylbenzene	-1.1	-1.3	(22)
37	methylaniline	-1.2	-1.1	(22)
38	dimethylaniline	-1.2	-1.2	(22)
39	p-methylnitrobenzene	0.9	0.9	(20)
40	m-methylnitrobenzene	1.0	1.0	(20)
41	o-methylnitrobenzene	0.9	0.9	(20)
42	2,3-dimethylnitrobenzene	0.9	0.9	(20)
43	3,4-dimethylnitrobenzene	0.9	0.9	(21)
44	2,4-dimethylnitrobenzene	0.9	0.9	(21)
45	2,6-dimethylnitrobenzene	0.8	0.9	(21)
46	2,3-dichloronitrobenzene	1.3	1.3	(21)
47	3,4-dichloronitrobenzene	1.4	1.3	(20)
48	2-chloronitrobenzene	1.1	1.1	(20)
49	3-chloronitrobenzene	1.3	1.2	(20)
50	4-chloronitrobenzene	1.3	1.2	(20)
51	2-fluoronitrobenzene	1.1	1.1	(20)

Table 6 (cont'd)
Observed vs. SPARC-Calculated of Substituted Aromatics EA

#	Molecule	Observed	Calculated	Ref.
52	3-fluoronitrobenzene	1.2	1.1	(20)
53	4-fluoronitrobenzene	1.1	1.1	(20)
54	2-methyl-4-fluoronitrobenzene	1.0	1.0	(20)
55	2,4,6-trimethylnitrobenzene	0.7	0.8	(21)
56	p-dinitrobenzene	2.0	1.9	(20)
57	m-dinitrobenzene	1.6	1.9	(20)
58	o-dinitrobenzene	1.6	1.8	(20)
59	p-nitrobenzonitrile	1.7	1.5	(20)
60	m-nitrobenzonitrile	1.6	1.4	(20)
61	o-nitrobenzonitrile	1.6	1.4	(20)
62	4-cyanobenzonitrile	1.1	1.0	(23)
63	3-cyanobenzonitrile	0.9	1.0	(23)
64	2-cyanobenzonitrile	0.9	1.0	(23)
65	2,6-dichlorobenzonitrile	0.7	0.8	(20)
66	3,5-dinitrobenzonitrile	2.2	2.3	(23)
67	o-nitrobenzaldehyde	1.5	1.5	(20)
68	m-nitrobenzaldehyde	1.4	1.5	(20)
69	p-nitrobenzaldehyde	1.7	1.6	(20)
70	o-trifluoromethylnitrobenzene	1.3	1.2	(20)
71	m-trifluoromethylnitrobenzene	1.4	1.3	(20)
72	p-trifluoromethylnitrobenzene	1.5	1.3	(20)
73	o-trifluoromethylbenzonitrile	0.7	0.8	(20)
74	m-trifluoromethylbenzonitrile	0.7	0.8	(20)
75	p-trifluoromethylbenzonitrile	0.8	0.8	(20)
76	3,5-di-trifluoromethylbenzonitrile	1.1	1.1	(20)
77	3,5-di-trifluoromethylnitrobenzene	1.8	1.6	(20)
78	trifluromethylpentafluorobenzene	0.9	0.9	(20)
79	4,4-difluorobenzophenone	0.8	1.0	(21)
80	2-bromonitrobenzene	1.2	1.1	(20)
81	3-bromonitrobenzene	1.3	1.2	(20)
82	4-bromonitrobenzene	1.3	1.2	(20)
83	3-methoxynitrobenzene	1.0	1.0	(20)
84	4-methoxynitrobenzene	0.9	1.0	(20)
85	3-nitroaniline	0.9	0.9	(20)
86	3-nitro-dimethylaniline	1.0	0.8	(21)
87	4-methylbenzonitrile	0.3	0.5	(23)
88	3,5-dicyanobenzonitrile	1.2	1.4	(23)
89	2,4,5-tricyanobenzonitrile	2.0	1.9	(23)
90	2-fluorobenzaldehyde	0.6	0.7	(25)

Table 6 (cont'd)
Observed vs. SPARC-Calculated of Substituted Aromatics EA

# Molecule	Observed	Calculated	Ref.
91 3-fluorobenzaldehyde	0.6	0.7	(25)
92 4-fluorobenzaldehyde	0.5	0.7	(25)
93 2-fluoroacetophenone	0.5	0.4	(25)
94 3-fluoroacetophenone	0.6	0.4	(25)
95 4-fluoroacetophenone	0.4	0.4	(25)
96 2-trifluoroacetophenenone	0.6	0.7	(25)
97 3-trifluoroacetophenone	0.7	0.7	(25)
98 4-trifluoroacetophenone	0.6	0.7	(25)
99 3-chlorobenzaldehyde	0.6	0.5	(25)
100 4-chlorobenzaldehyde	0.6	0.5	(25)
101 3-methylbenzaldehyde	0.4	0.5	(25)
102 4-methylbenzaldehyde	0.4	0.5	(25)
103 3-methoxycetophenone	0.5	0.5	(25)
104 4-methylbenzophenone	0.6	0.7	(28)
105 4-methoxybenzophenone	0.6	0.8	(28)
106 4-bromobenzophenone	0.9	0.9	(28)
107 4-chlorobenzophenone	0.9	0.9	(21)
108 4-fluorobenzophenone	0.8	0.8	(21)
109 3-chlorobenzophenone	0.9	0.9	(28)
110 3-bromobenzophenone	0.9	0.9	(28)
111 2-hydroxyacetophenone	0.9	0.6	(21)
112 2-nitroacetophenone	1.4	1.5	(20)
113 3-nitroacetophenone	1.3	1.5	(20)
114 4-nitroacetophenone	1.5	1.5	(20)
115 4-cyanoacetophenon	1.1	1.0	(20)
116 3-cyanobenzaldehyde	1.0	1.0	(20)
117 4-cyanobenzaldehyde	1.2	1.1	(20)
118 penta-florochlorobenzene	0.9	0.8	(26)
119 pentafluorobenzonitrile	1.1	1.0	(26)
120 pentafluoronitrobenzene	1.5	1.4	(26)
121 pentafluorobenzaldehyde	1.2	1.1	(26)
122 o-nitrobiphenyl	1.2	1.0	(20)
123 m-nitrobiphenyl	1.1	1.0	(20)
124 p-nitrobiphenyl	1.1	0.9	(20)
125 hexafluorobenzene	0.6	0.7	(20)
126 perfluorobiphenyl	0.9	1.2	(20)
127 decafluorobenzophenone	1.6	1.9	(20)
128 1,3-dinitronaphthalene	1.8	2.0	(20)
129 1,5-dinitronaphthalene	1.8	2.1	(20)

321

The case for in-ring N is similar. The substitution of N instead of one of the carbons in a benzene ring will decrease the electron density in the π^*-type SOMO state, which is going to stabilize the LUMO state and thus increase the electron affinity for pyridine [29,30] by 0.50 eV. The LUMO energy for pyridine is still relatively high, so that its electron affinity is -0.7 (Table 7) and no stable negative ion is formed in the gas phase state. The LUMO energy for in-ring N molecules can be lowered and stabilized by expansion of the π conjugation. Similar to carbon-aromatic systems, the EA is seen to increase significantly in the order of pyridine, quinoline, and acridine. A positive EA is expected for quinoline and acridine.

Table 7
Observed vs. SPARC-Calculated EA for Aza-Substituted Polycyclic Aromatics

#	Molecule	Observed	Calculated	Ref.
1	pyridine	-0.6	-0.7	(22)
2	pyrazine	-0.3	-0.2	(19)
3	pyrimidine	-0.5	-0.3	(19)
4	s-triazine	> 0	0.17	(22)
5	quinoline	0.2	0.2	(19)
6	quinazoline	0.6	0.5	(32)
7	quinoxaline	0.7	0.5	(32)
8	cinnoline	0.7	0.5	(32)
9	1,5-naphthyridine	0.4	0.6	(19)
10	1,6-naphthyridine	0.3	0.6	(19)
11	1,7-naphthyridine	0.3	0.6	(19)
14	2,6-naphthyridine	0.4	0.5	(19)
13	isoquinoline	0.1	0.2	(19)
16	pyrido[2,3-b]pyrazine	1.0	0.9	(32)
17	acridine	0.9	1.0	(32)
18	phenazine	1.3	1.4	(32)
19	pyridine maleic anhydride	1.4	1.5	(24)
20	4-cyanopyridine	1.0	0.8	(23)
21	pentafluoropyridine	0.8	0.8	(26)

Benzoquinone has a high electron affinity (Table 8). Expansion of the π system from benzoquinone to naphthoquinone to anthraquinone leads to a decrease in the electron affinity. Both Hückel MO and STO-3G calculations [31] predict a lower LUMO energy for benzoquinone relative to naphthoquinone. This agrees with experimentally measured electron affinities and the order is opposite to that observed for benzene, naphthalene and anthracene. Introduction of electron withdrawing groups increases the electron affinity of benzoquinone compared to benzene. On the other hand, methyl substitution or alkyl groups substitution leads to a decrease in electron affinity.

p - benzoquinone   p - naphthaquinone   9,10 - anthraquinone

Table 8
Observed vs. Calculated EA for Substituted Benzo-, Naphtho-, and Anthraquinone

#	Molecule[a]	Observed	Calculated	Ref.
1	pBQ	1.9	1.8	(20)
2	acetyl-pBQ	2.0	2.0	(23)
3	2-methyl-pBQ	1.9	1.8	(20)
4	2,6-dimethyl-pBQ	1.8	1.7	(20)
5	2,5-dimethyl-pBQ	1.8	1.7	(20)
6	methyl-trichloro-pBQ	2.5	2.3	(21)
7	3,6-dimethyl,2,5-dichloro-pBQ	2.2	2.1	(21)
8	3,6-dimethyl,2-chloro-pBQ	2.0	1.9	(21)
9	chlorotrimethyl-pBQ	1.9	1.9	(21)
10	trifluoromethyl-pBQ	2.2	2.1	(23)
11	dibromo-dimthyl-pBQ	2.0	2.1	(23)
12	2-methyl,5-chloro-pBQ	2.1	2.0	(21)
13	tetra-methyl-pBQ	1.6	1.7	(20)
14	cyano-pBQ	2.2	2.4	(23)
15	2,3-cyano-pBQ	2.8	3.0	(23)
16	chloro-p-BQ	2.0	2.0	(23)
17	2,6-dichloro-pBQ	2.5	2.2	(20)
18	2,5-dichloro-pBQ	2.4	2.2	(20)
19	2,3-dichloro-pBQ	2.2	2.2	(23)
20	trichloro-pBQ	2.6	2.4	(21)
21	tetrachloro-pBQ	2.7	2.7	(20)
22	fluoro-pBQ	1.9	1.9	(23)
23	tetrfluoro-pBQ	2.5	2.3	(23)
24	2-phenyl-pBQ	2.0	1.9	(20)
25	dimethylamino-pBQ	1.6	1.7	(23)
26	2,6-dimethoxy-pBQ	1.7	1.7	(20)
27	2,3-dimethoxy-5-methyl-pBQ	1.9	1.7	(20)
28	2,3-dichloro-tert-butyl-pBQ	2.3	2.2	(20)
29	2-chloro-5-tert-butyl-pBQ	2.1	2.0	(20)

Table 8 (cont'd)
Observed vs. Calculated EA for Substituted Benzo-, Naphtho-, and Anthraquinone

#	Molecule[a]	Observed	Calculated	Ref.
30	2-isopropyl-5-methyl-pBQ	1.8	1.8	(20)
31	2-tert-butyl-pBQ	1.8	1.8	(20)
32	2,6-di-t-butyl-pBQ	1.9	1.8	(20)
33	2,5-di-t-butyl-pBQ	1.8	1.8	(20)
34	phenyl-pBQ	2.0	2.0	(23)
35	tetra-bromo-o-BQ	2.6	2.4	(23)
36	tetra-chloro-o-BQ	2.6	2.5	(23)
37	p-NpQ	1.8	1.7	(20)
38	amino-p-NpQ	1.7	1.8	(23)
39	2-methyl-NpQ	1.7	1.7	(20)
40	2,3-dichloro-NpQ	2.2	2.1	(20)
41	2-oh-NpQ	1.6	1.7	(23)
42	5-oh-NpQ	2.0	1.7	(23)
43	o-NpQ	1.7	1.7	(23)
44	9,10-AnQ	1.6	1.6	(20)
45	methyl-9,10-AnQ	1.5	1.6	(23)
46	2-chloro-9,10-AnQ	1.7	1.8	(20)
47	2-ethyl-9,10-AnQ	1.6	1.6	(20)
48	bromo-9,10-AnQ	1.6	1.8	(23)
49	dibromo-9,10-AnQ	1.7	1.9	(23)
50	dichloro-9,10-AnQ	2.2	2.0	(23)
51	2-ethyl-9,10-AnQ	1.6	1.7	(20)
52	2-t-butyl-9,10-AnQ	1.6	1.7	(20)
53	1,8-dihydroxy-9,10-AnQ	1.7	1.7	(23)

[a] pBQ. = 1,4-benzoquinone; NpQ. = 1,4-naphthoquinone; AnQ. = anthraquinone.

Maleic anhydride   Phthalic anhydride   Maleimide   Phthalimide   Cyclopentedione

Cyclic unsaturated dicarbonyls such as maleic, anhydrides, maleimides and cyclopentenedione form a long-lived negative ion in the gas phase [33]. Similar to benzo-, naphthoquinone, the extra electron in these systems enters the LUMO, which is a π^* orbital resulting from a combination of $\pi^*_{c\text{-}c}$ and π^*_{co}. Their LUMO energies are higher, which leads to lower EA for these compounds relative to the quinones. Thus, the electron affinity of maleic anhydride is lower than that for benzoquinone. The electron affinity of the oxy compounds is larger than that of the NH and CH_2 bridged structures. The EA decreases as the electronegativity of the bridging atom decreases,i.e., in the order of OH, NH_2, CH_2. Substitution by the methyl group destabilizes the LUMO's of the quinones and the anhydrides. The substitution by electron-withdrawing substituents leads to significant increase of electron affinity as shown in Table 9.

Table 9
Observed vs. Calculated EA for Cyclic Dicarbonyls and Their Substituents

#	Molecule	Observed	Calculated	Ref.
1	maleic acid	1.1	1.1	(24)
2	maleic-anhydride	1.4	1.3	(24)
3	phenylmaleic-anhydride	1.8	1.4	(24)
4	methyl-maleic-anhydride	1.3	1.3	(24)
5	dimethyl-maleic-anhydride	1.2	1.2	(24)
6	dichloro-maleic-anhydride	1.9	1.7	(24)
7	phthalic anhydride	1.2	1.3	(24)
8	3-methylphthalic anhydride	1.2	1.2	(24)
9	4-methylphthalic anhydride	1.2	1.2	(24)
10	3,6-dichlorophthalic anhydride	1.7	1.6	(24)
11	tetrachlorophthalic anhydride	2.0	1.9	(24)
12	3-nitrophthalic anhydride	2.0	2.2	(24)
13	4-nitrophthalic anhydride	2.1	2.3	(24)
14	tetrabromophthalic anhydride	1.7	1.8	(23)
15	maleimide	1.2	1.1	(24)
16	methylmalemide	1.1	1.1	(24)
17	ethylmalemide	1.1	1.1	(24)
18	phenyl-phthalimide	1.2	1.1	(24)
19	phthalimide	1.0	1.1	(24)
20	Furil	1.3	1.3	(24)
21	Cyclopentedione	1.1	1.1	(24)

The electron affinities for some alkene and cyclo alkenes are shown in Table 10. The SPARC-calculated EA for these compounds is close to 2.0 eV, the same as the electron affinity for ethylene. Methyl substitution effects on electron affinity are small. In general, EA will decrease by almost 0.04 eV per substituent. 1,2-Dicyanoethylene and tetracyanoethylene (TCNE) are compounds of high electron affinity that are often involved as electron acceptors in charge-transfer complexes [34]. The additional electron in the negative ion enters the LUMO, which is the π^* orbital of ethylene lowered by conjugation with the electron-withdrawing cyano groups.

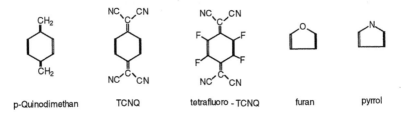

p-Quinodimethan TCNQ tetrafluoro - TCNQ furan pyrrol

p-Quinodimethane is expected to have a negative electron affinity similar to benzene [35]. Cyano substitution will lower the LUMO state substantially and increase the EA; hence, tetracyanoquinodimethane (TCNQ) electron affinity is high as 3 eV. Fluoro substitution in TCNQ will lower the LUMO state even more resulting in higher EA, e.g., tetrafluoro-TCNQ (Table 10). Pyrrole and furan both have a large negative electron affinity. The methoxy and the amine groups will raise the LUMO state lowering the EA for these compounds relative to the ethylene electron affinity.

Table 10
Observed vs. Calculated EA for Miscellaneous Compounds

#	Molecule	Observed	Calculated	Ref.
1	ethylene	-2.0	-2.0	(22)
2	propene	-2.0	-2.0	(22)
3	trans-butene	-2.1	-2.0	(22)
4	isobutene	-2.2	-2.0	(22)
5	trimethylethylene	-2.2	-2.1	(22)
6	tetramethylethylene	-2.3	-2.1	(22)
7	cyclohexene	-2.1	-2.0	(22)
8	1,4-cyclohexene	-1.8	-2.0	(22)
9	1,5-cyclohexene	-1.8	-2.0	(22)
10	pyrrole	-2.4	-2.6	(19)
11	furan	-2.4	-2.3	(19)
12	perflurormethylcyclohexane	1.1	1.1	(20)
13	nitromethane	0.5	0.6	(20)
14	tetranitromethane	1.6	1.6	(23)
15	trans-1,2-dicyanoethylene	1.2	1.0	(20)
16	tetracyanoquinodimethane	3.0	3.0	(23)
17	tetrafluoro-tetracyanoquinodimethane	3.2	3.3	(19)
18	tetracyanoethylene	3.1	3.1	(19)
19	alloxan	1.4	1.6	(19)

4. PHYSICAL PROPERTIES: ESTIMATION OF GAS CHROMATOGRAPHIC RETENTION TIMES

INTRODUCTION

Despite some limitations, Kov'ats index has found much greater usage than all other specialized retention specification schemes. The Kov'ats index is the only retention value in gas-liquid chromatography (GLC) in which two fundamental quantities, the relative retention and the specific retention volume are united [36]. Moreover, a series of explicit relationships between retention indices and a number of physicochemical quantities related to GLC have been developed. Also many different linear relationships between the Kov'ats index value for a molecule and other fundamental quantities such as carbon number, boiling point and refractive index have been derived [36,37].

The Kov'ats [38] index expresses the retention of a compound of interest relative to a homologous series of n-alkanes examined under the same isothermal conditions. The Kov'ats index for a particular compound of interest is defined as the carbon number (C_{eff}) multiplied by 100 of a hypothetical n-alkane having exactly the same net retention volume characteristics of the compound of interest measured under the same conditions :

$$K.I. = 100 * (\frac{\log V_{Nc} - \log V_{Nx}}{\log V_{Nc} - \log V_{N(c+1)}} + NC)$$

$$(20)$$

where K.I. is the Kov'ats Index of compound x, x is a compound with a retention between that of the first n-alkane and second n-alkane standard, C is the number of carbon atoms in the first n-alkane standard, C+1 is the number of carbon atoms in the second n-alkane standard, V_{Nx} is the net retention volume of compound X, V_{Nc} is the net retention volume of the first n-alkane standard, and $V_{N(c+1)}$ is the net retention of the second n-alkane standard.

Numerous investigators have attempted to calculate or predict Kov'ats Index using physicochemical descriptors like boiling point, density, dipole moment, etc. Unfortunately, all of the correlations of retention indices and the various physicochemical properties are either relatively limited in scope or their application is restricted to a particular chemical class. Other attempts to predict retention indices for a wide range of molecular structures using molecular bond length, molecular bond angle, topological indices [36,37,39], or other molecular characteristics have been only marginally successful. Most of these studies also were restricted to a particular class of molecules on a specific stationary liquid phase.

Despite all the attempts to predict Kov'ats index, no realistic scheme with widespread application for different classes of compounds for different polarity stationary liquid phases is available. Our goal is to develop mathematical models to calculate the Kovats index at any temperature based on a calculated Henry's constant for a wide range of different classes of compounds on different polar and non-polar stationary liquid phases.

4.1. SPARC Physical Models

For all physical processes (e.g., vapor pressure, activity coefficient, partition coefficient, etc.), SPARC uses one master equation to calculate characteristic process parameters:

$$\Delta G_{process} = \Delta G_{interaction} + \Delta G_{monmer} \qquad (21)$$

where ΔG(monomer) describes entropy changes associated with mixing, volume changes, or changes in internal (vibrational, rotational) energies going from the initial state to the final state. ΔG(monomer) depends only on the phase change involved and in the present application is presumed to depend only on solute/solvent volumes in each phase. ΔG(interaction) describes the change in the intermolecular interactions in the initial state and final state. For example, the interaction term for Henry's constant describes the difference in the intermolecular interactions in the gas phase versus those in the liquid phase. The interactions in the liquid phase are modeled explicitly, interactions in the gas phase are ignored, and molecular interactions in the crystalline phase are extrapolated from the subcooled liquid state using the melting point.

The intermolecular interactions in the liquid phase are expressed as a summation over all the intramolecular interaction forces between the molecules:

$$\Delta G_{interaction} = \Delta G_{dispersion} + \Delta G_{induction} + \Delta G_{dipole} + \Delta G_{H-bonding} \qquad (22)$$

Each of these interactions is expressed in terms of a limited set of molecular-level descriptors (density-based volume, molecular polarizability, molecular dipole, and H-bonding parameters) which in turn are calculated from molecular structure.

4.2. SPARC Molecular Descriptors

The computational approach for molecular-level descriptors is constitutive with the molecule in question being broken at each essential single bond and the property of interest being expressed as a linear combination of fragment contributions as

$$\chi°(molecule) = \sum_i (\chi_i° - A_i) \qquad (23)$$

where $\chi°_i$ are intrinsic fragment contributions (which in most cases are tabulated in SPARC databases) and A_i are adjustments relating to steric or electrometric perturbations from contiguous structural elements for the molecule in question in

328

the process model or medium involved. Both χ°_i and A_i are empirically trained either on direct measurements of the descriptor in question (e.g., liquid density based molecular volume) or on a directly related property (e.g., index of refraction, which can be related to polarizability) for which large reliable data sets exist.

4.2.1. Average Molecular Polarizability

In SPARC, fragment polarizability factored into atomic contributions, χ_j, and the polarizability of fragment ,i, is expressed as

$$\bar{\alpha}_i = \frac{1}{N_i} * [\sum_j \chi_j]^2 \tag{24}$$

where the summation is over all the atoms in fragment i, χ_j is the intrinsic atomic hybrid polarizability contribution, and N_i is the number of electrons in fragment i. The χ_j are empirically determined from measured polarizabilities and stored in the SPARC database (with exception of hydrogen, which is calculated from the measured polarizability of H_2).

The average molecular polarizability, α°, is expressed as

$$\bar{\alpha}^\circ = \sum_i (\bar{\alpha}_i - A_i) \tag{25}$$

where α_i is the polarizability of fragment i and A_i are the adjustments for the molecule in question. The only adjustment, A_i, currently implemented in SPARC is a 10% reduction in α_i for hydrocarbon fragments with an attached polar group or atom. The partition of polarizability into atomic contributions enables estimates to be made of molecular polarizabilities for any given molecular structure. The molecular polarizability can be calculated within less than 1% for a wide range of molecules. Figure 7 shows the observed versus SPARC-calculated refractive index at 25°C for alkane, alkene and aromatic systems. Examples of the calculation of molecular polarizability and index of refraction are given in the Appendix III.

4.2.2. Molecular Volume

The zero order density-based molecular volume is expressed as

$$V^\circ_{(25)} = \sum_i (V_i^{frag} - A_i) \tag{26}$$

where V^{frag} is the volume of the fragment and A_i is a correction to that volume based on both the number and size of fragments attached to it. The V^{frag} are determined empirically from the measured volume and then stored in the SPARC database. This zero order volume at 25°C is further adjusted for shrinkage resulting from dipole-dipole and H-bonding interactions:

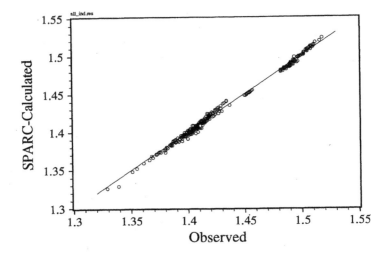

Figure 7.   Observed vs. SPARC-calculated refractive index at 25°C. The RMS deviation for this set was found to be equal to $3*10^{-3}$.

$$V_{25} = V_{25}^{o} + A_d \frac{\sum_i D_i^2}{V_{25}^{o}} + A_{HB} \frac{(\alpha_i \beta_i)_{max}}{V_{25}^{o}} \qquad (27)$$

where D_i is the dipole for the molecule, and α and β are the hydrogen bonding parameters of potential proton donor and proton acceptor sites within the molecule, respectively.

The volume at temperature T is then expressed as a polynomial expansion in (T-25) corrected as a function of hydrogen bonding (HB), dipole (D) and polarizability (P) interactions as

$$V_T = V_{25}[1 + f(P, D, HB) \sum_n a_n (T - 25)^n] \qquad (28)$$

where a_n are trainable parameters. The molecular volumes for a wide range of molecules can be calculated to better than 1%. Figure 8 shows the observed versus SPARC-calculated density based molecular volumes for polar and nonpolar compounds. Examples of the calculation of molecular volume are given in the Appendix III.

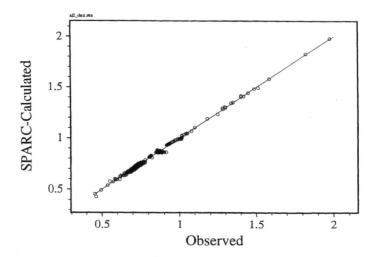

Figure 8   Observed vs. SPARC-calculated density at 25°C. The RMS deviation was found to less than $4*10^{-3}$.

4.3. Solute/Solvent Interactions

Models for self interactions between like molecules and between solvent-solute molecules have been developed to calculate physical properties. These interaction models build on the limited set of molecular-level descriptors (volume, polarizability, molecular dipole and H-bonding parameters) described above. These interaction models are dispersion, induction, dipole-dipole, and H-bonding. Dispersion interactions are present for all molecules including non-polar molecules. Induction interactions are present between two molecules when at least one of them has a permanent dipole moment. Dipole-dipole interactions exist when both molecules have dipole moments. H-bonding interactions exist when α_i β_j or α_j β_i products are non-zero.

4.3.1. Dispersion Interactions

In this review we present our calculation of the Kovats retention indices for alkanes, alkenes and aromatics on a non-polar stationary phase squalane. For this reason, we shall not discuss the interaction mechanisms for dipoles and hydrogen bonding.

Dispersion interactions occur between all molecules as a result of very rapidly varying dipoles formed between nuclei and electrons at zero-point motion of the molecules, acting upon the polarizability of other molecules to produce an induced dipole in the phase. The self interactions are expressed as

$$\Delta G_{ii}(disp) = \rho_{disp} (P_i^d)^2 V_i \tag{29}$$

and solvent-solute interactions are expressed as

$$\Delta G_{ij}(disp) = \rho_{disp} (P_i^d - P_j^d)^2 V_i \tag{30}$$

$$P_i^d(disp) = \frac{\alpha_i + A_{disp}}{V_i} \tag{31}$$

where i and j designate the solute and squalane molecule, respectively; P_i^d is the effective polarizability density of molecule i; ρ_{disp} is the susceptibility to dispersion; V_i and α_i are the molar volume and the average molecular polarizability described previously.

Dispersion is a short range interaction involving surface or near surface atoms and A_{disp} subtracts from the total polarizability, a portion of the contributions of sterically occluded atoms in the molecular lattice. Presently SPARC corrects for access judged to be less than that afforded by a linear array of atoms (i.e., for branched structures or rings small enough to prohibit intra penetration of the solvent). Branched (ternary or quaternary) atoms in an alkane structure will lose a small part of their intrinsic molecular polarizability depending on the size and number of appended groups, and the proximity of other branched carbons. Similarly, carbons in rings may lose their intrinsic polarizability contributions depending on ring sizes and the presence of a ring appendage.

4.4. Activity Coefficient Model

For a solute, i, in a liquid phase, j, at infinite dilution, SPARC expresses the activity coefficient as

$$- R T \log \gamma_{ij}^{\infty} = \Delta G_{interactions} + \Delta G_{monomer} \tag{32}$$

For the hydrocarbons in this study, the activity coefficient is given as

$$-RT \log \gamma_{ij}^{\infty} = \Delta G_{ijdisp} + RT \left(\log \frac{V_i}{V_j} + \frac{(\frac{V_i}{V_j} - 1)}{2.303} \right) \tag{33}$$

where the last term is the Flory-Huggins [40,41] excess entropy contribution of mixing in the liquid phase of placing a solute molecule in the solvent. When the solute and solvent have the same volume, the Flory-Huggins term will go to zero. Table 11 shows the observed versus SPARC-calculated activity coefficients in squalane. It should be noted that the negative log values are a consequence of the large Flory-Huggins contributions.

Table 11
Observed vs. SPARC-Calculated Values for the Log Activity Coefficient in Squalane

Molecule	Observed	Calculated
pentane	-0.24	-0.20
hexane	-0.19	-0.18
heptane	-0.15	-0.16
octane	-0.15	-0.15
nonane	-0.17	-0.14
2-methylpentane	-0.19	-0.16
2,4 dimethylpentane	-0.14	-0.13
2,5 dimethylhexane	-0.11	-0.12
2,3,4-trimethylpentane	-0.17	-0.14
cyclohexane	-0.28	-0.30
ethylcyclohexane	-0.23	-0.22
benzene	-----	-0.15
toluene	-----	-0.16
1,3 dimethylbenzene	-----	-0.14
1,4 dimethylbenzene	-----	-0.14
1,3,5 trimethylbenzene	-----	-0.11

4.5. Vapor Pressure Model
The vapor pressure P_i of a solute, i, is expressed as

$$- 2.303 \ RT \log P_i = \Delta G_{i_{disp}} - 2.303 \ RT \ (\log (T) + C) \qquad (34)$$

where $RT(\log (T) + C)$ describes the change in the entropic contributions [42] associated with the volume changes between the liquid and the gas phases. Figure 9 shows the observed versus the SPARC-calculated values for the vapor pressure for various molecular structures at 25°C. See Appendix III for sample vapor pressure calculations.

4.6. Henry's Constant
For a solutions that are so dilute that each solute molecule is surrounded only by solvent molecules, small changes in the solute concentration will not affect the composition of the nearest neighbor molecules. In this case, the intermolecular

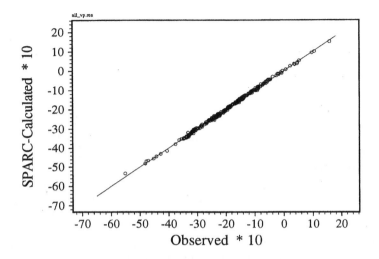

Figure 9.   Observed vs. calculated values for the vapor pressure at 25°C. The RMS deviation error for logP was 0.038.

interactions the solute molecule experiences will not change with concentration and the vapor pressure will be proportional to the mole fraction of the solute and Henry's constant may be expressed as

$$H_x = P_i^{\circ} \lambda_{ij}^{\infty} \tag{35}$$

where P_i° is the vapor pressure of pure solute i and λ_{ij}^{∞} is the activity coefficient of solute i in the squalane liquid phase (j) at infinite dilution. SPARC vapor pressure and activity coefficient models are used to calculate the Henry's constant for a solute in a squalane liquid phase.   Henry's constant can be related to the net retention volume, V_N, by

$$H_i = \frac{RT\,V_L}{M\,V_{V_N}} \qquad where \qquad K_i = \frac{V_N}{V_L} \tag{36}$$

where M is the molecular weight of the solvent, and V_L is the volume of the stationary phase.

Substituting in equation 20, we get

$$K.I = 100 \times \left(\frac{\log H_{Nx} - \log H_{Nz}}{\log H_{N(z+1)} - \log H_{Nz}} + NC \right) \qquad (37)$$

where H_{Nx}, H_{Nz}, and $H_{N(z+1)}$ are Henry's constant for a compound x, first n-alkane standard, and second n-alkane standard, respectively.

4.7. Calculation of Kovats Indices

Retention indices may be reproduced within a laboratory using modern instrumentation with considerable precision over finite time periods. Reproducibility of 0.1 units was reported by Schomburg and Dielmann [43] in 1973. However, squalane columns produced reproducible results for only for a few hours and, therefore, needed to be continually replaced. For routine operation, a reproducibility of about 1 Kovats Index unit might be expected with a squalane liquid phase.

Unfortunately, inter-laboratory reproducibility remains unsatisfactory, except for a few cases. The actual discrepancies between experimental values of retention indices for identical compounds obtained at different laboratories in routine analysis is assumed to be up to ± 10 Kovats units or even more [37].

4.8. Unified Retention Index

The unified retention index developed by Dimov [44,45] has been used to explain the variations in the retention index of simple hydrocarbons on squalane. The temperature dependence of retention index is well known, the function dI/dT being hyperbolic. A statistical treatment using simple regression analysis of the data allows computation of the unified retention index (UI_T) as

$$UI_T = UI_o + \left(\frac{dUI}{dT} \right) T \qquad (38)$$

where UI_o is the Kovats Index at 0° and dUI/dT is the temperature dependence where -dUI/dT is the slope of the plotted data. The UI_T is a statistically obtained value and, hence, it is more reliable than any individual I_{exp} value. Also dUI/dT is a more reliable value than dI/dT for estimation of the temperature dependence of retention indices. The UI_T and dUI/dT served as the observed values for the optimization of our dispersion parameters for prediction of the retention indices in this study.

Figure 10 shows the observed [44,45] versus the SPARC-calculated Kovtas Index at 25°C. The RMS was less than 7 Kovats units, a value that approximates interlaboratory experimental error. We also calculated the Kovats index as a function of temperature using the SPARC temperature dependence models discussed below.

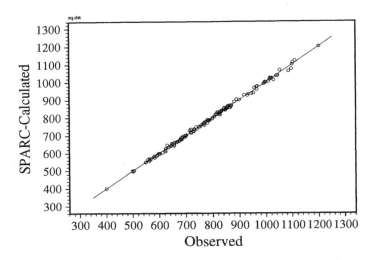

Figure 10. Observed vs. SPARC calculated values for the retention indices in squalane liquid phase at 25°C. The RMS deviation was less than 7 Kov'ats units.

SPARC calculates a physical property of interest at 25°C. In addition to the inherent temperature dependence described previously in equations 33 and 34 and the temperature dependence built into the volume calculator (equation 28), the susceptibility of dispersion at temperature T is modeled as a function of the polarizability density and the effective polarizability density, P'. In effect, this describes the small temperature dependence of enthalpy. For temperature, an "activity-driven" process, ρ_T, is given by:

$$\rho_T = [1 + (1 - \sum_{n}^{5} a_n (\frac{298.15}{T})^n) \times f(P, P')] \rho_{25} \tag{39}$$

where a_n are trainable parameters. These parameters were inferred from boiling point measurements at 1, 10, 30, 100, and 760 torr for more than 400 compounds spanning a range of over 700°C. Figure 11 and Table 12 show the observed versus SPARC calculated boiling points for wide range of molecules. The RMS deviation for this set was 3.4 degrees. Based on these temperature dependent models, we calculated the Kov'ats indices at 80°C as shown in Figure 12. The RMS deviation was less than 8 Kov'ats units.

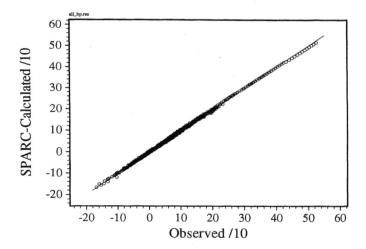

Figure 11.   Observed vs. SPARC calculated values for the boiling points at 1, 10, 100, 760 Torr. The RMS deviation for this set was 3.4 degrees.

Table 12
Observed vs. SPARC Calculated Boiling Points

Molecule	760 (Torr)		100 (Torr)		1 (Torr)	
	Obs	Cal	Obs	Cal	Obs	Cal
pentane	36.10	36.60	-12.6	-12.6	-76.6	-76.3
hexane	68.70	69.00	15.8	15.3	-54.0	-55.3
heptane	98.40	99.00	41.8	41.8	-33.2	-35.2
octan	125.7	126.8	65.7	66.8	-14.9	-15.7
nonane	150.8	152.6	87.9	90.6	3.60	3.00
2-methylpentane	60.30	55.80	8.10	5.00	-6.00	-6.40
2,4-dimethylpentane	80.50	76.40	25.4	23.4	-4.70	-5.04
2,5-dimethylhexane	109.1	106.0	50.5	45.5	-26.7	-31.4
2,3,4-trimethylhexane	139.0	137.0	76.0	76.0	-7.00	-9.00
cyclohexane	80.70	81.90	25.5	27.2	-47.0	-44.5
ethylcyclohexane	131.8	128.7	69.0	67.8	-14.4	-14.7
benzene	80.10	85.80	26.0	30.2	-45.0	-41.9
toluene	110.6	113.5	51.9	54.5	-26.1	-23.1
1,3 dimethylbenzene	139.1	138.5	76.8	76.5	-7.2	-5.90
1,4 dimethylbenzene	138.4	140.3	75.9	78.1	-8.1	-5.04
1,3,5 trimethylbenzene	146.7	162.2	99.8	97.7	11.6	10.9

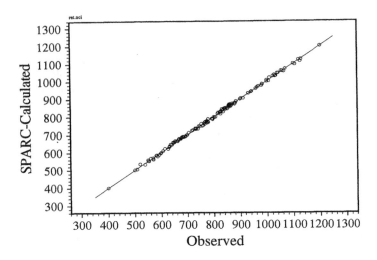

Figure 12.   Observed vs. SPARC-calculated values for the retention indices in squalane liquid phase at 80°C. The RMS deviation was less than 8 Kov'ats units.

5. SUMMARY

SPARC's chemical reactivity models predict pK_a as well as electron affinity values of a wide variety of structures. The calculated values are, with few exceptions, as reliable as the experimental measurements. The SPARC pK_a calculator has been tested on 4000 pK_a's for more than 3000 compounds *spanning a range of over 30 pK_a units*. The calculator can calculate multiple ionizations. For azo dyes, the calculator has calculated up to the 8th ionization constant. In addition, the user can ask for ionization steps that cannot be measured experimentally but which are useful in calculating zwitterionic equilibrium constants, and other important properties.

As a demonstration of the SPARC physical property calculator, we have presented the calculation of GC retention indices for a squalane liquid phase. This demonstrates the ability of the SPARC calculator to calculate molecular self interactions (vapor pressure) and solvent-solute interactions (activity coefficient) to generate other physical properties such as Henry's constant. The SPARC models can calculate physical properties as a function of temperature. This enables the calculation of vapor pressure at any temperature leading to the calculation of boiling point and boiling point as a function of pressure. By calculating vapor pressure and activity coefficient as a function of temperature, we were able to calculate Henry's constant as a function of temperature and therefore to calculate Kov'ats Index at any temperature.

However, the real test of SPARC does not lie in testing the prediction capability for pK_a's or electron affinity or in estimating many physical properties, but is determined by the extrapolatability of these models and "toolboxes" to other types of chemistry. The next area in estimating chemical reactivity will be the development of solvation models for ions. These models coupled with the gas phase electron affinity calculator and our neutral molecule solubility models will allow us to extend SPARC to the calculation of half-wave reduction potentials in any solvent. The present toolboxes will be used to develop gas and liquid diffusion models.

APPENDIX I

Distribution of NBMO electrons in odd alternate hydrocarbons.

Carbon atoms in an alternate conjugated network can be subdivided into two sets (one set starred and one set unstarred) such that no two starred or no two unstarred atoms are adjacent. Atoms belonging to the same set are said to be of equal parity. For odd alternate networks (i.e., containing an odd number of carbon atoms), atoms belonging to the smaller of the two sets will have vanishingly smaller NBMO amplitudes and will be assumed to be NBMO nodes. The existence of these alternating nodes provides an easily solvable set of equations for NBMO amplitudes. This enables the description of this "chemically significant" molecular orbital (MO) without determining the entire set of MOs (bonding, antibonding, etc.).

Consider the following conjugated network with an NBMO node at C_0:

$$\overset{*}{=}\overset{*}{C_1}\underset{\beta_{10}}{\overset{\beta_{10}}{—}}C_0\underset{\beta_{02}}{\overset{\beta_{02}}{—}}\overset{*}{C_2}=$$

with β_{03} connecting to $\overset{*}{C_3}$

I

The nodal condition at C_0 requires that

$$\beta_{10}a_1 + \beta_{02}a_2 + \beta_{03}a_3 = 0 \qquad (A1.01)$$

where the β's are π resonance integrals for the connecting bonds and the a's are NBMO amplitudes for the starred atoms. The NBMO charge density is given by $a_1 a_1^*$ or for real number coefficients, simply a_1^2, ($a_1^2 = b_1$). For an odd alternate system containing n carbons, there are $(n+1)/2$ AO coefficients to be determined.

The (n-1)/2 nodes provide (n-1)/2 equations of the type described above. These equations coupled with a normalization condition ($\Sigma_i b_i$ = number of NBMO electrons) provide solutions for each AO coefficient in terms of the β's. While conventional methods for solving simultaneous equations are straightforward, these procedures tend to be quite "cumbersome" for large n. Fortunately, for many conjugated networks, these coefficients can be inferred by inspection. Consider the simple allyl radical:

$$\overset{*}{\underset{\underset{-1(1)}{1}}{C}}\overset{\beta_{12}}{\rule{2cm}{0.4pt}}\underset{\underset{0(0)}{2}}{C}\overset{\beta_e}{=\!=\!=}\overset{*}{\underset{\underset{1(1)}{3}}{C}}$$

II

For the present, set all β = 1. If the amplitude a_3 is arbitrarily set a_3 = 1, the amplitude at atom 1 must be -1 to satisfy the nodal condition at atom 2. The actual NBMO amplitude will be given by Aa_1 when A is determined by the normalization condition. The charge densities (given above in parentheses) must sum to unity or A = $1/\sqrt{2}$. This inspection procedure can be extended to an expanded π network.

III

Again, if a_7 = 1, the other NBMO coefficients are determined easily by inspection; in this case A = $1/\sqrt{11}$. This inspection procedure will not always lead to complete specification of NBMO amplitudes but will always result in substantial reduction in "computational burden." Unfortunately, the use of the method requires considerable circumspection in choosing the proper reference or starting point for assigning relative NBMO amplitudes. Therefore, the method would be difficult to machine implement. These methods do, however, provide insight into distribution "patterns" that lead to an easily generalizable procedure for SPARC applications.

SPARC delineates conjugated networks into "rigid" π units (denoted R_π) connected by essential single bonds. For example, structure IV is fragmented into three R_π units plus an NBMO "source" unit (designated R_1):

340

-2(4)   -1(1)

-4(16)   4(16)   2(4)   1(1)

—C—C═══C—⟨⟩——⟨⟩

-2(4)   -1(1)

R_1 $R_{\pi 2}$ $R_{\pi 3}$ $R_{\pi 4}$

IV

Total NBMO charge density in each unit is denoted b_i, which can be expressed relative to the source unit b_1. For structure IV one can see by inspection,

$$b_1 = b_1 \tag{A1.02}$$
$$b_2 = b_1 \tag{A1.03}$$
$$b_3 = (3/4)b_2 = (3/4)b_1 \tag{A1.04}$$
$$b_4 = (1/4)b_3 = (1/4)(3/4)b_2 = (1/4)(3/4)b_1 \tag{A1.05}$$

The actual value of b_1 is then set by normalization. For this structure, if the NBMO contains one electron,

$$\Sigma_i\, b_i = 1 \text{ or } b_1 = 16/47$$

Although no formal proof is offered, the following recursive behavior can be inferred easily from the preceding discussions and examples. For a linear series of π units, the following recursion is realized.

$$b_1 = b_1 \tag{A1.06}$$
$$b_2 = f_2\, f_1^{\,o}\, b_1 \tag{A1.07}$$
$$b_3 = f_3\, f_2^{\,o}\, b_2 = f_3\, f_2^{\,o}\, f_2\, f_1^{\,o}\, b_1 \tag{A1.08}$$

$$b_m = b_1 \prod_{j=1}^{m} f_j\, f_{j-1}^{o} \qquad m \geq 1, \quad f_1\, f_0^{o} = 1 \tag{A1.09}$$

where f_j = total NBMO charge density in unit j per unit charge on the "outlet" atom of the previous unit (j-1); f_j^x = the fraction of NBMO charge in unit j residing on carbon x; x = 0 denotes the "outlet" atom connecting the j and (j+1) units. From structure IV, one can see that for ethylene units in a conjugated chain, $f_j = f_j^o = 1$, whereas for benzene units $f_j = 3/4$ and $f_j^x = 1/3$ for starred atoms. The total charge in a series containing n units is:

$$\sum_{m=1}^{n} b_m = b_1 \sum_{m=1}^{n} \prod_{j=1}^{m} f_j \, f_{j-1}^o \quad \text{(A1.10)}$$

and the fraction of total charge on the "x" carbon of unit "k" (denoted F_K^x) is given by:

$$F_k^x = \frac{f_k^x \displaystyle\prod_{j=1}^{k} f_j f_{j-1}^o}{\displaystyle\sum_{m=1}^{n} \prod_{j=1}^{m} f_j f_{j-1}^o} \quad \text{(A1.11)}$$

This applies also for branched series such as:

V

These network equations are better represented if the denominator is expanded:

$$\sum_{m=1}^{n} \prod_{j=1}^{m} f_j f_{j-1}^o = 1 + f_1 f_2^o \{1 + f_2^o f_3 [1 + f_3^o f_4 - (1 + f_{n-1}^o f_n)]\} \quad \text{(A1.12)}$$

For branched network V

$$F_k^x = \frac{f_k^x \displaystyle\prod_{j=1}^{k} f_j f_{j-1}^o}{1 + f_1^o f_2 [1 + f_2^o f_3 (1 + f_3^o f_4) + f_{2'}^o f_{3'} (1 + f_{3'}^o f_{4'})]} \quad \text{(A1.13)}$$

Where $f_{2'}^o$ and f_2^o denote the charge fraction in R_2 on the "outlet" carbons for the primed and unprimed branches, respectively.

Network Extensions: If an additional conjugated unit "n" is attached at the "y" carbon of unit "L" of an existing network, then

$$F_k^x = (F_k^x)_o \frac{1}{1+(F_L^y)_o f_n} \qquad k \neq n \tag{A1.14}$$

$$F_n^x = (F_k^y)_o \frac{f_n f_n^x}{1+(F_L^y)_o f_n} \tag{A1.15}$$

where the subscript "o" denotes charge distribution in the original network, prior to addition of unit n.

Multiple addition: If an additional unit m is connected at position z of unit p, the new distribution is given by

$$F_k^x = (F_k^x)_o \frac{1}{1+(F_L^y)_o f_n + (F_p^z)_o f_m} \qquad k \neq n,m \tag{A1.16}$$

$$F_n^x = \frac{(F_L^y)_o f_n f_n^x}{1+(F_L^y)_o f_n + (F_p^z)_o f_m} \tag{A1.17}$$

$$F_m^x = \frac{(F_p^z)_o f_m f_m^x}{1+(F_L^y)_o f_n + (F_p^z)_o f_m} \tag{A1.18}$$

One exceptional case that requires a slight modification to these NBMO network distribution functions involves loops or cross conjugations. For example, in the structure:

charge in $R_{\pi 3}$ can "arrive" via two paths (from R_1). These situations are "flagged" and the charge propagation modified to reflect NBMO *amplitude* additivity at the point of loop juncture.

$$b_3 = [\ (b_2 f_2^o f_3)^{\frac{1}{2}} + (b_2 f_{2'}^o f_{3'})^{\frac{1}{2}}\ ]^2 \qquad\qquad \text{(A1.19)}$$

NBMO propagation prior to this juncture point is as described previously. Before proceeding, the constraint on β_π should be relaxed. Within a given π unit, all bonds are described by a single β_π, either β_a for aromatic ring systems or β_e for ethylenic systems. A key element in describing NBMO "conduction" is the resonance integrals for the essential single bonds linking the π units (β_{ij}). Without retracting all the derivation steps, these resonance integrals are inserted in the recursion relationship. For example:

$$b_2 = \gamma_{12} f_2 f_1^o\, b_1 \qquad where \qquad \gamma_{12} = (\frac{\beta_{12}}{\beta_2})^2 \qquad\qquad \text{(A1.20)}$$

β_{12} describes the essential single bond linking π units 1 and 2, β_2 is the "internal" π resonance integral for unit 2, either β_a or β_e. These resonance integrals can be interjected thought the NBMO distribution algorithms by replacing f_i with $\gamma_{i,\,i-1}$, f_i.

APPENDIX II

Resonance Integrals: essential single bonds.

The π resonance integrals β_{ij} for essential single bonds govern π coupling between R_π units. The magnitude of β_{ij} reflects the extent of π orbital overlap between units $R_{\pi i}$ and $R_{\pi j}$. It is convenient to distinguish the electrometric verses geometric contribution to β_{ij}

$$\beta_{ij} = \beta_{ij}^e\, \cos\theta_{ij} \qquad\qquad \text{(A2.01)}$$

where Θ_{ij} is the dihedral angle describing alignment and β_{ij}^e is the electrometric contribution realized at $\Theta_{ij} = 0$. As described in Appendix I, β_{ij} enters into NBMO charge distribution as γ_{ij} for distribution of NBMO charge from unit i to j.

$$\gamma_{ij} = [\frac{\beta_{ij}}{\beta_j}]^2 = \gamma_{ij}^e\, P_{ij} \qquad\qquad \text{(A2.02)}$$

where $P_{ij} = \cos^2\Theta$; β_j is the characteristic resonance integral for R_π (i.e., $\beta_e \approx$ 28,000 cm^{-1} for ethylenic units or $\beta_a \approx$ 23,000 cm^{-1} for aromatic systems).

The description of π electron properties in any conjugated network of R_π units requires the estimation of γ_{ij} for all essential single bonds. The computational approach for γ_{ij} is a perturbation method similar to that used for other chemical properties. That is, the computation does not derive from "first principles" but extrapolates from measured values for reference compounds. Ethylene serves as a reference R_π unit for ethylenic (denoted e) systems and benzene is the reference for aromatic systems (denoted a). Reference parameters and calibration compounds (given in parentheses) are:

$(\gamma_{aa})_{ref}$	=	0.89 (biphenyl)
$(P_{aa})_{ref}$	=	0.87 (biphenyl)
$(\gamma_{ee})_{ref}$	=	0.47 (butadien)
$(P_{ee})_{ref}$	=	0.99 (butadien)
$(\gamma_{ae})_{ref}$	=	0.57 (styrene)
$(\gamma_{ea})_{ref}$	=	0.84 (styrene)
$(P_{ae})_{ref}$	=	0.93 (styrene)
$(P_{ea})_{ref}$	=	0.93 (styrene)

Substituents (S- or S+) are assumed to be ethylenic (type e) for reference parameter selection. Using Equation A2.02, one can show that the corresponding $(\beta^e_{ij})_{ref}$ are not highly variable, ranging from 19,000 cm^{-1} for $(\beta^e_{ee})_{ref}$ to 21,000 cm^{-1} for $(\beta^e_{aa})_{ref}$.

Computation of γ^e_{ij}.   For a specific compound, adjustments to $(\gamma^e_{ij})_{ref}$ are made for extended π conjugation (additional R_π units conjugated with $R_{\pi i}$ or $R_{\pi j}$) and (2) the presence of electron withdrawing (S+) or electron donating (S-) substituents.

Adjustments for extended conjugation are based on the π localization(L) energies $[(LE_\pi)_i$ and $(LE_\pi)_j$] of each coupled unit, determined at the point(s)-of-attachment (atoms i and j, respectively). A good approximation for π localization energy at an atom can be derived from a simple PMO algorithm [11,12] that involves the extraction of the atom-in-question and computation of the NBMO amplitude in the *fragment* at the point(s)-of-fracture.

For an ethylenic unit,

$$(LE_\pi)_i \approx 2\,\beta_e\,a_i \qquad (A2.03)$$

or if an atom in an aromatic ring, is extracted,

$$(LE_\pi)_i \approx 2\beta_a\,\Sigma_i a_i \qquad (A2.04)$$

where the summation is over the point(s)-of-fracture in the ring. The NBMO amplitude component is commonly termed the π reactivity number N (i.e., $N_i = a_i$ or $\Sigma_i a_i$). It is assumed that the γ^e_{ij} for an essential single bond should be inversely correlated with the sum of localization energies of the coupled R_π units, determined at the point(s)-of-attachment, or that

$$\gamma_{ij}^e \approx (\gamma_{ij}^e)_{ref} \frac{N_i^o + N_j^o}{N_i + N_j} \tag{A2.05}$$

where $N_i(N_i^o)$ and $N_j(N_j^o)$ are the reactivity numbers of each respective unit in the presence (absence) of extended conjugation.

For 1,3,5 hexatriene,

$$C_1 \!\!=\!\! C_2 \overset{\gamma_{12}}{\rule{1cm}{0.4pt}} C_3 \!\!=\!\! C_4 \overset{\gamma_{23}}{\rule{1cm}{0.4pt}} C_5 \!\!=\!\! C_6$$
$$R_{\pi_1} \qquad\qquad R_{\pi_2} \qquad\qquad R_{\pi_3}$$

$$\gamma_{12}^e \approx (\gamma_{12}^e)_{ref} \frac{N_1^o + N_2^o}{N_1 + N_2} \tag{A2.05}$$

$N_1(N_2)$ is determined by extracting carbons $C_2(C_3)$ and computing the NBMO amplitude $a_1(a_4)$ at carbons C_1 and C_4 in the respective fragments. The NBMO charge distribution is computed as described in Appendix I with the γ_{ij} for essential single bonds set at $(\gamma_{ij}^o)_{ref}$. Also, any heteroatoms within the π network are replaced by carbons.

For the above example:
$N_1 = a_1 = 1$

$$N_2 = a_4 [\frac{1}{1 + f_2^o f_3 \gamma_{23}}]^{1/2} = [\frac{1}{1 + 1.0 \times 1.0 \times 0.47}]^{1/2} = 0.82 \tag{A2.07}$$

Also, for the butadiene reference:

$N_1^o = N_2^o = 1$, therefore

$$\gamma_{12}^e = 0.47 [\frac{1 + 1}{1 + 0.82}] = 0.52$$

It is interesting to note that in the limit of a polyene of infinite length, the γ_{ij}^e will upgrade to a maximal value of

$$(\gamma_{ee}^e)_{max} = 0.47 \left[\frac{1}{\sum\limits_{n=0}^{\infty} (0.47)^n} \right]^{-1/2} = 0.65$$

or bond alternation will sustain.

If both $R_{\pi i}$ and $R_{\pi j}$ contain highly electronegative heteroatoms (O, N or S) located at sites of nonzero NBMO charge (in the calculation of reactivity numbers, N_i and N_j), a downgrade factor is applied based on the fractional amount of NBMO charge at the heteroatom location (q_H). The reduction factor $(1-q_{Hi})(1-q_{Hj})$ is applied in conjunction with any upgrade due to extended conjugation of Equation A2.06, or

$$\gamma_{ij}^e \approx (\gamma_{ij}^e)_{ref} \frac{N_i^o + N_j^o}{N_i + N_j} (1 - q_{Hi})(1 - q_{Hj}) \qquad (A2.08)$$

If, for example, carbons C_1 and C_4 in the hexatriene are replaced by heteroatoms, $q_{Hi} = 1$, and $\gamma^o_{12} = 0$ respectively.

A final γ_{ij} adjustment relates to π electron delocalization resulting from conjugated pairs of substituents (i.e., S_, S+). The extent of coupling is a function of (1) the "degeneracy" of the source/sink pair (described by a degeneracy factor, S+) and (2) the "degeneracy" of the intervening π network [described by $D_{\pi ij}$, where i(j) denote the positions of the S-(S+) substituents, respectively].

The substituent factor S±($0 \le S\pm \le 1$) describes the relative energy of a nonbonded electron in the "source" (S-) versus the "sink" (S+), with the degenerate or equi-energy case S± = 1. Examples of degenerate substituent pairs are (-O', =O) and (-NH$_2$, =NH$_2^+$). For some commonly occurring, nearly degenerate pairs, S+ is currently determined from spectral data and stored in SPARC. The π network "degeneracy" factor, $D_{\pi ij}$, is determined by replacing both S- and S+ by surrogates of CH$_2$'. $D_{\pi ij}$ is given by:

$$D_{\pi ij} = q_j/q_i \qquad (A2.09)$$

where q_i and q_j are NBMO charges in the surrogate source (q_i) and sink (q_j) boxes, respectively (computed with all intervening γ_{ij} set equal to 1).

The "strength" of conjugation of S- and S+ substituents is a function of the product of S±D$_{\pi ij}$. If S± D$_{\pi ij}$ = 1, two equi-energy resonance structures exist, and the nonbonded electron(s) in question distribute equally in the source and sink boxes. If S± D$_{\pi ij}$ <(>) 1, the nonbonded electron "favors" the source (sink) box, respectively.

Besides resulting in a shift of π electron(s) between source and sink boxes, resonant substituents can affect π electron redistribution within the intervening π electron network. In the limit of complete degeneracy ($S \pm D_{\pi ij} = 1$), bond alternation is lost with all $\gamma^e_{ij} = 1$. In fact as $S \pm D_{\pi ij}$ approaches unity, the actual definition of essential single bonds is "violated," although the PMO constructs for describing π electron properties remain valid. The adjustment of γ^e_{ij} for conjugated substituents is given by:

$$\gamma^e_{ij} = (1 - \gamma^{eu}_{ij}) \, S \mp D_{\pi ij} + \gamma^{eu}_{ij} \qquad (A2.10)$$

where γ^{eu}_{ij} is the upgraded γ^e_{ij} from Eqn. A2.08 (i.e., upgraded for extended conjugation and electron withdrawing heteroatoms). This "resonance" upgrade is applied to *all* "essential single bonds" in the intervening π network between S- and S+.

Computation of P_{ij}. A simple rule-based topological procedure is used to adjust $(\Theta_{ij})_{ref}$, Eqn. A2.03, taking into account steric twisting and orientational (i.e., H-bond) interactions. An "upgraded" Θ^u_{ij} is determined for each essential single bond.

$$\Theta^u_{ij} = (\Theta_{ij})_{ref} + \Sigma \, \Theta_{ij(steric)} - \Theta_{ij(H-B)} \qquad (A2.11)$$

$$\text{where } 0° \le \Theta^u_{ij} \le 90° \quad , \quad p^u_{ij} = \cos^2 \Theta^u_{ij}$$

The steric contributions (expressed as angle-of-twist) derive from substituents or rings bonded to $R_{\pi i}$ or $R_{\pi j}$ at positions α, β ortho or perri to the bond in question. Steric factors for specific structural elements are currently inferred from spectral data and stored in SPARC. The H-bonding orientational factor derives from intramolecular H-bonding between $R_{\pi i}$ and $R_{\pi j}$. Angle factors for commonly occurring H-bonding pairs are inferred from spectral data and stored in SPARC.

A final P_{ij} adjustment accounts for change in π character (relative to the reference compound) in the essential single bond as described by Eqn. A2.08 or A2.10.

$$P_{ij} = P^u_{ij} + (1 - P^u_{ij}) \frac{\gamma^e_{ij} - (\gamma^e_{ij})_{ref}}{(\frac{\beta_e}{\beta_j})^2 - (\gamma^e_{ij})_{ref}} \qquad (A2.12)$$

where β_j is $\beta_a(\beta_e)$ if $R_{\pi j}$ is aromatic (ethylenic). One can see that as $\beta^e_{ij} \to \beta_e$, $P_{ij} \to 1$ regardless of any steric factors.

For essential single bonds in ring-fixed geometries, Θ_{ij} is instantiated directly from rule bases and not modeled as described previously.

APPENDIX III

Sample Calculations

The following calculations will demonstrate the SPARC approach to calculating bulk polarizability, liquid density based volume, refractive index, effective molecular polarizability, vapor pressure and activity coefficient. Since the bulk of the molecules in this study are straight chain and branched alkanes, the two molecules chosen for the sample calculations are n-pentane and 2,5 dimethylhexane.

Polarizability

The fragments for polarizability calculations are CH_x units for alkanes. Only two intrinsic atomic polarizabilities are needed, $\chi(sp3\text{-}C) = 1.25$ and $\chi(H\text{-}C) = 0.314$. From Eqn. 24 α_i for the carbon is $(1.25*1.25/6) = 0.260$ and α_i for each hydrogen is $(0.314*0.314)/1 = 0.0986$. There are no corrections A_i (Eqn. 25) for the alkanes since there are no connected polar groups.

The bulk molecular polarizability for the alkanes, (C_nH_{2n+2}), from Eqn. 25 can be written as $\alpha = 4*(n*0.260 + (2n+2)*0.0986)$ where the units are $\text{Å}^3/\text{molecule}$. For n-pentane this yields 9.94 and for 2,5- dimethylhexane the value is 15.43.

Volume

For straight chain and branched alkanes the fragment values V^{frag} in Eqn. 26 are all that of $CH_x = 52.945$. The corrections A_i are due to the type, number and size of the substituents. For alkanes the type is always CH_x. The correction for this type is -19.2618. The correction for occluded volume from branching is 0 for branching less than 2, 3.31 for 2 branching and 8.42 for 3 branching. The size correction is 6.8 times the sum of the cone volumes. See Table A3.01.

Index of Refraction

Index of refraction is a good way to check the polarizability density for the molecule. The polarizability and volume can be related to the index of refraction using the Lorentz-Lorenz equation. For our units of cm^3/mole for volume and $\text{Å}^3/\text{molecule}$ for polarizability the Lorentz-Lorenz equation can be written as

$$\frac{n^2 - 1}{n^2 + 2} = \frac{0.6023 * 4\,\Pi\,P}{3\,V} \tag{A3.01}$$

where n is the index of refraction, P is the molecular polarizability and V is the liquid density based volume. Eqn. A3.01 leads to a calculated index of refraction of 1.354 (obs 1.358) for n-pentane and a calculated index of refraction of 1.388 (obs 1.392) for 2,5 dimethylhexane.

Table A3.01
Sample Calculation of the Volume for n-Pentane and 2,5 Dimethylhexane

	V^{frag}	Subtr-type	Branch	Size	Frag-Total
n-pentane					
C	52.945	-19.262	0	-6.8(0.113)	32.92
|					
C	52.945	-19.262*2	3.3	-6.8(0.05+0.107)	16.66
|					
C	52.945	-19.262*2	3.3	-6.8(0.091*2)	16.50
|					
C	52.945	-19.262*2	3.3	-6.8(0.05+0.107)	16.66
|					
C	52.945	-19.262	0	-6.8(0.113)	32.92
				Sum 115.6	
				Obs 116.1	
2,5 dimethylhexane					
C	52.945	-19.262	0	-6.8(0.213)	32.23
|					
C-C	52.945	-19.262*3	8.4	-6.8(0.05*2+0.11)	2.160
|	52.945	-19.262	0	-6.8(0.213)	32.23
C	52.945	-19.262*2	3.3	-6.8(0.131+0.126)	15.98
|					
C	52.945	-19.262*2	3.3	-6.8(0.131+0.126)	15.98
|					
C-C	52.945	-19.262*3	8.4	-6.8(0.05*2+0.11)	2.160
|	52.945	-19.262	0	-6.8(0.213)	32.23
C	52.945	-19.262	0	-6.8(0.213)	32.23
				Sum 165.2	
				Obs 166.1	

Effective Polarizability

Dispersion is a short range interaction involving surface or near surface atoms and A_{disp} in Eqn. 31 subtracts from the total polarizability, a portion of the contributions of sterically occluded atoms in the molecular lattice. Presently SPARC corrects for access judged to be less than afforded by a linear array of atoms. Branched (ternary or quaternary) atoms in an alkane structure will lose a small part of their intrinsic molecular polarizability depending on the size and number of appended groups, and the proximity of other branched carbons. For n-pentane there are no corrections so that the effective polarizability equals the calculated molecular polarizability of 9.94. For 2,5 dimethylhexane the polarizability of the 2 and 5 atoms are reduced by 1.72 times the sum of the sizes (see volume above) of the fragments attached to 2 and 5. Each atom is reduced by $1.72(0.05+0.109+0.05) = 0.36$. The effective polarizability for 2,5 dimethylhexane is calculated to be $15.43 - 2*1.72*(0.209) = 14.71$.

Vapor Pressure

Once the molecular polarizability and volume are known we can use Eqn. 29 to calculate the dispersion interactions. ϱ_{disp} is -2.571 (where this number has subsumed in it -2.303*RT) and the polarizability densities for n-pentane and 2,5 dimethylhexane are $9.94/115.6 = 0.086$ and $14.71/165.2 = 0.0891$, respectively. The dispersion contribution to the vapor pressure is calculated to be $-2.57*0.086*0.086*115.6 = -2.20$ for n-pentane and $-2.57*0.0891*0.0891*165.2 = -3.37$ for 2,5 dimethylhexane. The volume entropy terms (log(T) + C) are $\log(298) - 0.457 = 2.02$ at room temperature. Log(P) for n-pentane is then calculated to be $-2.20 + 2.02 = -0.18$. The observed vapor pressure for n-pentane is -0.17. Log(P) for 2,5 dimethylhexane is $-3.37 + 2.02 = -1.35$. The observed vapor pressure is -1.38.

Activity Coefficient

In order to calculate the infinite dilution activity coefficient for n-pentane and 2,5 dimethylhexane in squalane, we need the effective polarizability and volume for squalane. SPARC calculates the molecular polarizability of squalane to be 55.70, the effective polarizability to be 53.22 and the volume to be 530.0. The polarizability density of squalane is then calculated to be $53.22/530 = 0.10$. From Eqn. 30 the dispersion contribution to the activity coefficient can be calculated. For n-pentane this is $-2.57*(0.086-0.10)^2*115.6 = -0.06$ and for 2,5 dimethylhexane the value is $-2.57*(0.089-0.10)^2*165.2 = -0.05$. The volume entropy terms from Eqn. 34 are 0.26 and 0.17 for pentane and 2,5 dimethylhexane, respectively. The logs of the infinite dilution activity coefficients are -(-0.06 + 0.26) or -0.20 and -(-0.05 + 0.17) or -0.12 for n-pentane and 2,5 dimethylhexane, respectively.

ACKNOWLEDGMENT

The research described in this document was funded in part by the U. S. Environmental Protection Agency under Cooperative Agreement No. 815415 with the University of Georgia.

REFERENCES

1.   S. W. Karickhoff, V. K. McDaniel, C. M. Melton, A. N. Vellino, D. E. Nute, and L. A. Carreira, Unpublished report.

2.   L. P. Hammett, *Physical Organic Chemistry*, 2nd ed. McGraw Hill, New York, NY., 1970.

3.   J. E. Lemer and E. Grunwald. *Rates of Equilibria of Organic Reactions.*, John Wiley & Sons, New York, NY., 1965.

4.   Thomas H. Lowry and Kathleen S. Richardson, *Mechanism and Theory in Organic Chemistry*. 3ed ed. Harper & Row, New York, NY, 1987.

5.   R. W. Taft, *Progress in Organic Chemistry*, Vol.16, John Wiley & Sons, New York, NY., 1987.

6.   S. W. Karickhoff, V. K. McDaniel, C. M. Melton, A. N. Vellino, D. E. Nute, and L. A. Carreira, *Environ. Toxicol. Chem.*, 10 (1991) 1405 .

7.   S. H. Hilal, L. A. Carreira, C. M. Melton, G. L. Baughman and S. W. Karickhoff, *J. Phys. Org. Chem.*, In Press.

8.   S. H. Hilal, L. A. Carreira, C. M. Melton and S. W. Karickhoff, *Quant. Struct. Act. Relat.*, 12 (1993) 389 .

9.   S. H. Hilal, L. A. Carreira, C. M. Melton and S. W. Karickhoff, *J. Chromatogr.*, 662 (1994) 269.

10.   S. W. Karickhoff, L. A. Carreira, and S. H. Hilal, In preparation.

11.   M. J. S. Dewar and R. C. Doughetry, *The PMO Theory of Organic Chemistry*. Plenum Press, New York, NY., 1975.

12.   M. J. S. Dewar, *The Molecular Orbital Theory of Organic Chemistry*, McGraw Hill, New York, NY., 1969.

13.   E. P. Serjeant and B. Dempsey, *Ionization Constants of Organic Acids in Aqueous Solution*, Pergamon Press, Oxford, 1979.

14.   D. D. Perrin, *Dissociation Constants of Organic Bases in Aqueous Solution*, Butterworth & Co, London, 1965.

15.   D. D. Perrin, *Dissociation Constants of Organic Bases in Aqueous Solution: Supplement.*, Butterworth & Co, London, 1972.

16.   S. H. Hilal, L. A. Carreira, and S. W. Karickhoff, In preparation.

17.   A. V. Phelps and J. L. Pack, *Phys. Rev. Lett.*, 16 (1961) 111.

352

18.   Christodoulides, A. A., McCorkle, D. L. and Christophorou L. G., *Electron-molecule interactions and their applications*; Vol.2, Academic, New York, NY. 1984.

19   J. M. Younkin, L. J. Smith and R. N. Compton, *Theort. Chim. Acta.* 41 (1976) 157.

20.   S. Chowdhury and P. Kebarle, *Chem. Rev.*, 87 (1987) 518.

21.   R. T. McIver and E. K. Fukuda, *J. Am. Chem. Soc.*, 107 (1985) 2291.

22.   K. D. Jordan, P. D. Burrow, *Acc. Chem. Res.*, 11 (1978) 341.

23.   E. C. M. Chen and W. E. Wentworth, *J. Chem. Phys.*, 63 (1975) 3183.

24.   P. Kebarle and G. Paul, *J. Am. Chem Soc.*, 111 (1989) 464.

25.   L. W. Kao, R. S. Becker and W. E. Wentworth, *J. Phys Chem.*, 79 (1975) 1161.

26.   P. Kebarle and G. W. Dillow, *J. Am. Chem. Soc.*, 111 (1989) 5592.

27.   E. C. M. Chen and W. E. Wentworth, *J. Phys. Chem.*, 71 (1967) 1929 .

28.   E. C. M. Chen and W. E. Wentworth, *J. Chromatogr.*, 217 (1981) 151 .

29.   K. D. Jordan, P. D. Burrow, *J. Chem. Phys.*, 11 (1978) 341 .

30.   I. Nenner and G. J. Schultz, *J. Chem. Phys.*, 62 (1975) 1747 .

31.   M. D. Rozeboom, I. S. Tegino-Larsen and N. K, Horik, *J. Org. Chem.*, 46 (1981) 2338.

32   P. Kebarle and G. W. Dillow, *Can. J. Chem.*, 67 (1989) 1628.

33.   L. G. Christophorou, *Adv. Elec. Elec. Phys.*, 46 (1978) 55.

34.   R. Foster, *Organic Charge-Transfer Complexes*. Academic, New York, NY., 1969.

35.   A. Streitwieser, *Molecular Orbital Theory for Organic Chemistry*. Willey, New. York, NY., 1966.

36.   G. Tarjan, I. Timar, J. M. Takacs, S. Y. Meszaros, Sz. Nyiredy, M. V. Budahegyl, E. R. Lombosi and T. S. Lombosi, *J. Chromatogr.*, 271 (1982) 213.

37. J. K. Haken and M. B. Evans, *J. Chromatogr.*, 472 (1989) 93.

38. E.sz. Kovats, *Adv. Chrommatogr.*, 1 (1965) 31A.

39. L. S. Anker and P. C. Jurs, *Anal. Chem.*, 62 (1990) 2676.

40. P. J. Flory, *J. Chem. Phys.*, 10 (1942) 51.

41. M. L. Huggins, *J. Am. Chem. Soc.*, 64 (1942) 1712.

42. K. A. Sharp, A. Nicholls, R. Friedman and B. Honig, *Biochemistry.*, 30 (1991) 9686.

43. G. Schomburg and G. Dielmann, *J. Chromatogr. Sci.*, 11 (1973) 151.

44. N. Dimov, *J. Chromatogr.*, 347 (1985) 366.

45. D. Papazova and N. Dimov, *J. Chromatogr.*, 356 (1986) 320.

INDEX

368

Printed and bound by CPI Group (UK) Ltd, Croydon, CR0 4YY

03/10/2024

01040333-0004